Theory and Practical Exercises of System Dynamics

Book based on Vensim 9

Juan Martin Garcia is a renowned expert in System Dynamics and System Thinking, Ph.D. Industrial Engineer (Spain) and Postgraduate Diploma in Business Dynamics at the Sloan School of Management of the Massachusetts Institute of Technology (USA).

He has extensive industrial experience in engineering and manufacturing, and 20 years' experience teaching building simulation models in large companies and universities, now teaches the online courses of Vensim in http://vensim.com/vensim-online-courses/.

Theory and Practical Exercises of System Dynamics
Juan Martín García   2023   Spain
info@atc-innova.com

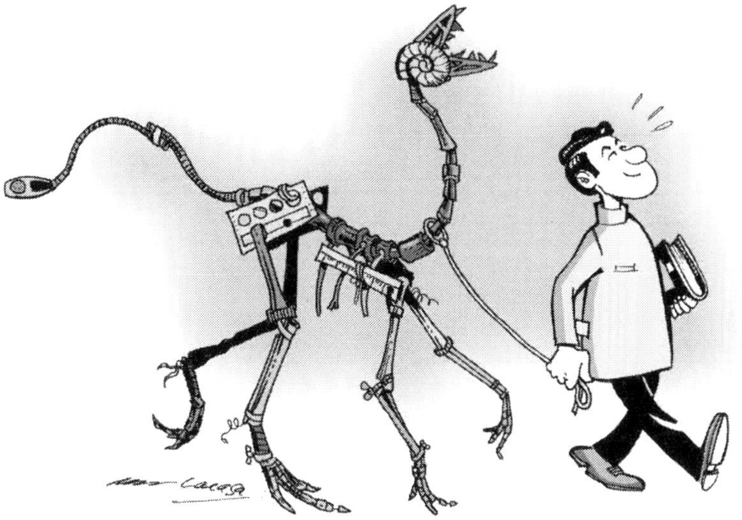

CONTENT

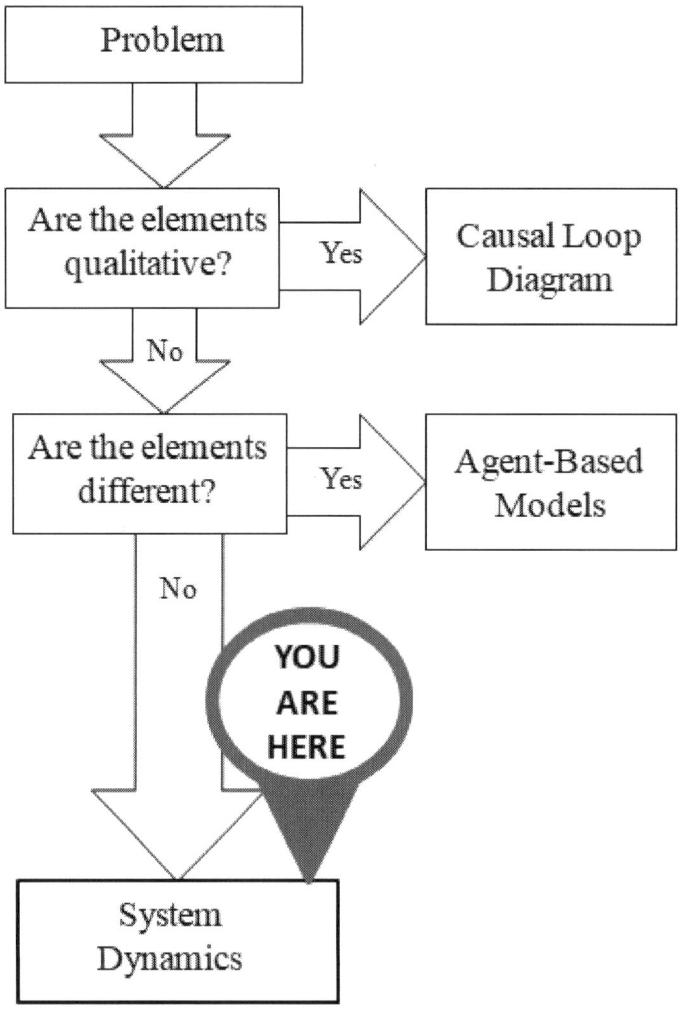

SHORTCUT

Start doing the cases 1 and 19 to discover the features of the software; later continue with cases 10 and 14 to see the functions, delays and tables. Continue with cases 2 and 3 to discover complex equations and policy simulations. Complete this trip with cases 20 to learn the use of qualitative variables and 21 with complex diagrams.

PREFACE

As the complexity of our world increases systems thinking is emerging as a critical factor for success, and even survival. How then can people become skilled systems thinkers? The most effective learning experiences combine experience with reflection, theory with practice.

Traditionally, theory was taught in school and university, and experience was gained in life outside those walls. But in the world of complex dynamic systems such as a business, society, or ecosystem, everyday experience fails because the time horizon and scope of the systems is so vast—we never experience the majority of the effects of our decisions. And without relevant experience, theory is uninteresting to students.

The old ways of learning fail. When experiments in the real world are impossible, simulation becomes the main way we can learn effectively about the dynamics of complex systems. For this reason I'm pleased to introduce Juan Martin Garcia's book, 'Theory and Practical Exercises of System Dynamics'. Juan combines theory and practice, experience and opportunities for reflection, so that newcomers to the field can learn for themselves how complex dynamic systems work. The examples span a range of important economic and social issues, from the aging of the population in developed economies to the course of contagious diseases to the accumulation of pollutants in the environment; everyone will find some examples here of direct personal interest.

The modeling exercises guide the learner through the process of building a working simulation; students will not only learn about the issues addressed, and in the use of state of the art simulation software, but will develop skill in the modeling process.

Juan has written a delightful first introduction to the field of system dynamics and complexity, and provides a much-needed addition to the literature available.

John D. Sterman
Cambridge, Massachusetts

FOREWORD

This can be a good scenario to fight against some myths circulating around the System Dynamics. Every single person not familiar with this methodology might have the perception that behind these words the most complex equations are hidden, all of them full of strange Greek symbols, differentials and integrals. The situation can even get worse as you have to deal with statistical data and try to decipher its blurry meaning. Eventually, you wait for complex software requiring a vast knowledge of programming.

There is no doubt that these myths, such as good guardians of knowledge as they are, easily throw weak minds out. But actually, myths are more a product of the imagination and the illiterate than the deliberate intent to hide a valuable knowledge source. In order to defeat them, it is necessary to mention that System Dynamics is applied to daily issues, that is to say, the real world. We will rarely find complex mathematical formulations in it but an amazing simplicity after carefully analyzing any side. Complexity is more apparent in the real world and often is the result of our ignorance. For example, integral equations symbolically represent an accumulation of material, energy or information. With regard to statistics, we will make use of them in special circumstances, more as a complement than a core point in our reason. Finally, concerning the alleged software complexity, we will prove how easy it can be. So, forget your fears and let's travel to Knowledgeland.

Once myths are gone, we have to ask the big question - What will System Dynamics do for us? It is possible to find thousands of answers to this question in several books but first, let us use a simple analogy. If you had to calculate the total amount of seats in a theatre with 30 rows and 30 seats per row, you could count them one by one as far as 899. In the end, good advice would be to count them again to be sure of the result. However, you can also use a methodology called 'multiplication' and by taking the numeric expression '30x30', you will obtain the result in a faster and more reliable way - 900. System Dynamics is a similar method in terms of offering the same fast and reliable results as opposed to other traditional ways to perceive reality. Hence, this methodology allows us to deal with the analysis of the apparent complexity found in economic, environmental, social or just mechanical issues.

The main purpose of this book is to serve as a reference for students studying this subject, guiding them in their first steps. Initially, we offer a brief perspective regarding this methodology and its basic concepts. Afterwards, a completion of basic exercises can be used in order to make the learning process faster. Just by following the directions, it will be realized how easy it is to satisfactorily catch on to this methodology.

This book does not aim to substitute the teacher's task. A teacher will solve the trickiest questions that arise. The Annex contains the most frequently asked questions (FAQ's) formulated around this topic, followed by accurate answers resulting in the readers having a better understanding to this methodology.

Practical exercises are provided to offer a cross-section of basic examples about what is necessary to know. They are not supplied to provide current research topics or a guide for projects in any particular area. Readers concerned with that will have to look over those models that match their discipline. To facilitate this task, some of the prolific references existing about models can be browsed on the Internet.

System Dynamics can be applied to daily matters. To the question 'When is it appropriate to use the System Dynamics?' The answer will be 'Every time some kind of feedback exists'. Such a reality is common in life (corporations, environmental and social topics) and it is the explanation of these phenomena that cannot be solved by one's intuition alone.

Exercises can be organized in many ways. Topics have been classified to make it easy to keep on track, firstly, environmental sciences, secondly, managing and social studies and finally, mechanic systems patterns. There is no need to do all the exercises to become competent in this methodology. You may choose those you consider the most significant or interesting.

In the book, problems and solutions are together, in contrast to other books where solutions are published in the final pages. This facilitates learning by doing different models. Simulation model design is a handcrafted job. There is a need to know how the technique works. Once acquired, the creator can freely decide the structure and not follow universal rules.

The concern about this methodology, that is, that all the studies consist of computer models, is far from being true. The application of this modus operandi aspires to obtain the simplest solutions to complex matters. Consequently, simulation models production is justified just a few times. Learning how to produce them is very useful to foster theoretical concepts and might be more effective in elucidating suitable responses. As a consequence, it is recommended not to start working on the exercises, particularly because the application of this methodology does not imply to create always a computer model.

Finally, this book seeks to be a careful and precise work which helps to avoid the typical doubts implicit in any learning process. Being conscious about the difficulties that come from learning with a book, we have tried to make a clear and attractive text. In spite of all the sincere efforts to accomplish what was previously explained, here is the final advice - find a teacher for this subject, whenever possible.

Juan Martín García
Spain

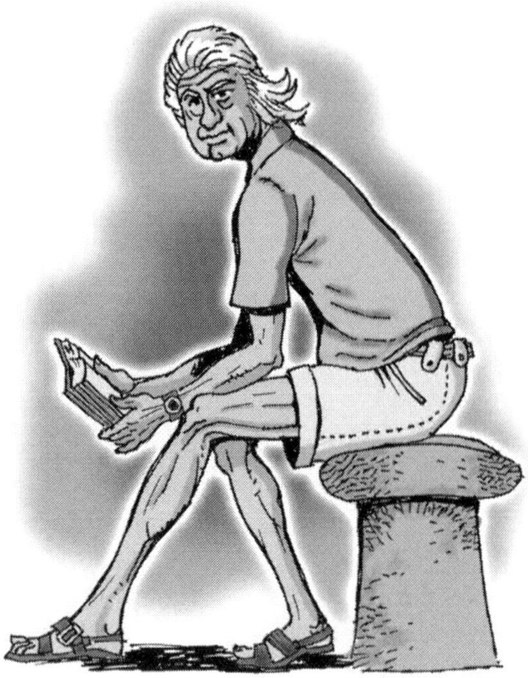

1. SYSTEM DYNAMICS

In the following pages we are going to describe approach to interpret the reality. It is likely that what we might call *the correct* or *the best* way to observe the reality does not exist, given that it is impossible to point at a certain direction as the best or more adequate.

However, this is a new concept for many people. In my opinion, it is a useful way to deal with the problems and challenges we are facing at the turn of the millennium - hunger, poverty, the degradation of the environment, war, etc. It would seem that the traditional ways of dealing with such issues haven't improved matters significantly.

The viewpoint presented here has several names. Here, we will call it 'System Dynamics'. I'm aware that 'system' could mean various things, and I will attempt to clarify my interpretation of the word through discussion and examples further on.

As an introduction, we will look at the characteristics of models that represent the world as a whole - as a global system.

We will then describe the characteristics of the situation of the world today from this viewpoint.

1.1. System Dynamics

We are increasingly aware that we live in a very complex and constantly changing reality and that it is more so every year. In order to make the decisions that are continually asked of us, we use mental models. However, these models don't always bring us closer to solving the problem, as the solution may be, as Jay Forrester calls it, *counter-intuitive*, even in the simplest of cases.

For instance, during a visit to the Science Museum with our children, we may have to explain why the hole in a water tank which is nearest the ground spouts water further than the spout above it. We may also have to explain why the image in a magnifying glass is inverted after a certain distance, instead of growing indefinitely.

As Ludwig von Bertalanffy notes: for those who wish to study science, and only science, any posterior question makes no sense. '*Quod non est in formula non est in mundo*'. Such is the only legitimate position of science. Despite this, if we wish to further our understanding, there is only one analogy that can explain that which is irrelevant to the physicist, the analogy of the only reality we know directly - the reality of our immediate experience.

All interpretations of reality, to use Kant's expression, are an adventure of reason. There are therefore only two alternatives to choose from - either we reject all interpretations of the essence of things or, (if we do attempt an interpretation) we must remain conscious of its analogous nature, since we don't have the slightest proof that true reality is the same as that of our own internal experience.

When faced with the common occurrence of a reality with a limited number of parameters, especially when these are quantifiable, we employ formal models which allow us to act with a reasonably high probability of succeeding. However, when faced with complex situations with an uncertain number of parameters that are difficult to quantify, we can resort to less formal models that provide a more structured view of the problem, its key aspects, and possible solutions.

Lynda M. Applegate states that computers today are designed to treat information sequentially, instruction after instruction. This works well if the problem can be structured and divided into a series of stages. It doesn't work well with complicated, unstructured tasks which require intuition, creativity and discernment.

The main application of System Dynamics is in this kind of complex and loosely defined environments, where a human being's decisions tend to be guided by logic. We must remember that science is currently based on measurable and reproducible phenomena. As specialists in marketing know, people also behave according to certain rules which are fairly easy to measure and reproduce, for example, market law (more demand pushes up prices, etc.).

With reference to this, Javier Aracil states in his book *Introducción a la Dinámica de Sistemas* (Introduction to System Dynamics) that computer models can provide information not attainable via mental models. They can show the dynamic consequences of interactions between components of a given system. When assessing the consequences of certain actions, the use of mental models means running the risk of obtaining erroneous conclusions. Intuition isn't reliable when the problem is complex. One possible reason for this is that we tend to think in terms of one-way cause-effect relationships, forgetting the structural feedback which almost certainly exists in such a system. When preparing a computer model, we must consider each step separately. The mental image we have of the system must be developed and expressed in a language that can be used to program the computer. Normally, any consistent and explicit mental image of any system can be expressed in this way. The mental images that we have of real systems are the result of experiences and observations. The explicit formulation of these experiences in a computer program forces us to examine, formalise and focus our mental images, thus providing us with a greater understanding through several perspectives.

Mathematical models, which are programmable, are explicitly expressed. The mathematical language used to describe the model leaves no room for ambiguity. A System Dynamics model is more explicit than a mental model and can therefore be expressed without ambiguity. The hypotheses upon which the model is built and the relationships between its constituent elements are present in complete clarity and are subject to discussion and revision. For this reason, a model's forecasts for the future can be studied in a completely precise way.

It is important to differentiate between the following two kinds of models: predictive models which are designed to offer precise information of the future status of the modelled system, as opposed to management models which are basically designed to decide whether option X is better than option Y. Management models don't require as much precision, since comparisons are equally useful. System Dynamics models are of the latter type.

As explained above, I understand the word '*system*' to mean a set of independent elements that interact with each other in a stable way. The first step towards understanding the behaviour of a system would be to define its constituent elements, and their possible interaction. The notion of Aristotle that the whole is more than the sum of its parts takes on a special meaning here.

The standpoint of System Dynamics is radically different to other existing techniques for the construction of socio-economic system models such as econometrics. Econometric techniques, which are based on behaviourism, use empirical data such as statistical calculus in order to determine the meaning and correlation between the various factors involved. The model is developed from the historical evolution of variables that are declared independent, and statistics is applied in order to determine the parameters of the system of equations that link them to other independent variables. These techniques can establish the behaviour of the system without the need for information regarding its internal functioning. This is how stock market models analyse the upward and downward trends in the values of shares, the rising and falling cycles, etc. They are designed in order to minimise the risk of losses, etc. They don't attempt to gain any detailed knowledge of the internal workings of the firms, as the value of a given company rises and falls according to its new products, new competitors, etc.

The basic objective of System Dynamics is different. It aims to gain understanding of the structural causes of a system's behaviour. This implies increased knowledge of the role of each element of the system, in order to assess how different actions on different parts of the system accentuate or attenuate its behavioural tendencies.

One characteristic that sets it apart from other methods is that it doesn't aim to give a detailed forecast of the future. By using the model to study the system and test different policies, we will deepen our knowledge of the real world, assessing the consistency of our hypotheses and the effectiveness of each policy.

Another important characteristic is its long-term perspective, that is, that the period studied is long enough for all significant aspects of the system to evolve freely. Only with a sufficiently broad time scale can the fundamental behaviour of a system be observed. We mustn't forget that the results of certain policies are sometimes not the most appropriate, for example, if the time horizon of the decision-making process was too short, or if there was a lack of perspective when the problem was addressed. In these cases, it would be useful to know the long-term consequences of actions taken in the present, and this can be more tangibly attained if we use a suitable model.

The long-term development will be understood only if the main causes of any possible changes are identified. This process is facilitated if the appropriate variables are chosen. Ideally, the limits of the system should include the whole set of mechanisms that are responsible for any important alterations in the main system variables over a broad time horizon.

System Dynamics allows the construction of models after a careful analysis has been conducted of the elements of a system. This analysis allows the internal logic of the model to be extracted.

Knowledge may then be gained of the long-term evolution of the system. It should be noted that the adjustment of the model according to historical data is of secondary importance, the analysis of the internal logic and the structural relationships within the model being the key issues involved in its construction.

(Note: All teaching material, and that includes this text, should be objective. This text aims to be so, but the author admits he hasn't always succeeded. For this he must apologise. Readers are invited to make their own assumptions as to what is an exposure of methodology, and what amounts to personal opinion.)

1.2. Identifying the Problem

What is the problem?

We are going to learn a method for constructing simulation models that help us determine the best solution for a given problem. These are therefore management models, not predictive models.

Firstly, we have to identify the problem clearly and give a precise description of the aims of the study. It may be obvious, but it is very important that the definition of the problem be correct, since all further steps depend on this. This is also very useful when establishing the amount of time and money that will be spent creating the model.

Once the core of the problem is defined, a description must be completed, based on the knowledge of experts on the subject, basic documentation, etc. The result of this phase should be a preliminary perception of the elements that have a bearing on the problem, the hypothetical relationships between them, and their historical behaviour.

The historical reference of a system is a record of the historical behaviour of the main elements that are believed to influence the problem. Where possible, they should be quantified. This is the graphical and numerical representation of the verbal description of the problem.

It's a good idea to ask ourselves whether it is necessary to construct a simulation model in order to find an efficient solution to the problem. This is an important question.

The construction of a model is a long and costly process. It can't be justified if there are other more simple ways of obtaining the same results. There are essentially two other ways - statistics and intuition.

- Statistics, or numerical calculus methods, are very useful for solving problems where there is an abundance of historical data or when we can assume reality will remain stable. For example, if you want to find out how many cars will drive past your house today, all you need is sufficient historical data and assuming the street hasn't changed, you'll get a good approximation.
- Intuition has got you where you are today, so don't underestimate it. For many problems, intuition provides the right answers, drawing on our experience and knowledge. Intuition is cheap and fast. Keep using it as often as possible.

Only when we can't apply one of these two options with certainty must we resort to constructing a simulation model.

Once the problem is defined, we will see that there are many directly or indirectly related aspects, or elements, which are also interrelated. They needn't be clearly or obviously interrelated. These elements constitute the system. We will now study reality as a system.

1.3. Defining the System

What is a system?

A system is a set of interrelated elements, where any change in any element affects the set as a whole. Only elements directly or indirectly related to the problem form the system under study here. In order to study a system, we must know the elements that make it up, and the relationships between them.

When we analyse a system we usually focus merely on the characteristics of its constituent elements. However, in order to understand the functioning of a complex system, we must focus also on the relationships that exist between the elements which form the system.

It is impossible to understand the essence of a symphony orchestra by merely observing the musicians and their instruments. It is the coordination that exists between them that produces beautiful music. The human body, a forest, a country, or the ecosystem of a coral reef are all examples of systems that are far more than the sum of their parts.

An ancient Sufi saying can illustrate this: *You can think, because you understand 'one', and you can understand 'two', which is 'one' plus 'one'. However, you must also understand 'plus'.* For example, in a traffic problem, many related elements converge - the number of inhabitants, the number of cars, the price of petrol, parking spaces, alternative transport, etc., and it is often easier and more effective to attempt to solve the relationships between the elements (*'plus'*) than the elements themselves.

A good method to begin defining a system is to write the main problem down in the middle of a blank page, surrounding it with the directly related elements. The elements which affect the main problem indirectly go around the appropriate direct elements. This will be the system that we will study in order to consider possible solutions to a given problem.

1.4. The Boundaries of a System

Where does the system end?

We have all heard the theory that a butterfly fluttering its wings in China could cause a tornado in the Caribbean. In our study, however, we will include only elements with a reasonable influence on the behaviour of a system. We mustn't lose sight of the objective, that is, to propose practical action towards effectively solving the problem at hand.

The system must contain as few elements as possible while providing a simulation that will truly allow us to decide which of the possible courses of action studied are the most effective solution to the problem. The models are generally small to begin with, with few elements. They are then expanded and perfected. Later on, elements which don't play a decisive part in the problem are eliminated. During the construction of a model, there are several extension and simplification phases in which elements are added and subtracted.

We can't ignore the relationship between the consumption of petrol and lung health. When we analyse the carbon combustion process in an electric power plant, we can see that, apart from energy, the following is produced - ash, suspended particles, SO_2, CO_2, etc. We can also see that there is no barrier between the desired product (electricity) and the by-products. Sometimes, the so-called *side-effects* are as real and as important as the *main effects*. The beauty of a system in nature is that the waste produced by one process serves to feed the next. Perhaps this is the model to follow for industrial design in the future.

The final size of the model must be such that its main aspects can be explained in ten minutes. Any model larger than this will fail.

1.5. Causal Loop Diagram CLD

How do we represent the system?

The set of elements that bear relationship to the problem and that account for the observed behaviour, along with the relationships that exist between these constituent elements (which often involve feedback) form the system. The causal diagram represents the key elements of the system and the relationships between them.

As discussed above, it's important to draft versions that will bring us increasingly closer to the final complex model. The minimum set of elements and relationships that serves to reproduce the historical reference of a system is that which forms the basic structure of the system.

Once the variables of a system and the hypothetical relationships between these variables are known, we can move on to produce a graphical representation. This diagram shows the relationships as arrows between the variables. These arrows are marked with a sign (+ or -) which indicates the kind of influence one variable exerts over the other. A '+' means a change in the influencing variable will produce a change of the same direction in the target variable. A '-' means the effect will be the opposite.

So, when an increase in A results in an increase in B, or a fall in A causes a fall in B, this is a positive relationship, as shown below:

<u>Positive does not mean good</u>

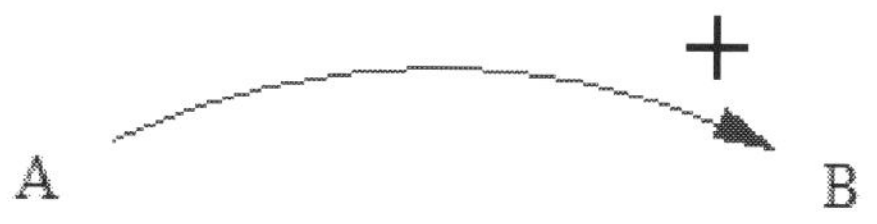

When an increase in A results in a fall in B, or a fall in A causes an increase in B, this is a negative relationship, which is expressed as follows:

<u>Negative does not mean bad</u>

1.6. Feedback loops

What is a feedback loop?

A closed chain of relationships is called a loop, or a feedback loop. When we turn on the tap to fill a glass with water, the amount of water in the glass increases. The amount of water in the glass, however, also has an effect on the speed at which it is filled. We fill it more slowly when it is fuller. Therefore, a loop exists.

The system formed by us, the tap and the glass is a negative loop, because it is designed to achieve a goal (fill the glass without spilling). Negative loops act as stabilising elements in systems designed to reach a given goal, like when a thermostat in a heating system guides the temperature towards the level specified by the user.

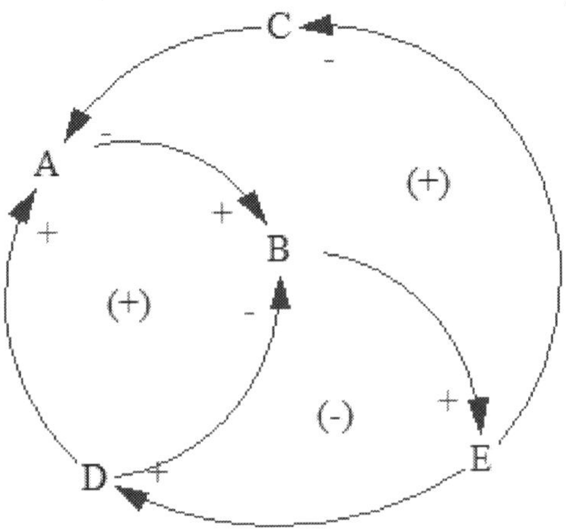

When we construct a model, loops appear. For example, those formed by ABEDA, DBED and ABECA in the following causal diagram. Loops are defined as 'positive' when the number of negative relationships is even. If the number of negative relationships is odd, the loop is 'negative' (just as -3 multiplied by +3 gives -9).

Negative loops tend to stabilise the model, while positive loops tend to destabilise it, independently of the basic problem at hand, because in essence a positive loop is: $x_{n+1}=f(x_n*x_n)=f(x_n^2)$ and a negative one is $x_{n+1}=f(x_n*(1-x_n))$ that is, the state of a variable x in period n + 1 is a function of the state it had in period n, exponentially amplified if it is a positive loop , or not, if it is a negative loop.

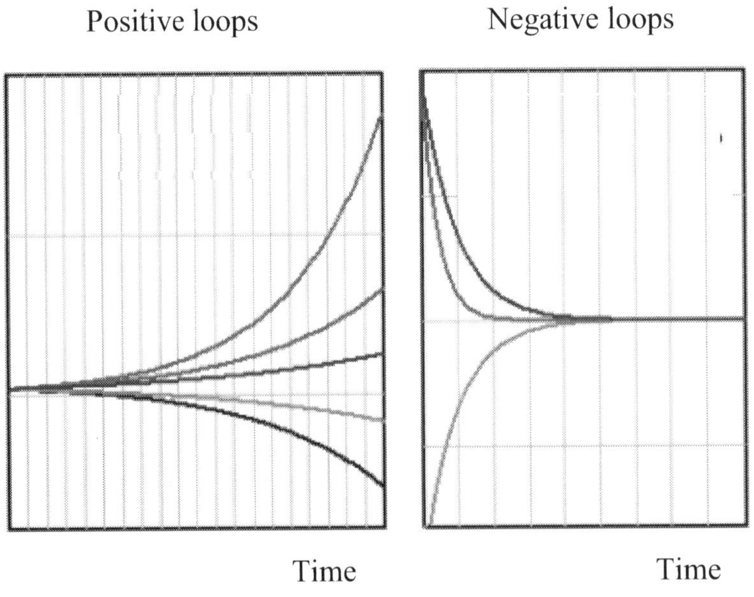

In order to attain a goal, huge efforts are sometimes made in the wrong directions. This is especially true in the personal, social, business and ecological fields. In an attempt to avoid this, Jay Forrester proposed a set of guidelines for the business world that can easily be extrapolated to other areas.

1) Whatever the problem is that has arisen, it is necessary to know the inner workings of the system, how it takes its decisions, how it operates. Don't be led astray by indications that point towards momentary or superficial factors, however visible they may be.

2) Often a small change in one or very few policies can solve the problem easily and definitively.

3) The key factors tend to be ruled out or judged to be unrelated to the problem at hand. They are rarely an object of attention or discussion and when they're identified, nobody can believe that they are related to the problem.

4) If somebody happens to have already identified a key factor, it is not unusual for action to have been taken in the wrong direction, thus seriously magnifying the problem.

Models enable us to conduct sensitivity studies and see which of the system's elements can have a decisive bearing on its behaviour. In other words, they enable us to identify the key factors. However, that doesn't mean we can't advance without their help.

The peculiarity of these key factors is that they are located in unexpected points or aspects that provoke counter-productive actions. This is difficult to illustrate with a causal diagram. The phenomenon seems to be attributable to the difficulty in interpreting the behaviour of a system that is already defined, rather than to any specific structure, as the effect of the interrelationships is beyond our capacity for analysis. (For me, this means that the system has more than three loops).

This inability to perceive and interpret the nature of the system and the identity of its key factors make for counter-intuitive behaviour by the system, with the result that our actions are in the wrong direction. Let us take a look at some examples.

a) A car engine manufacturing firm suffered a constant loss of market share. Every four years there was a major loss of customers who seldom came back. According to the firm's analyses, the problem lay in their policy on stocks of finished products. The company was reluctant, due to the high financial cost, to keep a large number of engines in stock waiting for orders to arrive. The policy was to keep stocks of finished products low. This policy saved a great deal of money but whenever there was an upturn in the economic cycle, the firm was overwhelmed with orders that they were forced to attend to with long delays. The customers then went to the competitors who supplied the engines more quickly. The firm responded to the loss of sales with a programme of cost-cutting measures, including further reductions in stocks of finished products.

b) Dairy farms are steadily disappearing. Measures are proposed to combat this, including tax cuts, soft loans and subsidies. There is plenty of incentive for anybody wanting to start up a small farm. However, the main reason why farms close is expansion. Farmers try to increase their income by producing more milk. When all the farmers do this, the same the market is flooded with milk and prices fall (as there is no intervention or guaranteed price. If there were, the burden would be shifted to the external factor). When the prices have dropped, each farmer has to produce more milk in order to maintain his or her earnings. Some manage to do so and others don't, and of the latter, those that are in the weakest position give up farming.

c) One of the key factors in any economy is the useful life of the installed capital. The best way to encourage the sustained growth of the economy is to stretch this useful life as long as possible. Yet the policy that is practised is one of accelerated obsolescence, or priority is given to replacing existing equipment with machinery designed to provide short-term economic growth.

d) The right way to revitalise the economy of a city and ease the problem of depressed areas occupied by people without economic resources isn't to build more subsidised housing. The solution is to demolish the abandoned factories and houses and create space to set up new businesses, thus allowing the balance between jobs and population to restore itself.

Ideally, we'd have a set of simple rules to find the key factors and know in which direction to act. It is not always possible to find these points by simply observing the system and this is where computer simulation models really come into their own.

1.9. Classification of Systems

1.9.1. Stable and Unstable Systems

A system is stable when it consists of or is dominated by a negative loop, and is unstable when the loop is positive. That is, when the dominant loop contains an odd number of negative relationships, we have a negative loop, and the system will be stable. The basic structure of stable systems is as follows:

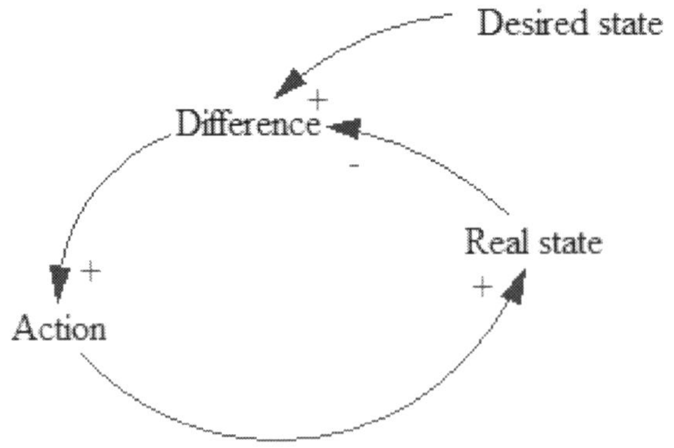

Here we can see that the system has a 'desired state' and a 'real state'. These two states are compared ('difference'), and on the basis of this value, the system takes 'action' to move the 'real state' towards the 'desired state'.

In this case the initial parameters are of relatively little importance, since the system will act according to the environmental conditions it encounters, so if it is hungry, it will look for food and when it finds it, it will deal with its next objective, and so on.

It is important to note that in stable systems the structure that generates the behaviour is always the same, that is, there is an odd number of negative relationships and the loop is negative.

This means that the system permanently compares its real state with the desired state and when there is a difference, it takes action to bring its real state closer to the desired one. Once these two states coincide, any change in the real state will result in action (proportional to the difference) to regain the desired state.

This is how we usually find systems. By the time we get close to them, they are in a position of stability. If a system is unstable, we are unlikely to be able to study it as it will have disintegrated before we can analyse it. However, if we are designing a totally new system, we should take the trouble to find out whether it is going to be stable and if we are designing a change in a stable system, we have to ensure that we are not changing it into an unstable one.

Examples of systems that are not in an optimum situation but carry on over the years – i.e., stable systems – can be found in many fields, for example, government, workers and bosses together produce inflation that is harmful to all. Rich countries and poor countries trade with raw materials, each with a different political and economic objective, and the result is permanent price instability.

Let us suppose that the government intervenes in the system with a particular policy that puts the state of the system where it wants it. This will cause major discrepancies between the other elements of the system which will intensify their efforts until, if they succeed, the system is back very close to the initial position after each element has made a huge effort. For example, I think in the work that has gone into improving the traffic in the city Barcelona over the last 10 years. The traffic improved for a few years after the opening of the Ring Roads, but now we are faced with the same problems as before – except that they affect many more cars.

The most effective way of combating the natural resistance of the system is to persuade each element to change its objectives in the direction in which we want to lead the system. Then the efforts of all the elements will be directed towards the same goal and the effort will be minimum for all as they won't have to resist the tide going the other way. When this can be achieved the results are spectacular. The most common examples of this are the mobilisation of the economy in wartime and the recovery after wars or natural disasters.

A less warlike example can be found in the birth rate policy in Sweden in the 1930s, when the birth rate fell below the rate of natural replacement. The government made a careful assessment of its objectives and those of the population and found that an agreement could be reached on the basis of the principle that the important thing isn't the size of the population but its quality. Every child should be wanted and loved, preferably in a strong, stable family, and have access to excellent education and health care. The Swedish government and citizens agreed on this philosophy. The policies that were introduced included contraception and abortion, education on sex and the family, unhindered divorce, free gynaecological care, aid for families with children in the form of toys, clothes, etc., rather than cash, and increased spending on education and health. Some of these policies seemed strange in a country with such a low birth rate, yet they were introduced, and since then the birth rate has risen, fallen and risen again.

Some systems lack feedback and the models we build must show that fact. For instance, if we know the initial parameters of a clam (type, weight, etc.) and we control the environmental conditions in which it will live, we can safely predict its weight after 6 months. There's a 'transfer function' between the start and end values and we have to find it, but that's all.

Other examples: God is someone who gets his real state to coincide with his desired state instantly. Suicide is the response of those who perceive that they will never get their real state to coincide with their desired state and that therefore, all action is pointless. Please note: The more intelligent a system is (i.e., the clearer its vision of its objectives), the more stable it will be. This is applicable to people.

1.9.2. Hyperstable Systems

When a system consists of several negative loops, any action taken to modify one of its elements is offset not only by the loop in which that element is located but also by the whole set of negative loops which act to support it, thus superstabilising the system.

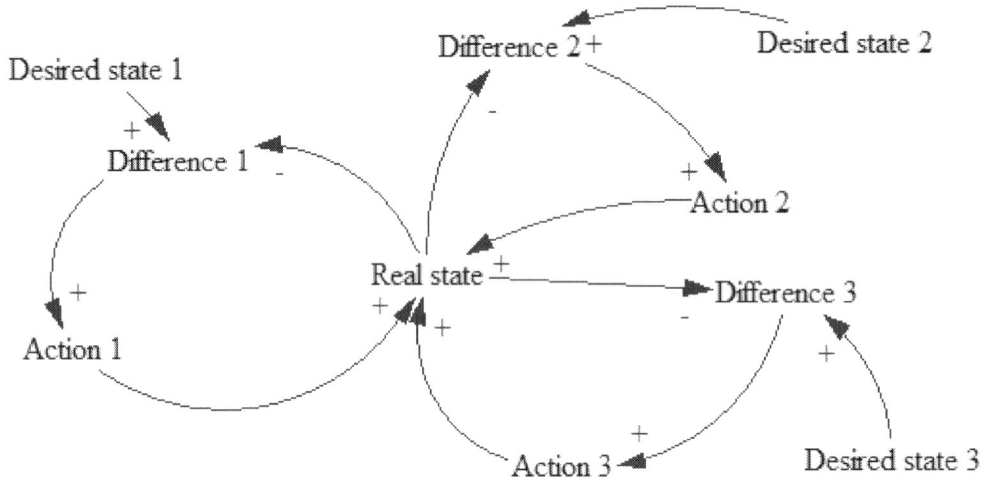

An analysis of the system can be helpful. Any complex system, whether social or ecological, is made up of hundreds of elements. Each element is only linked to a limited number of variables which are important to it, and which it permanently compares with its objectives. If there is a discrepancy between the state of these variables and its objectives, the element acts in a particular way to modify the system. The greater the discrepancy, the more energetic the action taken by the element on the system.

The combined action of all the elements that attempt to fit the system to their objectives leads the system to a position that none of the elements actually wants but in which all of them find the smallest gap between the parameters that are meaningful for them on the one hand and their objectives on the other.

Why do many problems persist despite continual efforts to solve them?

As we have just seen, systems base their stability on the actions of all its elements in pursuit of different objectives, trying to get the rest of the system as close as possible to its desired position.

From this moment on, if an element of the system or an external agent attempt to modify its stability, the other elements will take action to go back to the initial situation, thus neutralising the action that altered its stability.

So the answer is simple - systems resist any change we try to introduce because its present configuration is the result of many previous attempts like ours (unsuccessful ones, otherwise the system would be different today) and an internal structure that renders it stable and capable of neutralising changes in its surroundings, such as the one we made with our action.

The system achieves this as a whole, by rapidly adjusting the internal relationships between its elements in such a way that each continues to pursue its own goal and together they neutralise the action exerted on them from outside.

1.9.3. Oscillating Systems

We will see later in the case studies that for a system to display oscillating behaviour it has to have at least two stocks, which are elements of the system that produce accumulations.

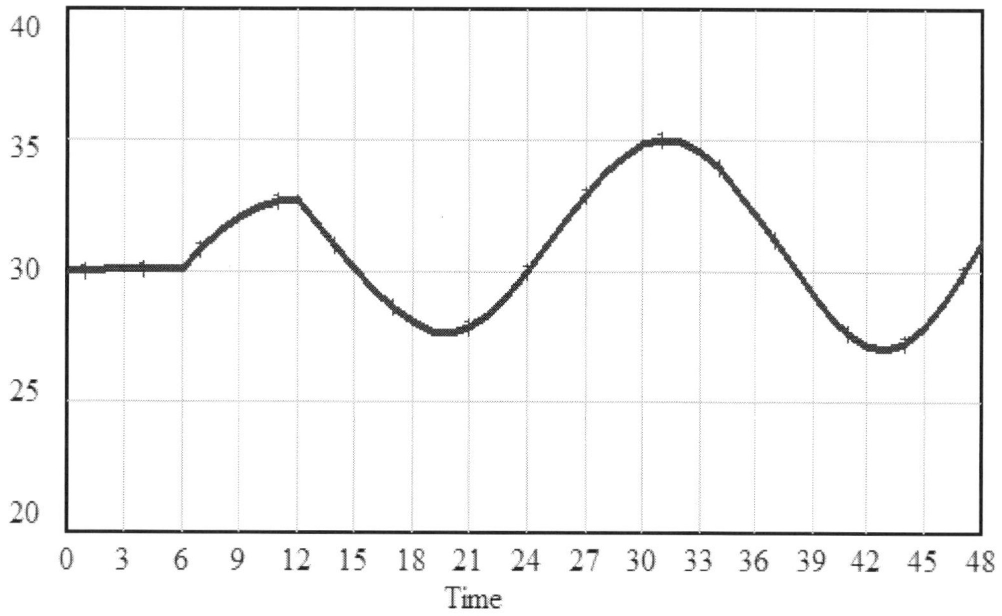

1.9.4. Sigmoidal Systems

These are systems containing a positive loop that acts as the dominant feature at the beginning, causing the system to undergo an exponential take-off. Subsequently, control of the system is taken over by a negative loop that cancels out the effects of the earlier positive one and provides the system with stability, setting it to a particular value asymptotically.

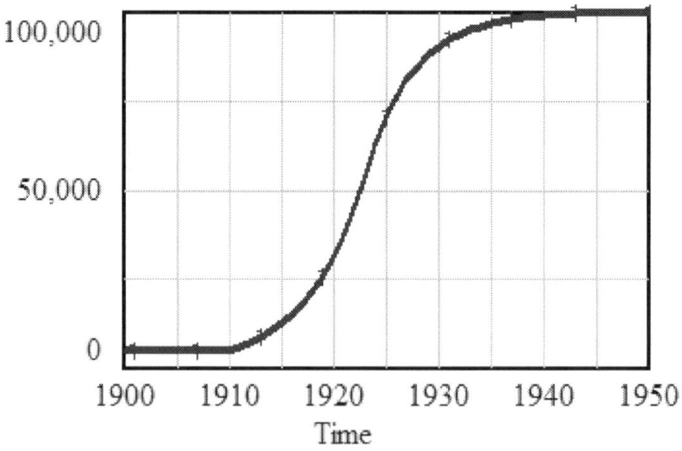

It is important to keep sight of the fact that in this case, we are dealing with the same system all the time, dominated by one part of it in one period, and by a different part later on. So in order to regulate its behaviour, we will have to find a way to play up or down the part of the system we are interested in. We also have to be aware that in the mid-term the negative loop will stabilise the system at its target value. All we can do is regulate the time scale and the way in which the system reaches its objective.

1.10. Generic Structures

In complex systems, we can observe the same structure, that is, desired state - real state - difference - action, over and over again in very different contexts. On top of this base structure, generic structures have been identified that tend to appear regardless of the object of study.

There is always the same 'intelligent' structure that seeks to bring the real state closer to the desired state.

1.10.1. Resistance to Change

When new managers joins a firm, usually with new objectives, they often find that its employees put up resistance to everything they propose – 'They have already tried that. That won't work here. Our customers like it the way it's always been. That proposal is very risky.' In short, the company acts as a system that has managed to survive innumerable economic crises in the past, and as a structure, is capable of neutralising any change, whether from inside or outside, due to the multiple relationships between its members. Each pursues a different objective, yet as a whole they have succeeded in endowing the firm with stability, although that doesn't mean its position is necessarily the most efficient. For this reason it is often wise for new managers to seek the commitment of the general manager for their new objectives as a way of achieving a certain amount of strength and aligning the other elements in the company towards these objectives.

Many systems are not only resistant to new policies designed to improve their state (greater productivity, lower costs, etc.) but also show a persistent tendency to worsen it, despite the efforts to improve the situation. Examples abound in the business world - productivity, market share, quality of service, etc. And on a personal level, we all know somebody with a tendency towards obesity in spite of repeated diets!

1.10.2. Erosion of Objectives

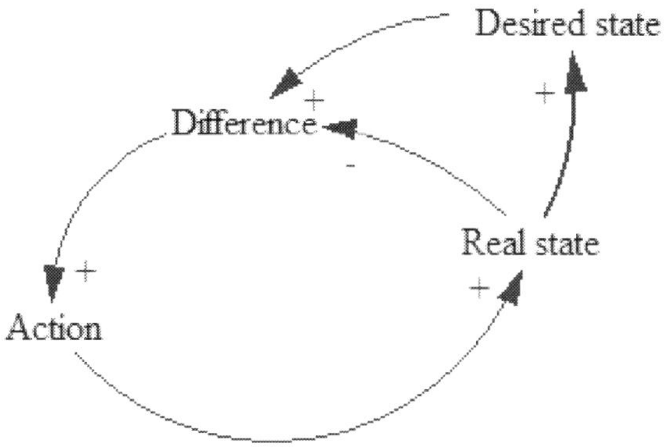

The action required to shift the real state towards the desired state always demands an effort and this effort in turn requires a consumption of time, energy, money, etc.

It is normal for the real state to 'contaminate' the desired state, that is, **for the system to try and avoid the consumption of energy required to take the action.** The desired state is initially reconsidered, since if it coincided with the real state, no action would be necessary. The diagram below shows this 'contamination'.

If contamination occurs, the desired state is modified until it is the same as the real state. The difference is then zero and therefore there is no action to be taken. And so the real state of the system doesn't change.

There are only two ways of avoiding this process:

1.- Find a 'hero' system. That is, convince the system that it doesn't matter how much effort is required to reach the desired state, it just has to be reached. (Personally, I can assure the student that this way of avoiding the contamination process doesn't tend to get results in the 21st century.)

2.- Get an 'external element' to serve as a reference or anchor for the desired state, so that it can't be altered by pressure exerted on the system, and so that the system has no capacity to alter the 'external element'.

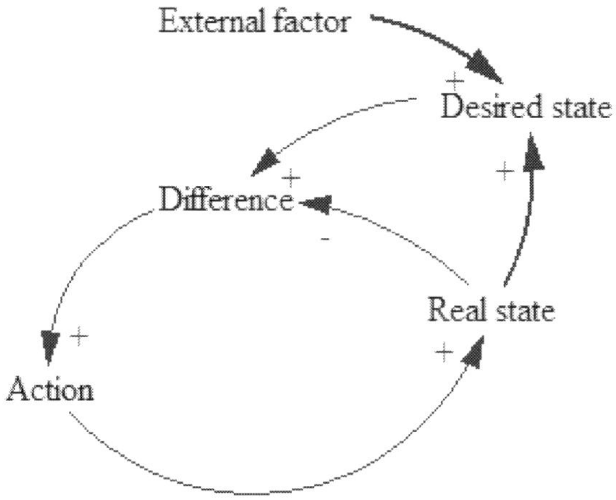

In Spain, when secondary school students consider the possibility of carrying on studying at a public university, they already have a fairly accurate idea of the minimum mark they have to get in their schoolwork and the entrance exams. Their desired state is that minimum entrance mark. It's not negotiable. Their real state tends to be a lower mark than the entrance mark in the early years of secondary education, so they perceive a difference, which leads them to take action (studying harder) in order to get their real state to match the desired state. If students know what it is they want to study, their family doesn't need to push them at all. The system isn't contaminated because the desired state (the minimum entrance mark) isn't alterable.

Later, when they start at university, if you ask, they'll say they want to be a great professional and that they're going to get an average mark of 10 in their degree. With the first exams come the first failures, which make them: 1) study harder than they'd anticipated, and 2) tone down (contaminate) their desired state, from the desired 10 to the non-negotiable minimum of 5.

The structure that brings about this behaviour is based on the idea that the system includes a particular objective (e.g., desired weight) that is compared with reality (real weight), and the discrepancy between these two values triggers an action which is proportional to the size of the gap. This is the usual pattern, seen up to now as a negative loop that tends to gradually pull the system towards its objective if it encounters some discrepancy. However, sometimes the state of the system can condition or modify the desired state, either because the real state is very long-lasting, because the action taken involves a great effort, or indeed for some other reason. The initial goal shifts towards the real state of the system.

This relieves the need to take action as the discrepancy has been reduced, not because the system has approached the objective but because the objective has approached the real state. As a result, the action taken is smaller.

In the case of the weight of obese people, this occurs when they accept that the target weight was too ambitious and that a more realistic target (a higher weight) is better. This argument serves as an excuse to follow a less strict diet. When they see their weight doesn't fall, they reconsider the ultimate target once again, and so on, until they think that actually their real weight is best, at which point they don't have to follow any sort of diet (this would have involved a sacrifice).

There are plenty of examples of this pathology in environmental pollution, law and order, traffic accidents, etc. In all of them, a poor performance becomes the standard in the face of the effort required to do something effective.

A system that bases its objectives on reality and intends no more than to improve on it is permanently drawn towards poor results. A system that gets its targets from outside itself is immune to this type of process.

It may seem paradoxical but if a student is convinced that he must pass all his subjects in July because his father has imposed it as an immovable objective, for whatever family reasons, it will be easier for him than if he himself had made that decision. If it's a personal decision it can be reconsidered when some of the subjects prove to be too difficult. He can accept to leave one or two for next time, which means less studying.

However, if the objective is non-negotiable, this risk doesn't arise, and he has to study as hard as necessary to reach the objective.

Economics provide any number of examples. In Spain, nobody remembers such low rates of inflation as there are now. Any government would be satisfied and would be happy to give up reducing inflation further, as that would mean taking very unpopular measures (a wage freeze for civil servants). If the target for inflation was in the hands of the government, corrective measures would have been less strict in the past and the present, since they would have meant less public spending and therefore lost votes. However, the target for inflation was imposed as a condition for entering the euro zone and as such was beyond the control of the government which pulled out all the stops and took all the unpopular measures they deemed necessary because there was a fixed goal with a deadline and it was non-negotiable.

The obvious antidote to this pathology is to fix absolute objectives for the system that are not based either on the past or the present situation and take corrective measures depending on the difference.

An absolute objective loses credibility if it is raised or lowered and it won't get it back. We see this sometimes when an objective is raised because the initial objective has been reached. When this happens, everybody expects the initial objective to be changed again (but this time downwards) when the results are lower than the initial objective.

1.10.3. Addiction

Sometimes the real state of the system matches the desired state not as the result of action but due to support from outside the system. This support may or may not be permanent and may or may not be of interest, but the net effect is to bring the real state into line with the desired state, resulting in zero difference and therefore action by the system is unnecessary.

This phenomenon occurs when there is an objective that serves as a point of comparison with the state of the system. On the basis of the discrepancy observed, corrective measures are taken proportionally but in this case, the action taken doesn't serve to bring the system's real state closer to its desired one but rather to create the perception that the real system is close to the desired one, whereas in fact this action has no such effect.

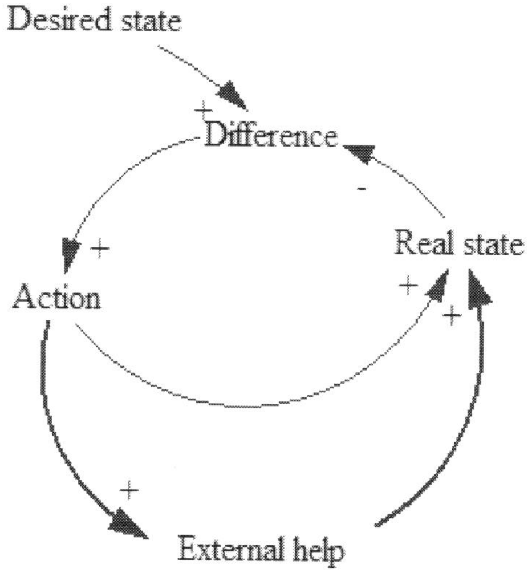

The lack of clear perception of the real system leads to a situation in which the necessary corrective measures aren't taken because the state of the system is perceived as being closer to the objective than it really is.

When the immediate or short-term effect of the action disappears, the problem (i.e., the discrepancy between the real state and the desired one) reappears, often with greater intensity, so the system reapplies some measure that appears to solve the problem whenever the effect of the previous measure starts to fade.

Alcohol, nicotine and caffeine are obvious examples of addictive substances. Another case that springs to mind is the use of pesticides, which eliminate, together with the pest in question, the natural control mechanisms. As a result, the pest will reappear as soon as the effect of the pesticide abates, but this time without any natural control.

In cases of addictive systems, it is difficult to find suitable policies, since the action taken offers apparent results in the short term, but once the process is rolling, it is difficult to stop. Obviously, the best approach is to be aware of these types of processes, in other words, to be wary of using measures that attack the symptoms but make the system worse when they are relaxed.

Once the addictive process has been started, you have to expect at least short-term difficulties if you plan to stop this process, be it physical pain for somebody who takes an addictive drug, rising petrol prices on inclusion of the associated environmental costs, or more pests and lower-quality food until such time as natural predators return.

Sometimes it is advisable to wean yourself off an addiction gradually. But it is always less costly to avoid the addictive process in the first place than to stop it later.

1.10.4. Shifting the Burden to the External Factor

As they get older and spend more and more time reading, some people gradually get poorer eyesight. In the end they can't read what is written on a blackboard, and can't renew their driving licence, so they get glasses or contact lenses.

Then, in one year their eyesight worsens as much as it did in the previous 30 years. So their glasses become a necessity not only to see at a distance but also to read a document. Apparently this happens because for years, the muscles around the eyes have been straining to compensate their poor vision and when this effort is no longer necessary, they cease to act and end up losing this ability totally. Before long, they need stronger lenses.

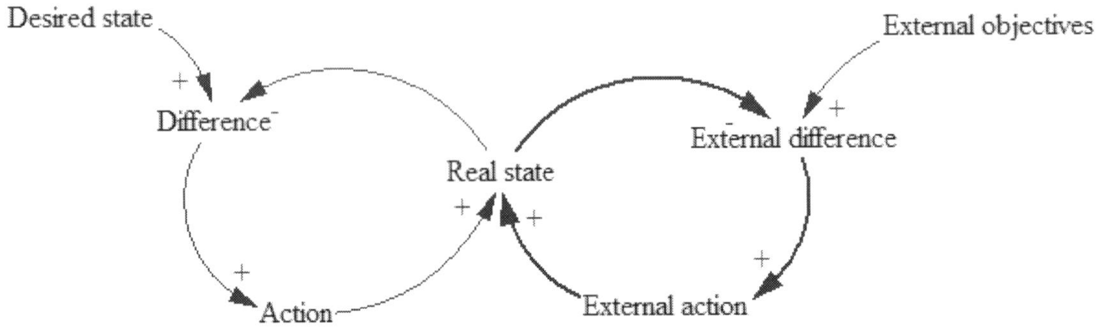

This is a classic example of shifting the burden to the external factor. In this sort of system an external force keeps the system in the desired situation. A well-intentioned, benevolent and very effective force decides to help us to get the system where we want it. This new mechanism works very well.

But with this process, through the active destruction of the impediments that redirected the system towards the desired position, or simply through atrophy, the original forces that worked to correct the position of the system are weakened. When the system moves away from the desired position the external factor makes an extra effort, which weakens the original forces still further. In the end the original system adopts a position of total dependence on the external factor, as its original corrective forces have disappeared completely and in most cases, irreversibly.

It is easy and fun to find other examples of shifting the burden to the external factor. Here's the start of a possible list.

Problem	System	External factor
Difficulty of calculating	Mental skill	Calculators
Care of the elderly	Family	Social Security
Lack of pasture in winter	Wild deer	Provision of fodder
Infections	Human body	Antibiotics

Seeking the aid of an external factor to get the system where we want it to be isn't in itself a bad thing. Usually, it is beneficial and enables the system to better tackle objectives. Yet the dynamics of the system can be problematical for two reasons. Firstly, the external factor that intervenes doesn't tend to perceive the consequences of its help on the elements of the system, particularly on those that performed the same task as itself. Secondly, the system doesn't stop to think that this help is temporary; they lose their long-term perspective and so become more vulnerable and dependent on the external factor.

The withdrawal of aid from a system that is being helped, whether it is the human body, a particular area of ecological value or a human community, doesn't tend to be easy and is often simply impossible. This process of withdrawing help without harming the system must be based on identifying the internal elements of the system that in their original state, took care of correcting the problem, strengthening these mechanisms and, as they begin to do their job, gradually withdrawing the help.

1.10.5. Short and Long-Term Effects

A rational analysis of the problem at hand based on our capacity for synthesis and our ability to imagine things seems to be a bad guide to finding the key factors. We generally pay attention to the components of the system and their behaviour in the short term, all on the basis of incomplete information. Consequently, firms reduce their stocks of finished products when sales are seen to slump, the government extends its tax reductions for small farmers, and policies are introduced to encourage firms to replace their machinery instead of maintaining it properly. They are all very reasonable policies. But there's still something inside us that just might make us realise that our customers' dissatisfaction with our long delivery schedules, or farmers' permanent concern with increasing their output, or the idea of replacing a machine that's productive, all means something, but we haven't given it the right interpretation.

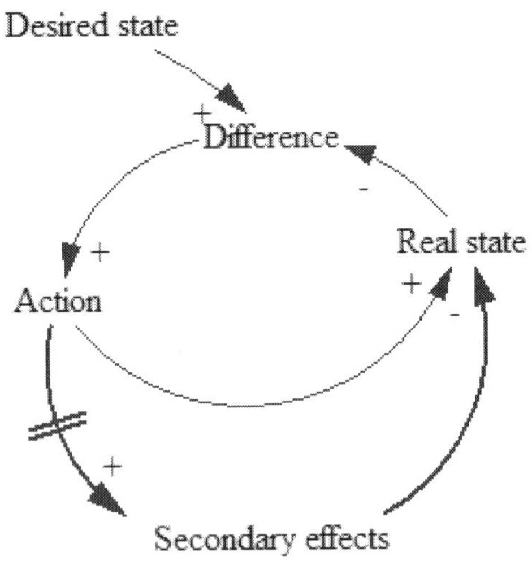

Finally, I'd like to say that in my opinion, we have the capacity to understand not only simple systems but complex ones too, and to find the key factors. What we don't appear to have is the capacity to articulate the arguments to convince others or even ourselves that what we're perceiving is right. We expect the solution to be closely related to the symptom; we expect long-term profit to start with short-term profit, or a strategy that's satisfactory for all the agents involved. Yet we know complex systems don't behave that way. So something inside us still insists somehow that maybe that simple, effective solution isn't the best. And then we carry on proposing policies that can't work, denying ourselves other simpler and more effective ones that could.

We try to compete instead of cooperating, and we try to reach the limits of the environment's capacity instead of admitting that we have already gone too far. The results are famine, war, pollution and depression. And right in front of us, within reach of our capacity for understanding, stand balance between countries, peace, equality and sustainable development.

1.11. World Models

For most people, especially engineers, world models conjure up enormous computers with huge amounts of information about any topic you can imagine. The first well-documented world simulation model was relatively simple. It was published in 1971 by Jay Forrester. Since then, the proliferation of world models has been massive. These models have been created in different parts of the world, using different techniques with very diverse goals in mind. Even so, making these models is clearly limited to the amount of information their designers can include. Each of these models uses just a minute amount of the information that is available about the world. Most of them focus on economic, demographic and agricultural production factors.

A small number deals with environmental problems and the use of natural resources. Almost none refer to war, politics, new ideas or natural disasters. Most assume technology doesn't change or changes automatically and exponentially without a price tag, allowing more and more production at an ever decreasing cost. Some models show the world as a simple unit and a continuous whole, while others divide it into 10 or 15 regions, and still others divide it into a hundred nations. Some use a time span to the year 2100 and others just project a few years into the future. Some, especially the former, raise heated discussions and others have been explicitly instructed to refute the conclusions of previous ones.

We are going to see some of the characteristics of these world models. They are frequently misunderstood or interpreted by an audience that is either quite gullible or quite sceptical about computer models.

1. There is quite an assortment of models. They have been made by people with defined political and cultural beliefs and they, therefore, usually tend to be quite slanted in different ways. There is no such thing as a model in the socio-economic, ecological or other realm which is 'objective'; in other words, neutral, and therefore, not influenced by those who have made it.

2. The models, especially those of the world, are tremendously complicated in what they refer to (detailed structures by age of the population, varied economic sectors, complex sketches about business and extensive classification of people by income) and amazingly simplistic in what they omit (weapons, age of infrastructures, motives, social norms, political structures, the starting point and destination of the flow of raw material, etc.).

3. A model is a list of mathematical equations that explicitly sums up a view of the world. It is bolstered by statistical parameters and logical consistency that is able to produce phrases such as: 'If all of these hypotheses are correct and there aren't any others which should be considered and they are still valid in the future, then the logical results are…' (The reader of this text can copy this just in case they need it someday).

The importance of these models perhaps resides not so much in them but in the fact that underlying them is the effort of people from various continents and ideologies interpreting the world from different points of view, given the limitations of traditional work tools. They have all had to observe the world as a whole and pondered the long-term implications of the ties that connect population, capital and the production of goods and services that link all nations. They have immersed themselves in global statistics and constructed a model that has portrayed the global situation generally and relatively consistently - each buyer needs a salesperson; each birth must eventually be linked to a death. Once capital is set, its original use for a tractor factory cannot be changed for a hospital.

Consequently, although some topics and specific details are given a different perspective, there are some common conclusions and sensations obtained when observing the world as a closed system in all the models. The model creators, who are generally hostile and critical toward each other, are surprised that their basic conclusions coincide.

The following points sum up these basic conclusions:

1. There is no physical or technical reason why physical or technical needs cannot be met in a reasonably predictable future. These needs are not met now because of political and social structures, as well as values, norms and world views, not because of physical limitations.

2. The population and consumption of physical resources cannot grow indefinitely on Earth.

3. Clearly and simply, information is incomplete concerning the degree to which the Earth can absorb and meet the needs of a growing population, as well as the capital and

residues this population generate. There is a lot of partial information, which optimists interpret optimistically and pessimists pessimistically.

4. The continuation of current national policies in the future will not bring us closer to a desirable state where human needs will be better met. The outcome will be an increase in the gap between rich and poor, problems with available natural resources, environmental destruction and worsened economic conditions.

5. In spite of these difficulties, current trends will not necessarily continue in the future. The world can begin a transition period in which it can direct itself toward a different future, not only quantitatively but qualitatively.

6. The exact nature of that future condition and whether it will be better or worse than the current one is not predetermined. It is the result of decisions and changes made (or not made) now.

7. When the problems are obvious to everyone, it is too late to do anything. That is why policies to change social processes should be implemented in previous stages. That is how they will have a major impact with the least consumption of resources.

8. Although technological progress is to be expected and will undoubtedly (?) be beneficial, no strictly technological change alone will lead to a better future. Social, economic and political restructuring leads to a better future in a more effective way.

9. The interdependence of people and nations through time and space is greater than what is commonly thought. Events that occur in one part of the world can have medium-term consequences that are impossible to know by intuition and probably impossible to predict, perhaps not even partially, with computer models.

10. Due to these interdependencies, simple actions that try to obtain a specific effect can be entirely counterproductive. Decisions must be made within the broadest context of analysis of feasible areas of knowledge.

11. A focus on cooperation to attain individual or nationwide goals is frequently more beneficial in the medium-term for all the parties involved than focuses or strategies based on confrontation.

12. Many plans, programs and agreements, particularly those which are international in scope, are based on suppositions about the state of the world that are inconsistent with physical reality. Too much time and effort is spent in designing and debating policies that are, in fact, simply impossible.

For the people who have devoted time and effort in creating these world models, the common conclusions are not, after all, surprising. They have acquired an intuitive view of how the complex system in which we live functions. And the points above are no more than the key to how a complex system works.

The final conclusion of global models is quite simple: The world is a complex, interconnected, and finite system with relationships among the ecological, societal, psychological and economic scopes. We tend to act as if this were not so, as if it were divisible, separable, simple and infinite. Our persistent problems directly originate from this lack of perception. No one wants or works to produce hunger, poverty, pollution or the extinction of species. Very few people are in favour of arms use, terrorism, alcoholism or inflation. These phenomena are produced by the current system as a whole, in spite of the effort made against them. In some cases, the policies used solve problems, but many problems historically defy any solution. Perhaps we need a new perspective to solve these problems.

1.12. Control Questionnaire

After reading this paragraph, it is advisable that the reader answer the following questionnaire to look at what is understood and what must be reviewed before continuing.

a. Give some examples of **SYSTEM**. Remember the definition of system as a set of interrelated elements such that one element affects the behaviour of the whole set. For example: a city.

Answer:

b. Name some **ELEMENTS** of the system. For example: persons, cars, pollution, streets, etc. Incorrect elements would be: the government, the city, Barcelona, colour, asphalt, etc. They are valid as elements of the system if we can notice when the element increases or decreases, improves or worsens, etc.

Answer:

c. Name the **UNITS** of the elements. For example: Persons: number of persons, Pollution: number of particles in suspension/m3, Streets: m2.

Answer:

d. Give some examples of **CAUSAL RELATIONSHIPS** indicating whether they are positive or negative. For example: The more rainfall, the more people with umbrellas (positive).

Answer:

e. Give some examples of **LOOPS**, which are formed where there is a 'closed circuit' between two or more elements of a system. For example: The warmer you are the fewer clothes you wear (negative) and the more clothes you wear, the warmer you are (positive).

Answer:

f. Give a system example that has **JUST ONE GOAL**, indicating the goal. For example: a mower; goal: to cut grass.

Answer:

g. Give a system example that has **SEVERAL GOALS**. For example: a company, where the businessperson has the following objectives: the most profit, increasing the number of clients and increasing product quality.

Answer:

h. Give a system example that has **GOAL EROSION**. Indicate some ways to avoid contamination (in other words, erosion) of the goal by the real situation by securing it to an external element. For example: students usually have an erosion of their initial goal of getting excellent grades. In this case, an external element that can prevent this erosion is the grades of a 'rival' student.

Answer:

i. Give a system example that shows **RESISTENCE TO CHANGE**. For example: We prefer to wear our old shoes because they are more comfortable than new ones.

Answer:

j. Give a system example indicating the **LIMITING ELEMENT** that prevents an action. For example, fire does not spread because there is no more wood left; the youngster does not study because there is no more paper; the car will stop when it runs out of petrol.

Answer:

k. Give system examples and some of its **KEY ELEMENTS**. For example, the amount of salt in food is a key factor for it to be edible since, if we put in too much salt, no one will be able to eat the food.

Answer:

2. STOCK AND FLOW DIAGRAM SFD

2.1. Stock and Flow Diagram

The Stock and Flow Diagram (SFD), also known as the Forrester Diagram, is a characteristic diagram in System Dynamics. It is a translation of the causal diagram to a terminology that makes it easy to write equations on a computer. Basically it is a reclassification of elements.

There are no strict rules about how to make this transformation but there is a generally accepted way of tackling the process. The steps to follow are:

1. Take a mental snapshot of the system and the elements included in it (people, sq kms, litres, animals, etc.). These are its **stocks.**
2. Find or create the elements that constitute the 'stock variation' (persons/day, litres/hour, etc.). These are the **flows.**
3. The remaining elements are the **auxiliary variables**.

This will do as a general guide. Later on you can make alterations and those stocks that are going to remain constant (e.g., sq m of a room) can be defined as an auxiliary variable instead of stock, which makes the diagram simpler.

That is the whole procedure. Now let us take a closer look at how these elements are represented.

The 'stocks' are those elements that show the state of the model at each instant - they present an accumulation and only vary as a function of other elements called 'flows'. The 'clouds' in a flow diagram are stocks with inexhaustible content. Stocks are represented by a rectangle.

e.g.: people, sq km, litres,...

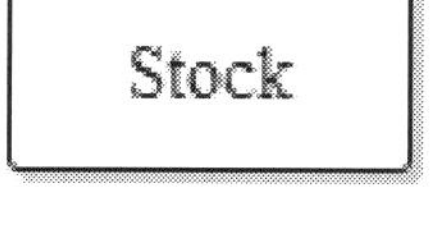

Flows are elements that can be defined as time functions. They can be regarded as representing decisions taken in the system which determine the variations of the stocks.

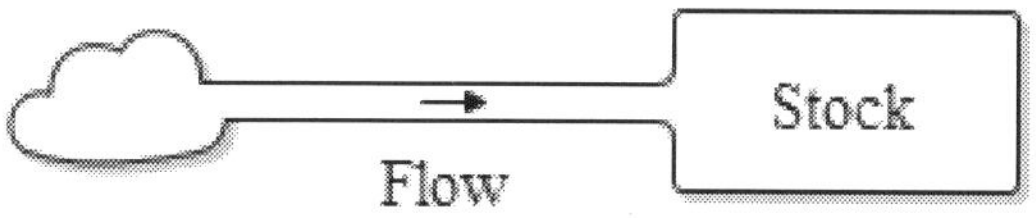

e.g.: people/day, sq km/year,...

'Auxiliary variables' and 'constants' are parameters and allow a better visualisation of the aspects that affect the behaviour of flows. e.g.: density, weight, ...

The physical magnitudes between flows and stocks are transmitted through what are known as 'material channels'. There are also 'information channels' which, as their name indicates, convey information that due to its nature, is not stored.

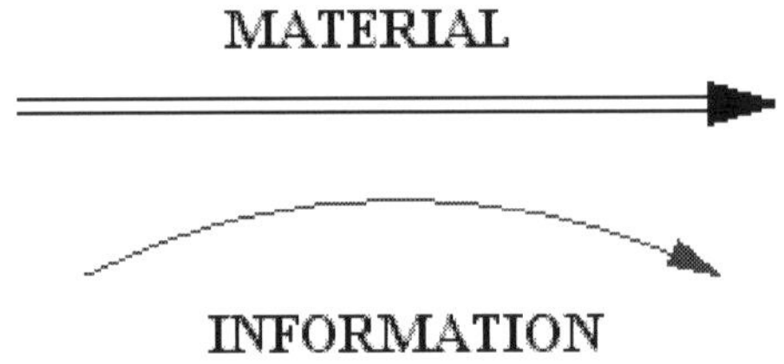

Finally, we should define the 'delays', which simulate the time taken to convey material or information. In socio-economic systems, there are often delays in the transmission of information and material and they have an important effect on the behaviour of the system.

For delays in material we have the functions DELAY1 and SMOOTH. For information delays we use DELAY3 and SMOOTH3. First-order delays with respect to a step input will produce an exponentially asymptotic curve, whereas a third-order delay leads to a sigmoidal curve. To some extent, information delays act as smoothing filters for the input variable.

2.2. Computer Simulation

In this step, we write the concise instructions (i.e., equations) for the computer to interpret our vision of the system. e.g.:

Food Generation Rate = (Food Capacity-Food)/Food Regeneration Time

There are a number of PC-compatible software packages available on the market that require no more than user-level knowledge and meet the needs of users (students, professionals, etc.) quite well. The most frequently used languages and programs are (in alphabetical order) Dynamo, Ithink, Powersim, Stella and Vensim.

Using these languages enables you to avoid having to formulate non-linear relationships mathematically by building Tables (see 'Case studies' below).

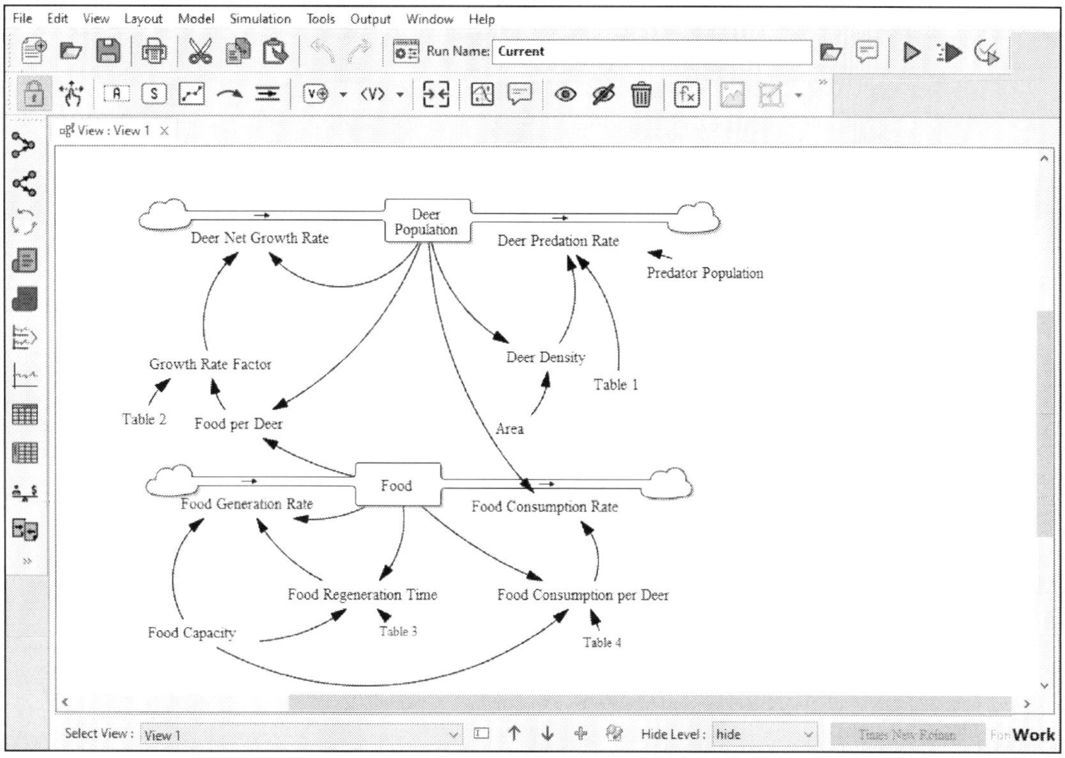

In this step, you have to give numerical values to system variables, functions and tables. This is one of the many ways in which System Dynamics differs from most traditional modelling methods. For example, in econometrics, most of the research effort is dedicated to determining the exact value of the parameters that characterise the system under study.

However, in System Dynamics the parameters are only calculated to the necessary degree of exactitude for the model to fulfil its purpose. Given that social systems tend to be fairly sensitive to changes in the values of the parameters, it isn't necessary to dedicate a great deal of time to calculating them exactly.

We can work from a set of approximate values with a view to obtaining an initial idea of the behaviour of the model. Later on we will use sensitivity analyses to identify the relatively small number of parameters whose values have a significant effect on the behaviour of the model or its responses to different policies. In this way we'll discover which parameters are to be calculated exactly.

The advantages this method provides as regards savings in time and effort are obvious.

2.3. Behaviour of the Model

Once the equations have been fed into the computer we can obtain an output in the form of the evolution over time of the parameters that we have given it.

We can also compare the behaviour of the model with reality, as the output provided by the model enables us to verify the truth of our hypotheses. On the basis of the difference between the model and reality, we have to reconsider our initial hypotheses and make the necessary adjustments to the model.

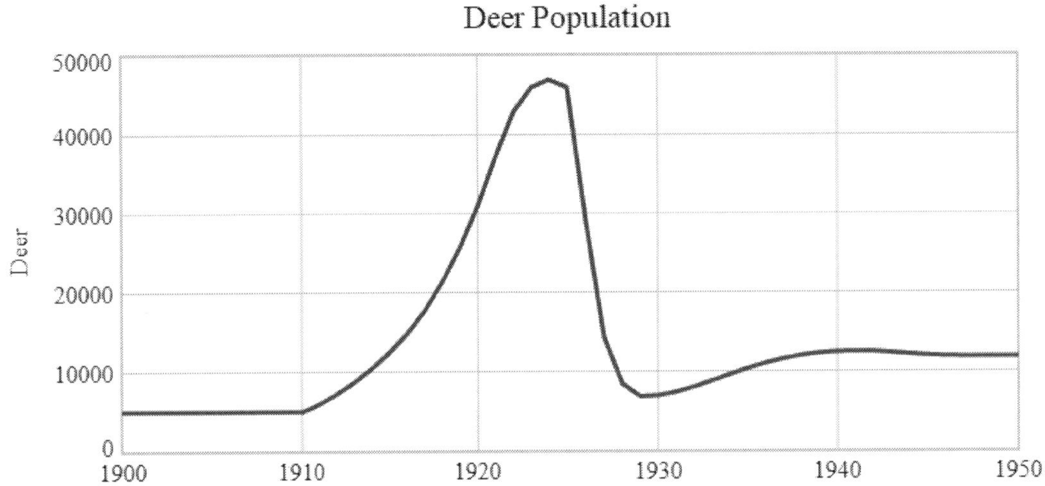

A dynamic system possesses a number of aspects that can be evaluated, including:

- Its ability to reproduce the historical data of the system being modelled, under normal and extreme conditions;
- The acceptability of the assumptions made when defining the model; and
- The plausibility of the numerical values adopted for the parameters of the model.

Of course, the first criterion will only be of importance when the others are also verified, as there will be an infinite number of models that are capable of suitably reproducing the historical data of the system without being related to the mechanisms that make up the structure of that system.

Our judgement as to how a model fulfils the above criteria should not be restricted to considering the quantitative information available, since most of the relevant knowledge about social systems is qualitative and held by experts in the field we are considering.

It is important to bear in mind that a model that passes all the evaluation tests is neither an incontestable description of reality nor the only model.

Working from a given set of initial conditions, we determine the evolution of each of the model's variables over the time horizon chosen for the evaluation and record this evolution by means of graphs. By comparing these graphs with their corresponding historical data, we can find out whether the main characteristics of the real behaviour are fulfilled.

This evaluation will then serve as the basis for perfecting the model, by correcting the defects that have been detected and introducing any improvements that are considered appropriate. This reformulated model is then used to carry out a new simulation followed by a further analysis and evaluation. This process is repeated until the model is considered to satisfy the set objectives to a sufficient degree, or until the result of the changes we could introduce would not be worth the effort.

As we said earlier, the first simulation is done with a set of parameter values and initial conditions that are not necessarily exact. Sometimes, in the absence of data, these values will be based on the opinions of experts in the field of the system being studied. As a result, they won't be very exact but often they will nonetheless be useful.

2.4. Analysis of the System

Finally, once the model yields an output that is coherent with the past and the present situation, we can simulate the impact of policies or decisions that will lead us to the solution of the problem.

It is also possible to locate key factors (leverage points), by means of sensitivity analyses.

The final model should be relatively simple, so we will have to make the necessary additions and any possible simplifications in order to make the resulting model understandable for the receiver and manageable for the user. On top of the great deal of time and effort required to obtain a complex model, it can be just as complicated as the reality it represents, which hinders communication and means that the model is of very little help in solving the problem at hand.

It should be stressed that although a System Dynamics model has the appearance of a complex set of mathematical equations that might look like it could make a perfect prediction of the future, this is not what we are after. What we are trying to do is increase our knowledge of the system we are studying and create a useful tool for analysing policies which should be judged on the basis of the overall trends they generate.

This difference between the appearance of the model and its intention tends to deceive users who often think they're dealing with a different sort of tool. This possibility recedes when the model is regarded as what it really is - the explicit representation of a mental model. This perception is easier to grasp if the creator of the model keeps the use of inaccessible symbols to a minimum, clarifies the mathematical equations with graphics that are easily understood by non-specialists, and accompanies them with a consistent justification, in other words, if transparency is one of the premises for the creation of the model.

Moreover, the limitations of the model must be made clear, particularly when it is used to predict the impact that different policies would have on the system. We must not forget that the evolution we obtain of the behaviour of the model is based on certain hypotheses regarding the present structure of the system and the circumstances that make up its environment. Therefore, the conclusions drawn will be valid only insofar as the fundamental structure of the system or the outside circumstances that might influence its sensitive parts remain unchanged, and only if we can assume that the model is a correct portrayal of the present.

Consequently, if a model is to continue to be useful over time it has to be periodically checked. In this way we can introduce any significant alterations that may have arisen in the system represented by the model.

Another factor to be borne in mind is the dose of subjectivity involved in the use of models. In the particular case of System Dynamics, the creation of a causal diagram representing the various interactions between variables is one of the most subjective stages, but not the only one. Subjectivity can also exist when quantifying and using the available data, when interpreting the results, and so on.

The intrinsic subjectivity of the modelling process isn't really a negative factor, since it is impossible to study a social system totally objectively.

2.5. Weaknesses of Models

As early as 1971, Van der Gritten pointed out a number of weak points in the models created using this methodology. These criticisms can be summed up as follows:

- Empirical content is lacking. In other words, the results of the model ought to be comparable with real data.
- Functional relationships include ideas and criteria that do not always have the support of theory, evidence or experience.
- The results are sensitive to variations in certain inputs and parameters.
- Many models are deterministic, in the sense that they don't include possible reactions to predicted future changes.

Although we could no doubt add one or two more, basically these are the main aspects of this type of model that are criticised.

Obviously, these methods will always attract criticism due to the nature of the problems that are tackled, unless first of all it is made clear that System Dynamics models are not predictive models, and that they're not designed to find exact values but comparative ones, that is, they enable us to compare different alternative policies on the basis of the scenario to which they lead.

2.6. Five of the Author's Experiences

System Dynamics is used in company operations, such as **Project Management**. The traditional Project Management tools make it possible to organize tasks that are to be done in a linear fashion but run into problems managing unforeseen events, sudden changes in planning or errors in tasks that have already been carried out.

System Dynamics does not try to substitute the classical PERT or Project Management in ordering tasks that make up a project but it can help to understand and prevent the usual problems that appear in project implementation, such as delivery delays, poor product quality or the increase in actual costs with respect to the budget.

System Dynamics is used not only in the design of large infrastructure projects, such as dams and roads, but in much smaller business project management, such as the change of factory location or the launching of a new product.

In the area of **Production Management**, System Dynamics allows us to better understand the cause of some problems such as production delays or the fluctuation in the number of pieces there are in the finished product warehouse and transparently simulate the impact of various ways of organizing production.

In this field, System Dynamics provides a dynamic view of the aspects that are involved in production and makes it possible to carry out simulations about the model to identify those key aspects.

Production Management models are apparently very complex because it is necessary to integrate many different factors, but actually the relationships that exist among the elements obey well-established rules and norms, which make the job of creating simulation models easy.

In **Process Management,** a System Dynamics model allows us to simulate the structure of each process and carry out sensitivity analyses of its various phases. The start-up of a new process frequently has many unforeseen events which are difficult to sort out when it is already running because any change influences other parts of the process. A Dynamic Systems simulation model helps to identify the impact of small random variations in the general evolution of the process. It can also help identify delayed areas in the process, so that the whole production process is organized taking them into account without creating false alarms because of them.

In the area of **Human Resource Management**, System Dynamics is also a very valuable contributor, since it makes it possible to analyse the influence in corporate development of non-quantifiable factors such as motivation, corporate goal policy or the employee training level in quite concrete problem analysis such as, according to my own experience, the number of work accidents.

This tool makes it possible to work with elements that cannot be quantified and organize them in a simulation model that allows corporate management to obtain conclusions that have a practical application.

In **Research and Knowledge Management** for new product development, System Dynamics simulation models are a much more efficient and economic tool than carrying out lab experiments. They should, therefore, always be a mandatory prior step to actual experiments. There are two important reasons for this. One is the expense, since any error in the lab experiment can be quite economically costly to solve. There are also the people and the facility's physical safety. Consequently, actual experiments must be used to confirm simulation model results.

Lastly, we can briefly mention the applications in **Business Administration and Management**, which range from real estate or financial investment research, organizational change management, knowledge management and new technology implementation to the design of quality control policies.

System Dynamics is also applied to the enhancement of management skills through Business Games, such as the well-known Beer Game, where participants can understand the significance of delays in the transmission of information and how our own mental model influences the perception of a situation.

Models created with System Dynamics are widely applicable at companies. Five tasks developed by the book's author are briefly described below.

Due to the limitations companies impose on the people who carry out these types of studies, the name of the company and key details of the work carried out are omitted.

A. A Gas Company's Distribution Management Model

A company involved in gas exploration, importation, distribution and exportation needed a management model to integrate all the aspects of its activity taking costs and relative product prices, as well as government norms, into account.

The management model had to identify policies that optimized economic results, evaluated the impact of government restrictions and **carry out simulations to determine policies to be implemented when there were sudden changes in supply sources.**

This work allowed the planning department's personnel to create simulation models with System Dynamics and make a preliminary version of the model that dealt with this specific purpose.

Specifically, the model made it possible to analyse the supply from its own exploration and that of imports, taking limitations, such as transportation capacity, into account. In addition, the model made it possible to evaluate demand with the details of its components, such as families, companies, transportation sector and export, taking into account existing prices in each of those markets and their future trends, as well as the influence of government norms that prioritize the distribution to specific sectors. Lastly, the model proved to be an excellent basis to carry out simulations about possible cuts in imports and determine their consequences and best alternatives in advance.

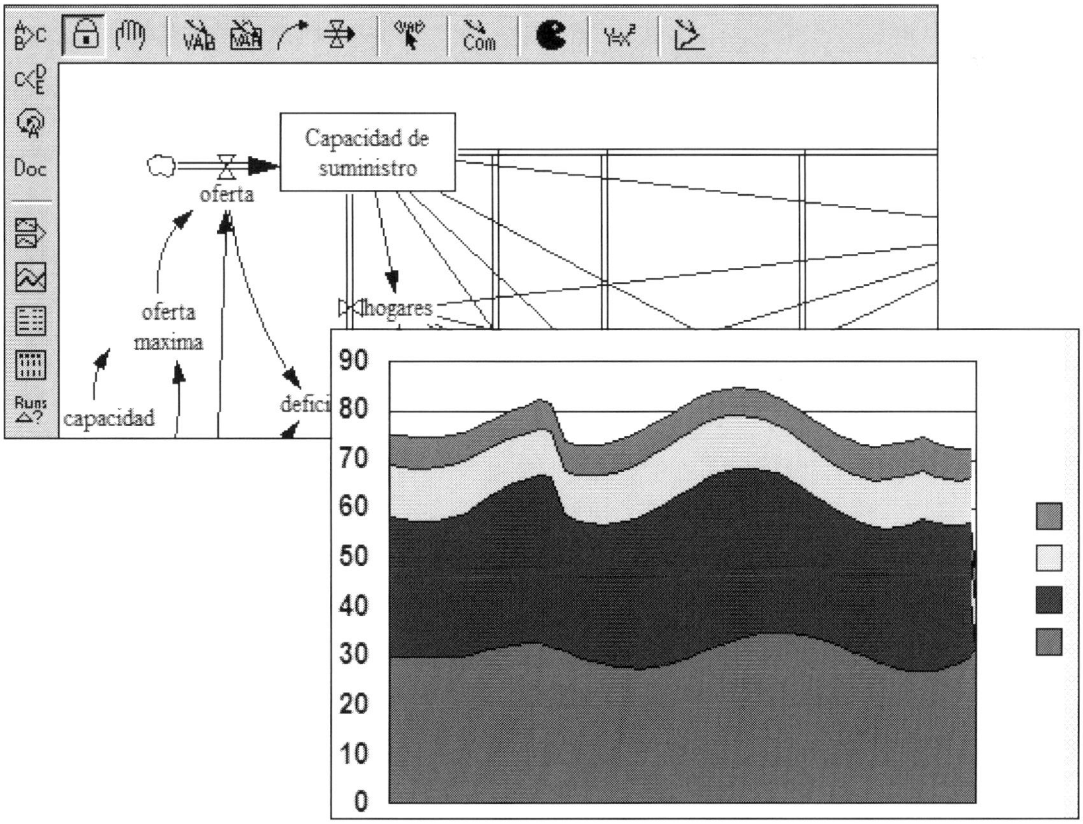

B. A Petrochemical Company's Analysis and Price Forecasting Model

The sales price of one of the main products of a multinational chemical company is determined to a great extent by the product's international pricing. **This pricing has had significant fluctuations in the last few years.** Since the product can only be stored for a short time, the company must try to regulate its production to obtain maximum production when prices are high and decrease production when prices are low, since there can even be losses during those periods.

For production planning, the company uses the services of an international consulting firm that carries out price forecasts. Nonetheless, the company wants to have its own simulation model to help it understand the international dynamics of price creation in order to plan its production more accurately.

Price fluctuations occur because buyers and sellers of the product plan production based on the price forecasts they receive. Buyers advance their purchases when they believe that prices will rise, causing a price hike and sellers advance their sales with discounts when they believe that prices will fall, which results in a price decrease. Therefore, any small fleeting change in price is amplified in ever-increasing fluctuations.

The model the company created preliminarily after a week of training its employees proved to be more reliable in historic price forecasting than those received from its regular external consultant.

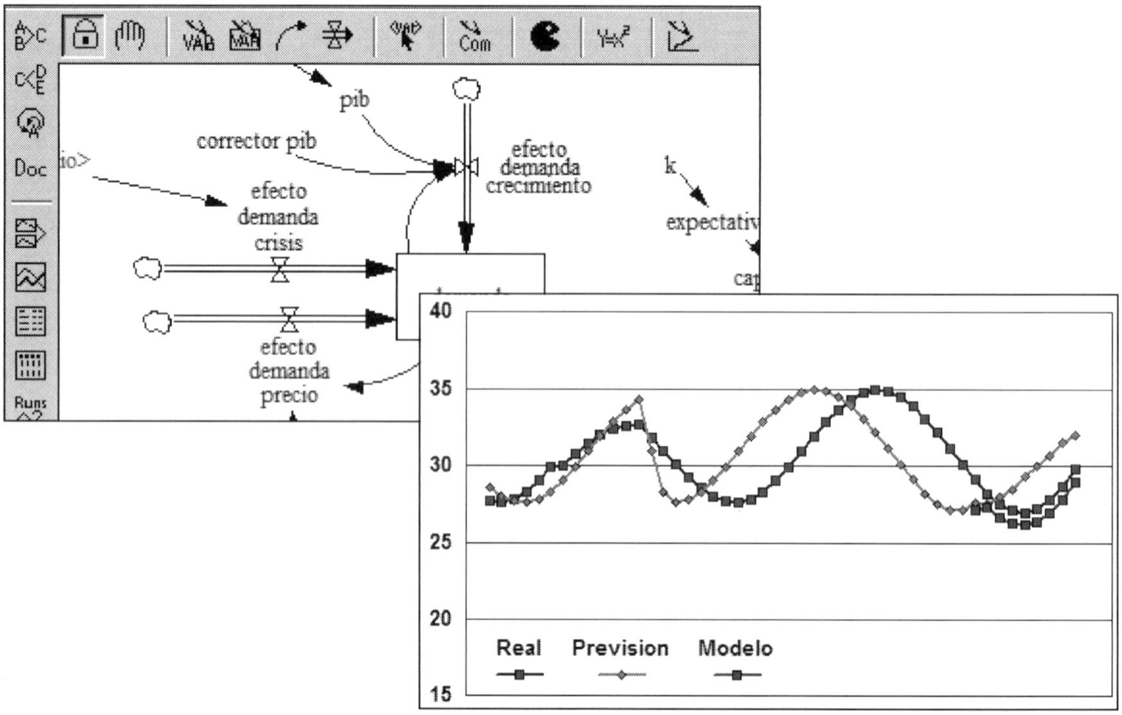

C. An Auditing Company's Human Resource Organization Model

The auditing company had noticed an ongoing significant difference between the number of clients and auditors available to attend them in the last few years, so that **there were periods in which clients could not be attended due to a lack of auditors and other periods in which auditors were idle due to a lack of clients.**

Company employees are divided into junior and senior auditors. Mandatory specialized auditor training requires a long trial period in which junior auditors cannot directly attend clients and only collaborate with senior auditors.

Upon completing a one-week training course about System Thinking concepts with the management team, it was able to compile the essential factors that had an influence on the problem at hand in a single causal diagram, according to the contributions each of them made.

The causal diagram served to facilitate a dialogue about the problem to explicitly illustrate the limitations of the employee training process, analyse the relationships between the departments that intervened, and raise the possibility of new hiring and personnel selection policies. In addition, some classical System Thinking behaviour patterns were identified that explained the observed behaviour. Finally, the key aspects that corporate policies had to focus on to solve the problem were identified.

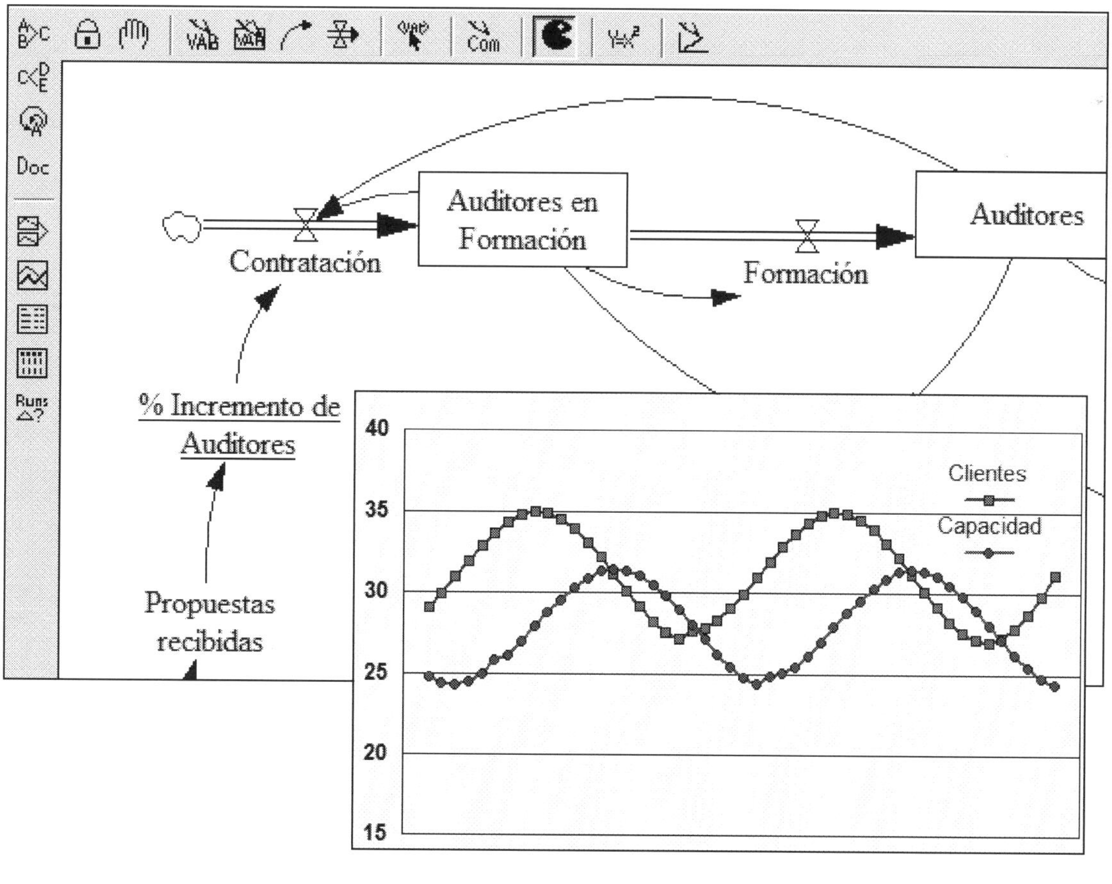

D. Cost Policy Management Model

The company was proud of the increasing reductions in cost it had attained in constructing new electric power plants. This was the result of an intense cost analysis policy and optimization of the construction process. Nonetheless, **the total construction cost of the new electric power plants had increased due to indemnity payments** to people affected by their construction. There is an exponential increase in this cost, so much so that the indemnity cost becomes greater than the cost of constructing the last power plants.

Because most of the elements that affect the indemnity cost are not quantifiable, the management team proposed studying a policy that would stabilize or reduce the indemnity costs based on System Thinking.

To carry out this analysis, a large group of middle managers from a variety of areas in the company took a training course on System Thinking and System Dynamics. Once the course was completed, the problem of the exponential increase of indemnity costs was analysed. Although the issue was complex, a consensus was finally reached about one key factor in which company policies had to coincide to control indemnity costs.

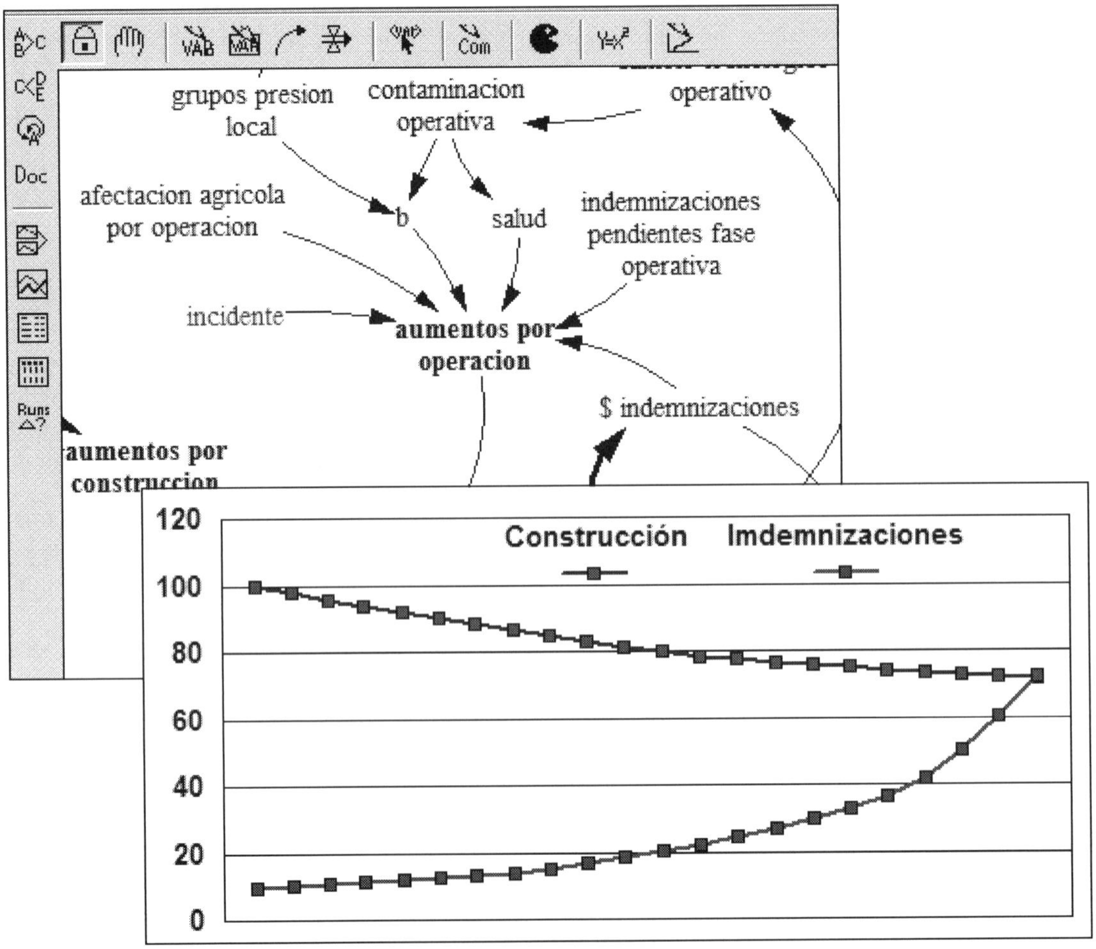

E. Work Accident Reduction Management Model

The company had carried out an intense work accident reduction campaign in the last few years that, together with an economic incentive policy, made it possible to drastically reduce serious accidents within a short time. Nevertheless, the indicator's quarterly follow-up that evaluates this showed a change in trend during the last year, since **a progressive increase of serious accidents had occurred.**

The indicator or ratio that rates serious work accidents considers situations that have required hospital care for the worker referenced at one million hours/man hours. This indicator has a quarterly follow-up and had gone from a value of 5 three years ago to a value of 0.2 the previous year. Since then, there had been a gradual increase to a value of 0.8 when the study was carried out. This increase had disconcerted the corporate managers who had followed the campaign, which included employee training and economic incentives.

An intensive System Thinking and System Dynamics training course was carried out for a large group of technical and management staff, since many factors are involved in the problem to be considered and it was necessary to get their points of view. Afterwards, the construction of a causal diagram that involved the main aspects of the problem was completed. In this process, they were able to incorporate the contributions of the implicated departments into one sketch, finally obtaining a clear vision of the structural causes that had produced the recent increase in the number of accidents and the essential factors to take into consideration in defining future work accident policies.

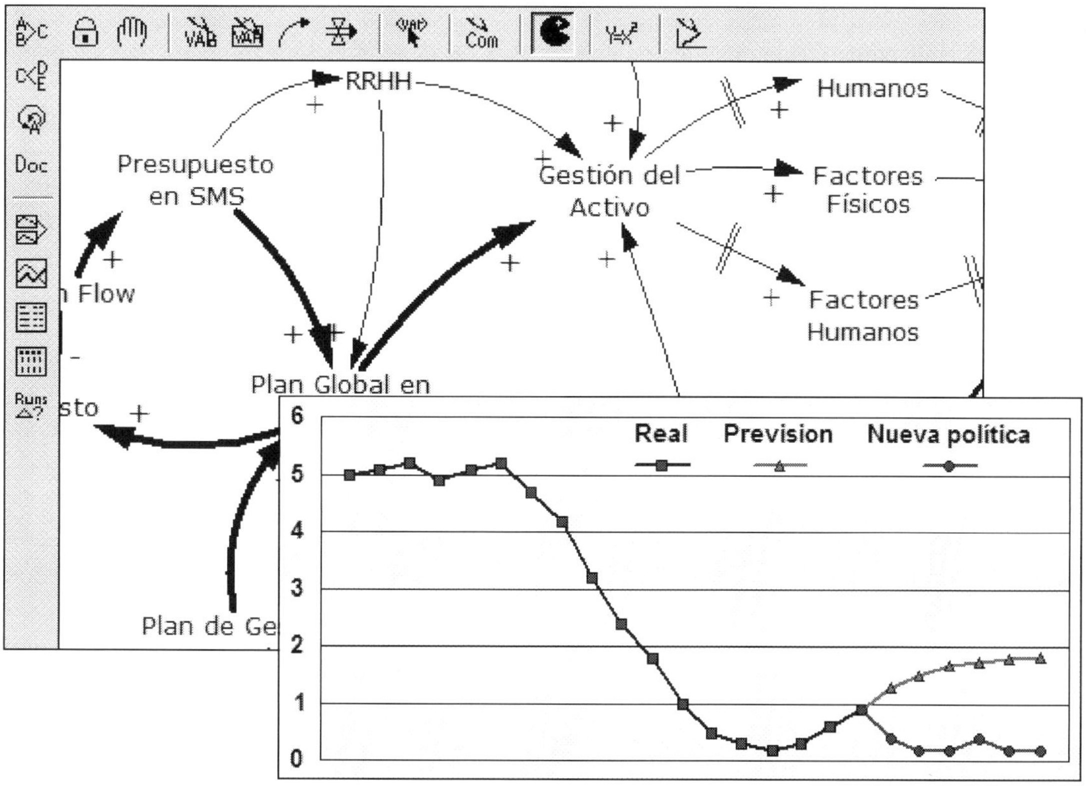

2.7. Control Questionnaire

a. Give some examples of **STOCKS**. These are elements where there is an accumulation, such as: persons, dogs, books, etc.

Answer:

b. Give some examples of **FLOWS**. These are those elements that make Stocks vary. Their units always have a time component, such as litres/year, persons/hour, etc. Note: Flows are not ratios, such as persons/hectare or price/kilo.

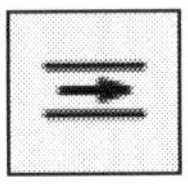

Answer:

c. Give some examples of **AUXILIARY VARIABLES**, which are all the elements that are not Stocks or Flows, such as the price of a kilo of bread or animals per acre.

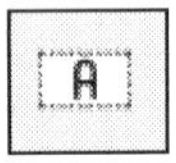

Answer:

3. PRACTICAL EXERCISES

Index to locate items

(number of the **exercises** in which you can find the item)

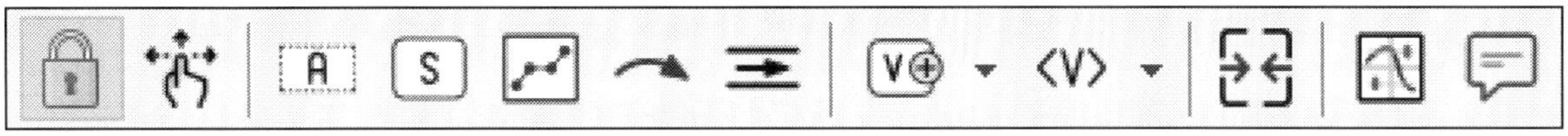

Simulate Setup: 8
SyntheSim: 21
Integration method: 26

Outputs
Output graphs: 1, 2, 8, 10, 14, 23, 26
Output tables: 9, 22
Causes-strip tool: 1
Line Markers: 17
X-Y graph: 26

 # Exercise

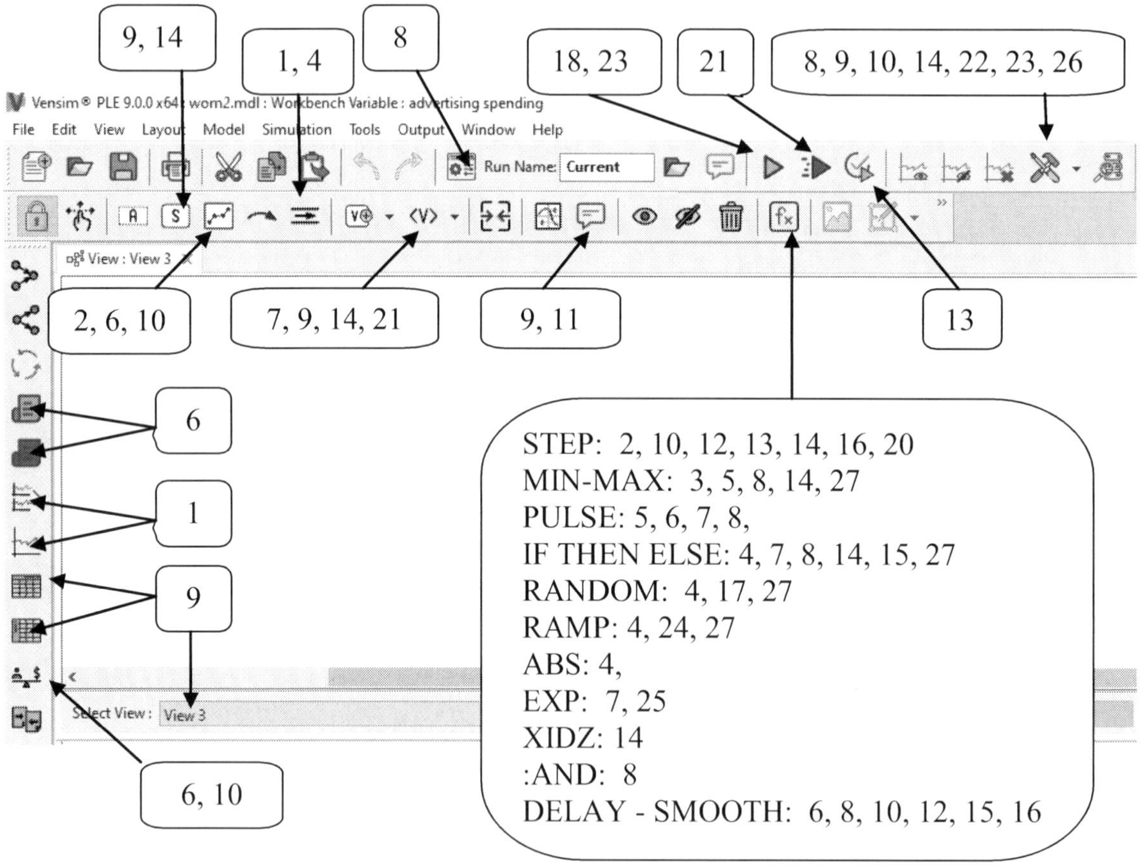

Environmental Area

3.1. Population growth

This chapter focuses on the mechanics of getting such a model up and running, and reviewing the results.

This chapter helps you get started by:
- Moving step-by-step through an example.
- Detailing many key functions and activities.
- Providing a starting point for further application and exploration.

This chapter describes a set of steps you can take to build and analyze a simple model, which lets you try out some of the most important features of Vensim.

The problem

The problem posed is to develop a model of population growth in a situation where there are no restrictions on growth. It is a representation of the natural rate of growth in a species population when there are no predators, and no restrictions on habitat, food supply or water supply.

The idea underlying the model is simple. Births and deaths are proportional to population, and population is the accumulation of births less deaths. Not surprisingly, this is a simple model to build and understand.

The model

This model features a simulation model of a rabbit population. The modelling process starts with sketching a model, then writing equations and specifying numerical quantities. Next, the model is simulated and simulation output saved to a dataset. Finally, the simulation data can be examined with analysis tools to discover the dynamic behaviour of variables in the model.

A system dynamics simulation model is solely determined by the equations that govern the relationships between different variables. The full equation listing completely describes a Vensim model. The structural diagram of a model is a picture of relationships between variables.

Graphic system dynamics models should be clearly presented to facilitate model building, analysis, and presentation.

Software download and install

Go to www.vensim.com – Download – Free Downloads and follow the instructions to download and install Vensim PLE in your computer.

Start

Click the New Model button,  or select the menu item File - New Model.

In the Time Bounds dialogue, leave INITIAL TIME at 0, type 100 for FINAL TIME. Click on the drop-down arrow for Units for Time, and select 'Month'. Click on OK (or click Enter).

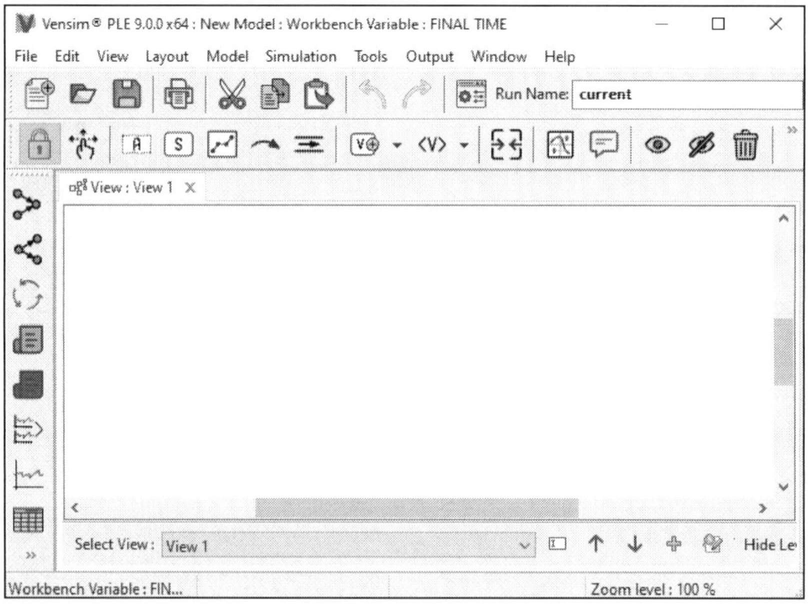

The sketch

Stocks are entered with the Stock sketch tool. Flows are entered with the Flow sketch tool. Vensim draws a rate with one arrowhead, which indicates the direction of flow.

Constants and Auxiliaries are entered with the Auxiliary sketch tool just as words, with no shape attached.

Select the Stock tool and click somewhere in the middle of the sketch. An editing box will appear. Type the word **Population**, and click the Enter key.

Select the Flow sketch tool. Click once (single click and release of the mouse button) about 5 cm. to the left of the Stock **Population**, then move the cursor on top of **Population** and click once again. Type the word **Births,** and click Enter.

Click once on the Stock **Population** then move the cursor about 5 cm. right and click again. Type the word **Deaths**, and click Enter.

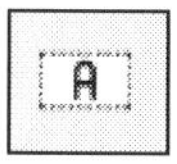

Select the Auxiliary sketch tool. Click on the sketch below **Births**, type **Birth rate** and click Enter.

Click on the sketch below **Deaths**, type **Average lifetime** and click Enter.

Select the Arrow sketch tool, click once on **Birth rate** then once on **Births**. Click once on **Average lifetime**, then once on **Deaths**.

Click once on **Population**, then once on the sketch a little below and left of **Population**, then once on the flow **Births**.

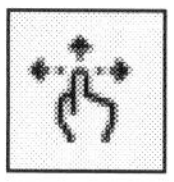

Click once on **Population**, then once on the sketch a little below and right of **Population**, then once on the Flow **Deaths**. Click on the small circles in the arrows to bend them.

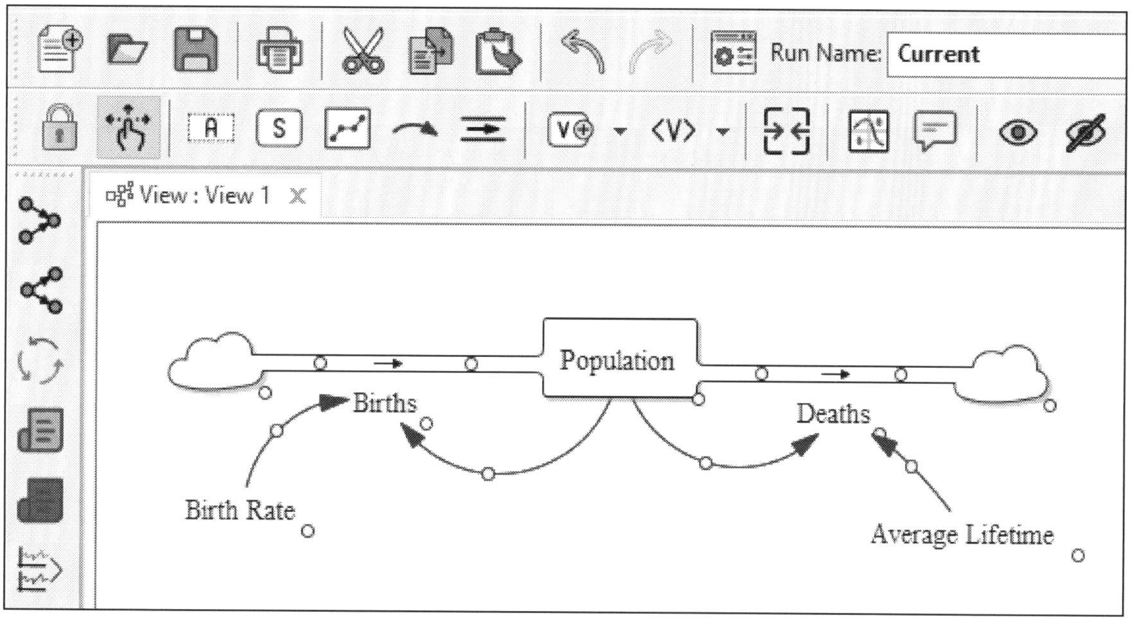

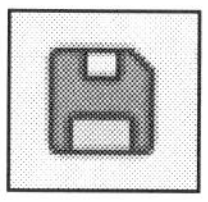

Click the Save button and save your model in the directory that you prefer. Name your model - population.mdl.

Select the Delete icon to delete everything you want to delete ... 

The structure of the **Population** model is now complete. A positive feedback loop from **Population** to **Births** increases **Population**, and a negative feedback loop from **Deaths** decreases **Population**.

Equations

In order to simulate, the model needs a set of equations that describe each relationship. These equations are simple algebraic expressions, defining one variable in terms of others that are causally connected. For example:

Births = Population * Birth rate

Looking at the sketch view, **Birth rate** has no causes - it is a Constant in the model. This Constant has a numerical value which we will fill in later.

Click on the Equations sketch tool. 

All the variables in the model will turn black. The highlights indicate which variables still require equations.

Click on the variable **Births**.

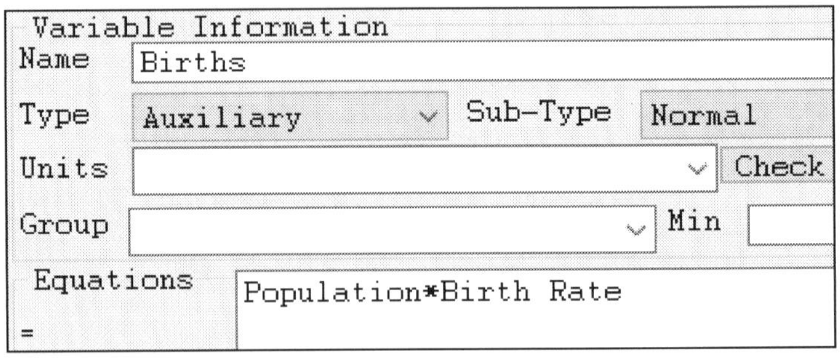

The Equation Editor opens. The top of the editor has the name of the variable we clicked on: **Births.** The drop-down list box on the left shows the type of variable: Auxiliary.

Vensim considers Rates and Auxiliaries to be the same Type of variable. The cursor is positioned in the equation editing box (next to the = sign). Complete the equation for **Births** as below (in the editing box)

Click on the variable **Population** in the Variables list (in the middle of the Equation Editor), then type the * symbol, then click on **Birth rate** in the Variables list. Click OK.

Click on **Population**.

The Equation Editor opens and is slightly different from what we saw with the variable **Births.**

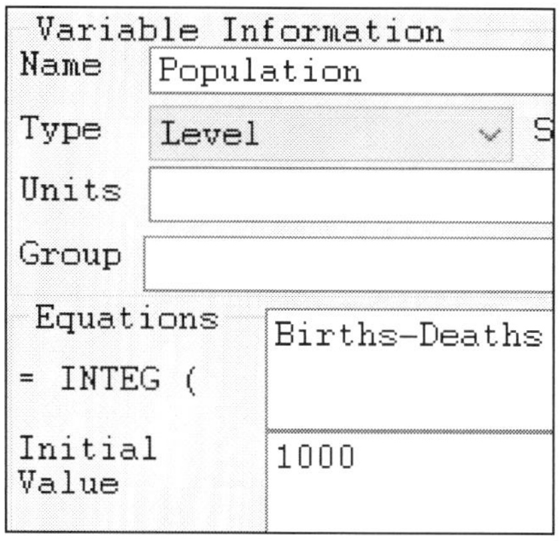

The drop-down list box on the left shows the type of variable - Level (Level is equivalent to Stock)

An equation is already present in the equation editing box. Because we connected Rates with the names **Births** and **Deaths** to the Stock, Vensim automatically enters the Rates to the Stock equation.

The Equation Editor for a Stock has an extra editing box to set the starting or initial value. The cursor is placed there. In the Initial Value editing box, type in 1000. This value is the initial number of rabbits at the start of the simulation.

Click on **Birth rate**. Type in the numbers 0.04 in the editing box.

Complete the remaining two equations as they are shown in the Equations listing below.

Average lifetime = 50
Birth rate = 0.04
Births = Population * Birth rate
Deaths = Population / Average lifetime
Population = Births - Deaths
Initial value: 1000

Save the model

Click the icon Save model and write the name Population. The model will be saved as Population.mdl

Questions

Please, try answer in a paper the following questions without using the model.

1. What evolution will the population take when the positive loop (Population-Births) prevails?
2. What evolution will the population take when the negative loop (Population-Deaths) prevails?

Later, you can compare your answers with the model behaviour.

Simulate

Click on the Simulate button. The model will simulate, showing a work-in-progress window until completion.

The results

This model has been designed to show the evolution in the rabbit population. Select the icon Move/Size, then double click on the Stock **Population** in the sketch. This selects it as the Workbench Variable. Check the Title Bar at the top of the Vensim window to see that **Population** is selected.

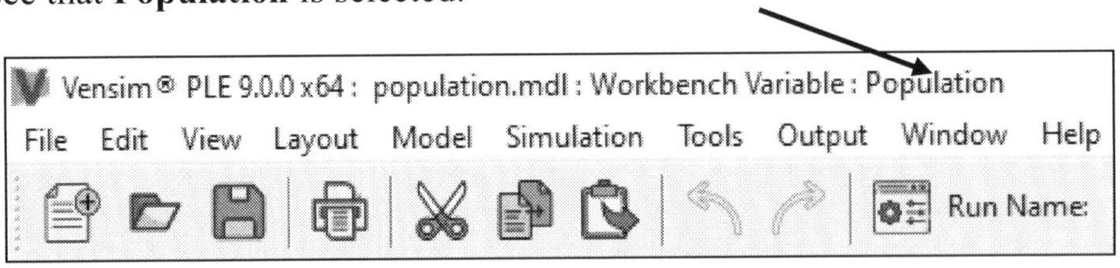

Click on the Graph tool. A graph of **Population** is generated:

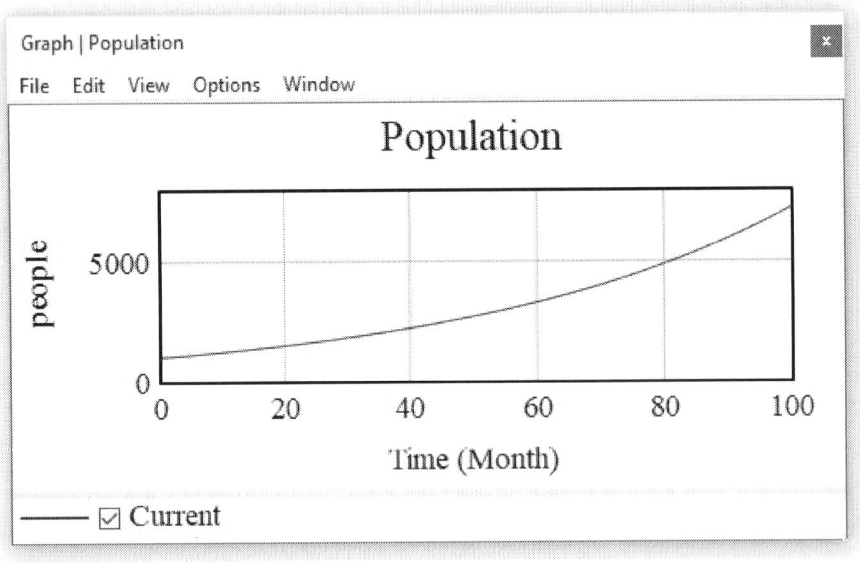

Now click on the Causes Strip tool. A strip graph is generated showing **Population** and its causes **Births** and **Deaths**.

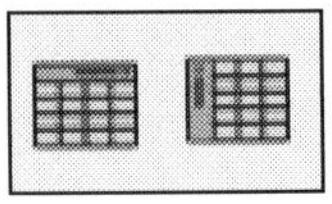

You can also display the results as numbers.

3.2. Modeling the ecology of a natural reserve

The Kaibab Plateau is a large, flat area of land located on the northern rim of the Grand Canyon. It has an area of about 1,000,000 acres. In 1907, President Theodore Roosevelt created the Grand Canyon National Game Preserve, which included the Kaibab Plateau. Deer hunting was prohibited. At the same time, a bounty was established to encourage the hunting of pumas and other natural deer predators. In 1910 nearly 500 pumas were trapped or shot. As a result of the reduction in the numbers of the Kaibab pumas and other natural enemies of the deer, the deer population began to grow quite rapidly. The deer herd increased from about 5,000 in 1907 to nearly 50,000 in 15 years.

As the deer population grew, Forest Service officials and other observers began to warn that the deer would exhaust the food supply on the plateau. Over the winters of 1924 and 1925, nearly 60 percent of the deer population on the plateau died. The deer population on the Kaibab continued to decline over the next fifteen years until it finally stabilized at about 10,000 in 1940.

Imagine you were an official of the National Forest Service in 1930 and you were interested in the fate of the deer population on the Kaibab Plateau. To examine some alternative approaches to the problem, you have decided to build a simulation model.

Your main concern is the growth and rapid decline of the deer population observed over the period from 1900 to 1930, and the future course of the population from 1930 to 1950. Thus, the time frame for the model you will build is the fifty years from 1900 to 1950, and the problematic behaviour (reference mode) is the behaviour of deer population.

Once you have developed an adequate model, you can use it to examine some alternative ways of controlling the size of the deer population on the plateau.

*The model discussed is based, in part, on the work of Professor Donella Meadows of Dartmouth College and Michael Goodman of Pugh-Roberts Associates, Inc. Historical information about the Kaibab Plateau is, in part, based on Edward J. Kormondy, Concepts of Ecology (Englewood Cliffs, N .J.: Prentice-Hall).

In building a model, it is often helpful to begin with a simple model and then expand it in several steps. We will start with a small model focusing on a few of the factors that influenced the size of the deer population on the Kaibab Plateau. Later, the model will be enlarged to include additional elements of the plateau ecology, and finally obtain a model that generates the behaviour.

To face the problem, we will build three models, (each of which relate to the previous) that we will call Deer 1, 2 and 3. The reason is that it is very difficult to create a model that reproduce the behaviour of stability of the number of deer up to 1907, growth up to 1924 and collapse in 1930.

Next steps:

Model 1 will reproduce the situation of stability in the number of deer that existed up to 1907.

Model 2 will be the same as Model 1 but reproducing a step in 1907 in which the pumas were eliminated in a few years, and we will improve it in other aspects.

Model 3 will be as Model 2 but adding the information that they facilitate - the deer began to die from hunger. It will be the base model on which we will be able to simulate different alternatives, the reintroduction of pumas, to give forage to the animals, to introduce hunters, etc.

Model 1

Objective: To reproduce the stable initial conditions of the population of deer, with the limited data that we have.

Frequently found in the description of any problem is a lot of unnecessary (for our analyses) data. On the other hand, it normally lacks some very useful information. We would spend a lot of time and money if we needed to go to the Grand Canyon - aerial trip included - to get some data in the field. So, let us do some reasonable hypotheses which allow us to work. Later, if we see that this hypothesis can significantly affect the actions proposed, we will study them with a specialist. Remember that the goal of the study is to END UP PROPOSING SOME ACTION, not to focus on the details.

So let us find some reasonable initial values, without deep knowledge of the subject we are studying, and without spending any time searching the data that has no influence in the simulation and on the final proposals.

They tell us that before white man arrived, the population of deer was balanced, which means that the number of births less the deaths by senility (we call it the vegetative increment) was equal to the number of deer hunted by the pumas.

Let us see what could be a reasonable vegetative increment:

From the whole population, only females have descendants. If every female had a fawn, the birth rate would be 0.5, but there are females too old and too young to have fawns, so the rate can be reduced 0.3.

In regards to deaths, the average life expectancy of an animal of this kind is not known, but it could be about 10 years (rate equal to 0.1). Therefore we can assume a rate of vegetative increment of 0.2 (i.e, 0.3-0.1).

This is to say that with a population of about 5,000 deer, every year it would increase by 5,000*(0.3-0.1) = 1,000 deer.

Therefore the quantity of deer hunted every year is also 1,000.

How many pumas were there? We don't know, but if we consider that the pumas should feed basically on animals small and easy to hunt, we can assume 2 deer a year for each puma is a good figure (perhaps even a little high). Therefore, 1,000 divided by 2 gives us initially 500 pumas.

The model should be built as described in the documentation with the simulation software and the user should try to understand its behaviour.

Open Vensim and go to File - New Model

define:

INITIAL TIME = 1900
FINAL TIME =1950
TIME STEP=1
Units=year

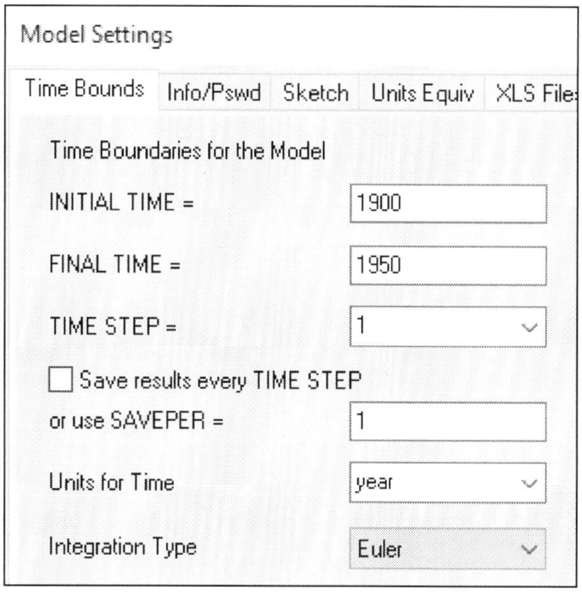

Start by drawing the Stock and then add the Flows.

Model 1

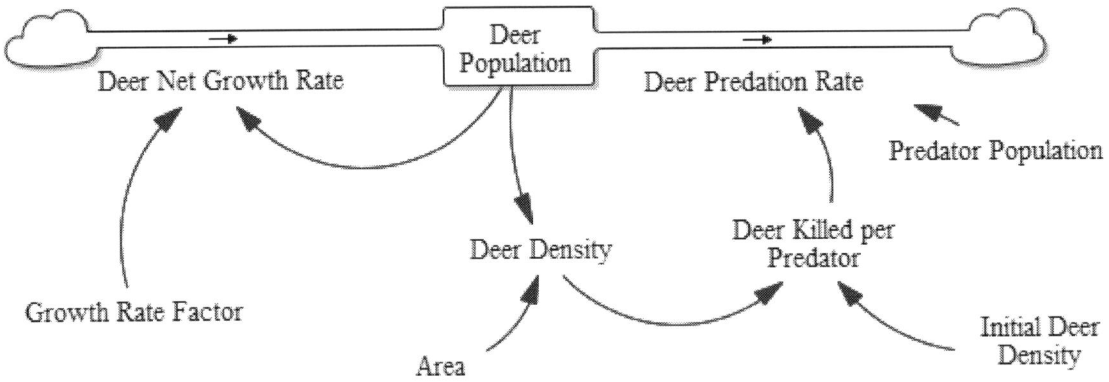

Equations 

(01) Area= 1000000
Units: acre

(02) Deer Killed per Predator = Deer Density/Initial Deer Density
Lookup: (0,0),(1,2),(2,4),(4,6),(20,6)
Units: Deer/(year*puma)
The point (0,0) means that if there are no deer, each predator will hunt 0 deer, which is obvious. The point (1,2) is based on the fact that in the stable scenario (Deer Density = Initial Deer Density) if there are 5000 deer, with a birth rate of 0.2, the natural increase is 5000 * 0.2 = 1000 deer. So if there are 500 predators, they each kill 1000/500 = 2 deer / year. Point (2,4) assumes that if the deer population doubles, the number of deer hunted per predator also doubles. In points (x, 6) it is assumed that, due to their small size, each predator can only eat and hunt one deer every 2 months, or 6 a year.

(03) Deer Net Growth Rate=Deer Population*Growth Rate Factor
Units: Deer/year

(04) Deer Population= +Deer Net Growth Rate-Deer Predation Rate
Initial value: 5000
Units: Deer
Initial amount of deer is 5,000

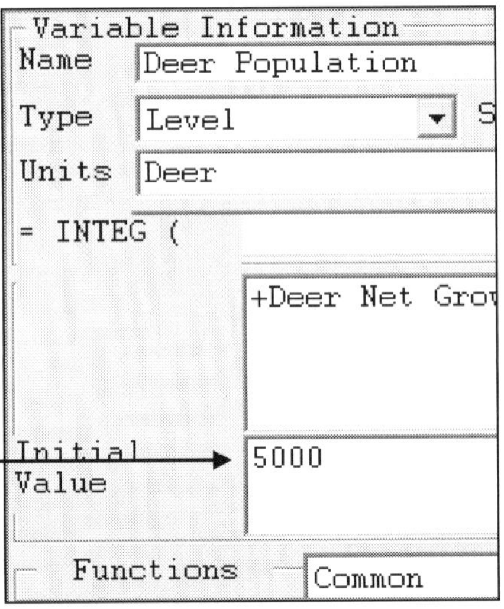

To define this variable, set Type=Auxiliary, Sub-Type = with Lookup, and As Graph=enter values in the table (see image below).

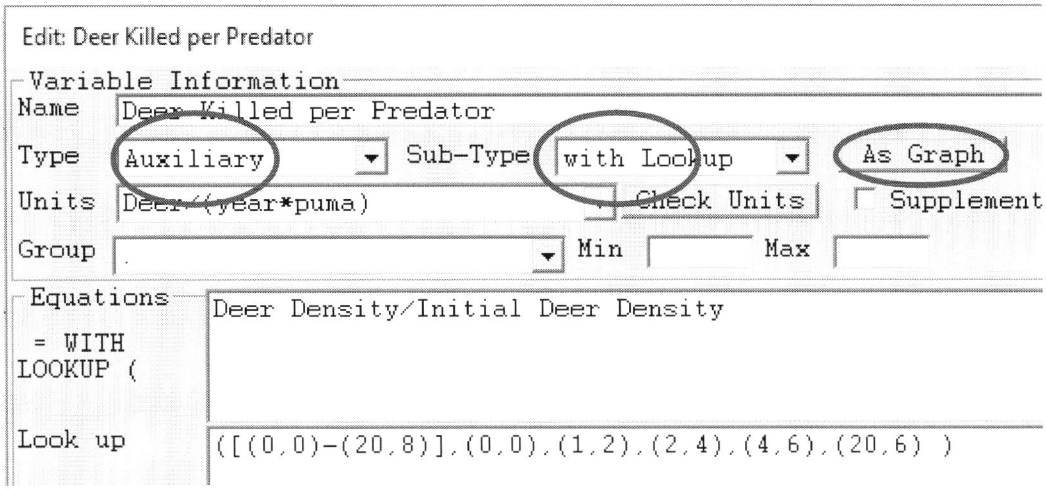

Figures in brackets do not affect model calculations.

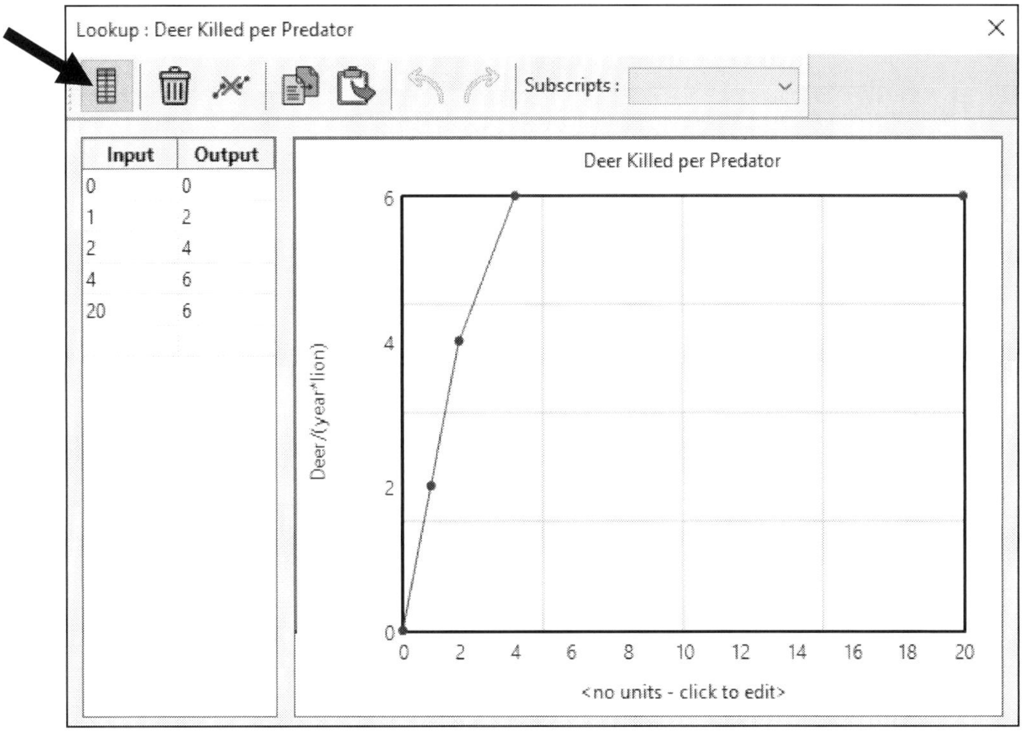

(05) Growth Rate Factor= 0.2
Units: 1/year
Here we work with the net growth rate of the deer population. This means births minus deaths. If every female had a fawn, the rate will be 0.50%. But there are females too old and too young to have fawns, so we reduce this amount to the 0.30%. Now we need to consider the life expectancy, say 10 years. This is minus 0.10% due to deaths. So we assume the Net Growth Rate Factor equal to 0.2

(06) Deer Predation Rate= Deer Killed per Predator*Predator Population
Units: Deer/year

(07) Initial Deer Density= 0.005
Units: Deer/acre
5,000 deer / 1,000,000 acres

(08) Predator Population= 500
Units: puma
We assume that Net Growth Rate was 1000 deer, so Deer Predation Rate must be also 1000 deer. In Lookup we define that with a density of 0.005 deer/acre are killed 2 deer/predator, so we have 1000 (deer) / 2 (deer/predator) = 500 predators.

(09) Deer Density=Deer Population / Area
Units: Deer/acre
This is the amount of acres per deer

Simulate

Display the result. All variables remain constant

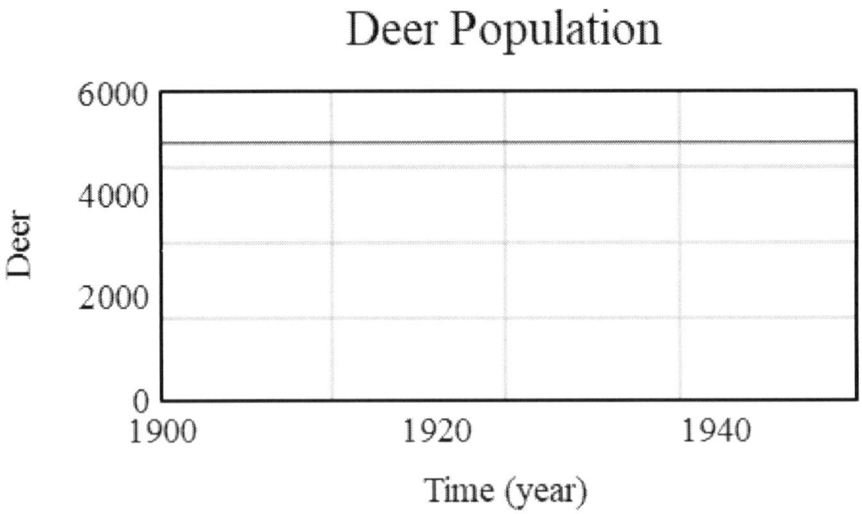

Click the icon to check the units before to continue.

Model 2

Now we will reproduce the elimination of the pumas in a few years. In the documentation of the model you can observe that it is implemented with the STEP function (Go to the Annex II to see the use of some functions as RAMP and PULSE).

We will also modify the growth rate factor that was constant at 0.2 and we will make it depend on the quantity of food per deer in such way that if they don't have enough food, the rate will be smaller. To implement this we will use a Table. We will assume that the quantity of grass is constant. The model should be built as described in the documentation with the simulation software and the user should try to understand its behaviour.

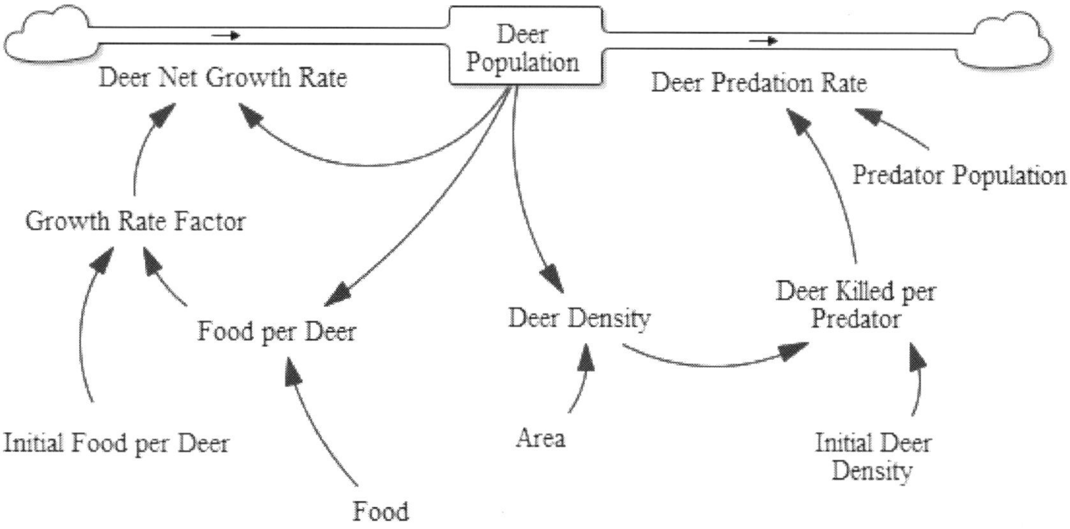

(01) Area= 1000000
Units: acre

(02) Deer Density=Deer Population / Area
Units: Deer/acre

(03) Deer Killed per Predator = Deer Density/Initial Deer Density
Lookup: (0,0),(1,2),(2,4),(4,6),(20,6)
Units: Deer/(Year*puma)
When density is equal to 0, the deer killed is 0. Point (0,0). When the Deer Density is equal to the Initial Deer Density each predator hunts 2 deer each year. Point (1,2).

(04) Deer Net Growth Rate= Deer Population*Growth Rate Factor
Units: Deer/Year

(05) Deer Population= +Deer Net Growth Rate-Deer Predation Rate
Initial value: 5000
Units: Deer

(06) Deer Predation Rate= Deer Killed per Predator*Predator Population
Units: Deer/Year

(07) Food= 100000
Units: ton
In this simulation we will take this element like a constant.

(08) Food per Deer= Food / Deer Population
Units: ton/Deer
Initially the amount is 100,000/5,000 = 20 tons per deer.

(09) Growth Rate Factor = Food per Deer/Initial Food per Deer
Lookup: (0,-0.6),(0.05,0),(0.1,0.2),(1,0.2),(2,0.2)
Units: 1/Year
Point (1, 0.2) is the equilibrium point (in which Food per Deer = Initial Food per Deer) and we assign the value 0.2 for the rate of increase (as in kaibab 1). Point (2, 0.2) we assume that although there is more food, the rate of increase will not increase. Point (0, -0.6) At this point we analyze an extreme scenario when Food per Deer = 0, there is no food, we assume that 60% of the deer will die, although we can also assume that they all die. The rest of the points are values that can be placed with a certain logic, but do not respond to additional information.

(10) Initial Deer Density= 0.005
Units: Deer/acre
5,000 deer / 1,000,000 acres

(11) Initial Food per Deer= 20
Units: ton/Deer

(12) Predator Population= 500-STEP(500,1910)
Units: puma
Here we are reproducing the hunting or fall of the pumas during 1910.

Behaviour

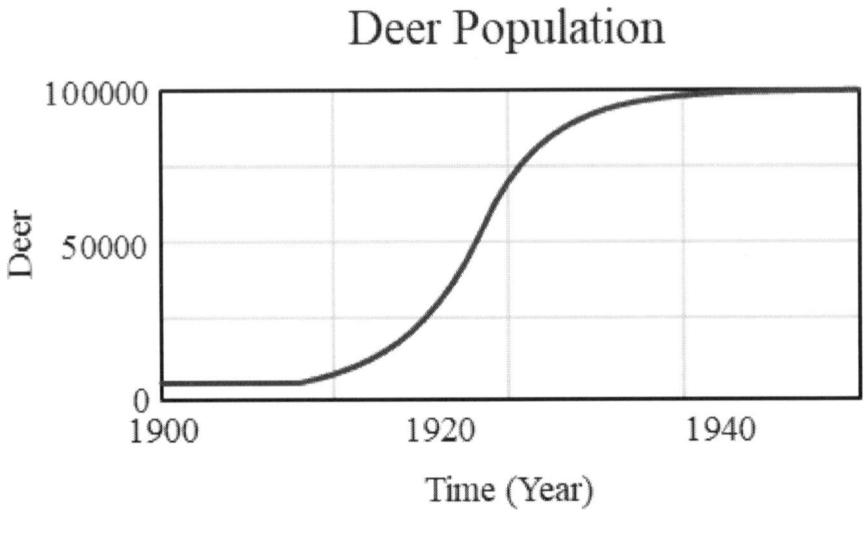

Model 3

In this model, the observed behaviour has been reproduced, including the fall in the number of deer after 1924, assuming the growth of grass was a constant and the model is based on of the number of deer that existed.

The model should be built as described in the documentation with the simulation software and the user should try to understand its behaviour.

Note: Remove the old auxiliary variable "Food".

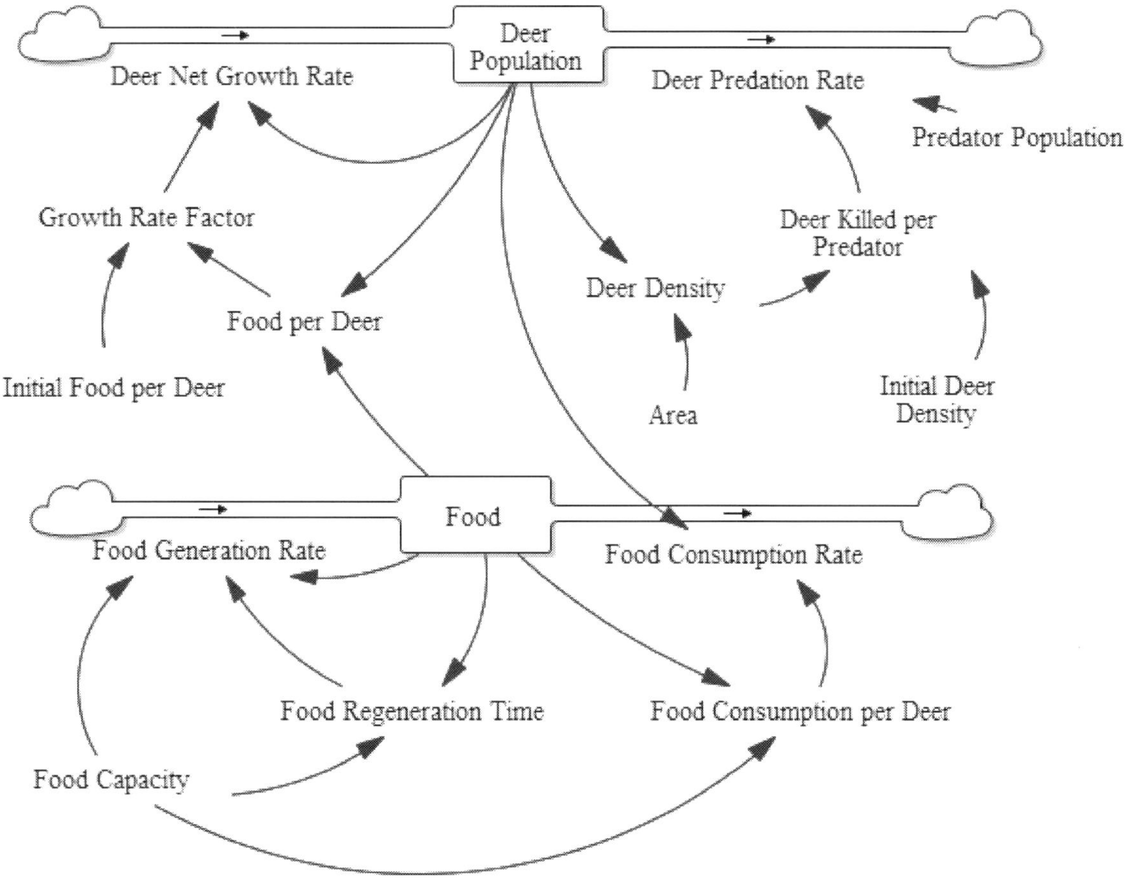

(01) Area= 1000000
Units: acre

(02) Deer Density= Deer Population / Area
Units: Deer/acre
This is the amount of acres per deer

(03) Deer Killed per Predator = Deer Density/Initial Deer Density
Lookup (0,0),(1,2),(2,4),(4,6),(20,6)
Units: Deer/(year*puma)

(04) Deer Net Growth Rate= Deer Population*Growth Rate Factor
Units: Deer/year

(05) Deer Population= +Deer Net Growth Rate-Deer Predation Rate
Initial value: 5000
Units: Deer

(06) Deer Predation Rate= Deer Killed per Predator*Predator Population
Units: Deer/year

(07) Food= Food Generation Rate-Food Consumption Rate
Initial value: 100000
Units: ton
Now Food depends from the Food Consumption Rate and from the Food Generation Rate.

(08) Food Capacity= 100000
Units: ton

(09) Food Consumption per Deer = Food/Food Capacity
Lookup: (0,0),(0.2,0.4),(0.4,0.8),(1,1)
Units: ton/Deer/year
Food Consumption per Deer will depend on the Food Capacity and present Food. We will assume that the ratio (Food/Food Capacity) will be the best way to determine the Food Consumption per Deer, using Table. When Food – Food Capacity we will assume that Food Consumption per Deer will be 1 ton. A lower ratio will produce a lower Food Consumption per Deer.

(10) Food Consumption Rate= Deer Population*Food Consumption per Deer
Units: ton/year

(11) Food Generation Rate=(Food Capacity-Food)/Food Regeneration Time
Units: ton/year

(12) Food per Deer=Food/Deer Population
Units: ton/Deer

(13) Food Regeneration Time = Food/Food Capacity
Lookup: (0,40),(0.5,1.5),(1,1)
Units: year
Food Regeneration Time will depend on the Food Capacity and present Food. We will assume that the ratio (Food/Food Capacity) will be the best way to obtain the Food Regeneration Time, using Table. When Food = Food Capacity we will assume that Food Regeneration time will be 1 year. A lower ratio will produce a higher Food Regeneration Time.

(14) Growth Rate Factor = Food per Deer/Initial Food per Deer
Lookup: (0,-0.6),(0.05,0),(0.1,0.2),(1,0.2),(2,0.2)
Units: 1/year

(15) Initial Deer Density= 0.005
Units: Deer/acre
5,000 deer / 1,000,000 acres

(16) Initial Food per Deer= 20
Units: ton/Deer

(17) Predator Population= 500-STEP(500,1910)
Units: puma

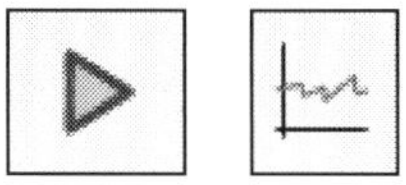 Simulate and display the graph

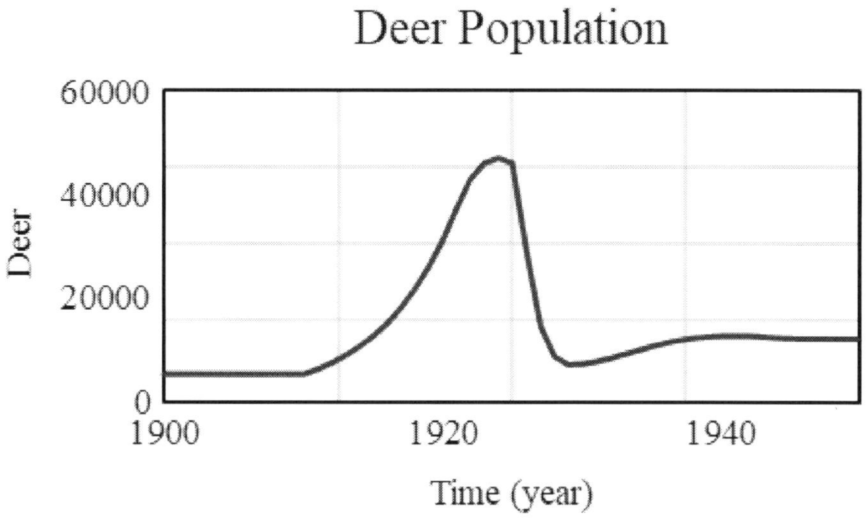

Now you can use this model to examine some alternative ways of controlling the size of the deer population on the plateau after 1930.

Prediction models need to 'tune' the model a lot based on the historical data, to improve their possibilities of success.

Here we are working with 'comparative models', to compare different alternatives or policies. Do you believe that the study will change its conclusions, if we dedicate more time (and money) to tune the model?

To see more than one variable together in the same screen, click the icon Control Panel, then choose the options: Custom Graphs – New. Select the variables that you like to see, OK, and Display

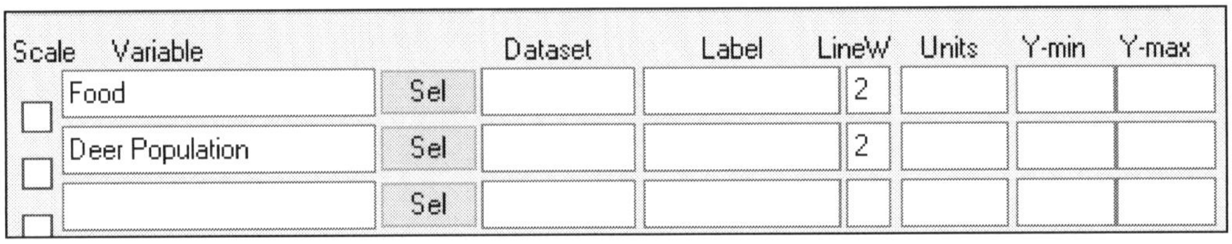

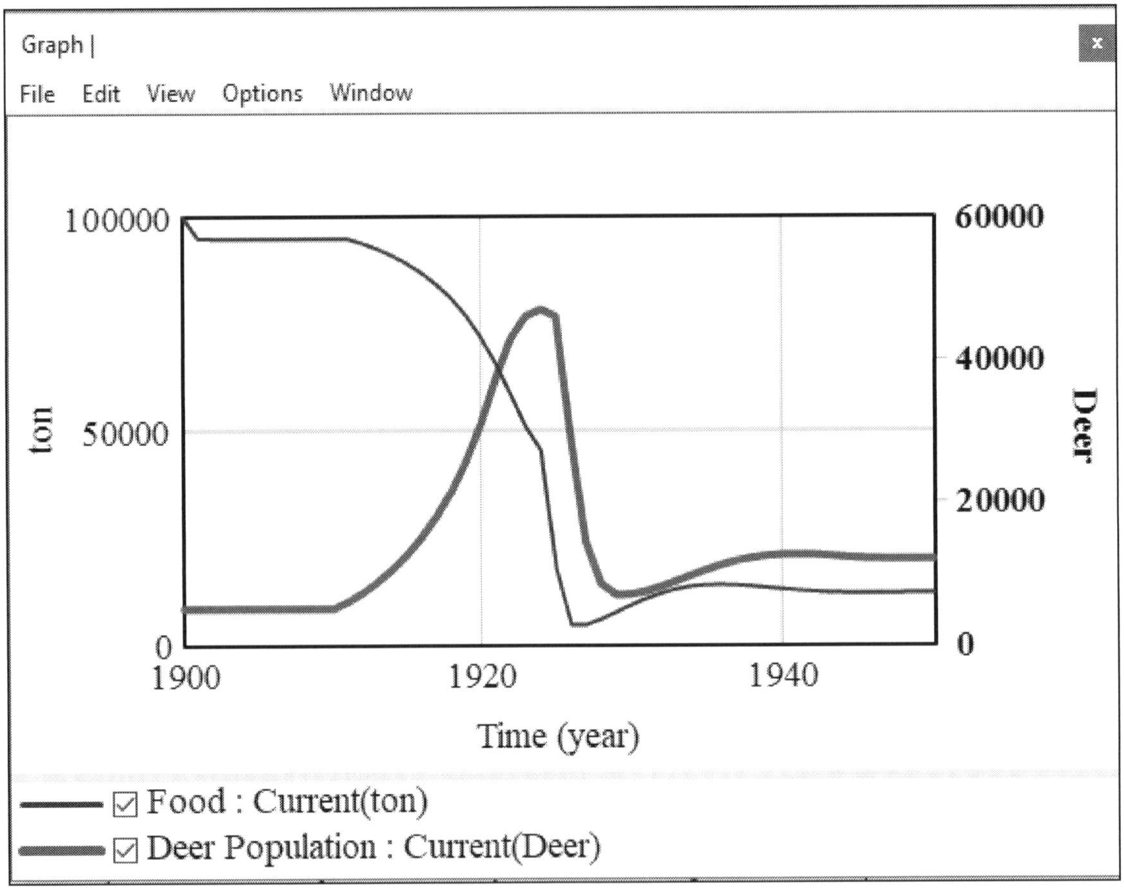

3.3. Effects of the intensive farming

One of the great mysteries of human history has been the sudden collapse of one of the main centers of Mayan civilization in Central America at a time when it was apparently at a peak of culture, architecture and population around A.D.800.

No one knows exactly why this society of several million people collapsed, but new research shows a gradually tightening squeeze between population and environment that may have been crucial to the fall.

Tropical environments are notoriously fragile. By understanding what the Mayans did to theirs, modern humans may get some useful guidance on how to treat tropical environments today.

Just before the final cataclysm, the new research suggests, the population in one area ranged from about 200 to 500 persons per square kilometer. This population-density almost certainly required advanced agriculture or large-scale trade.

Within two to four Mayan generations, which probably meant less than 100 years, the population dropped to what it had been almost 2,000 years before - 20 or less per square kilometer and sometimes far below even that sparse population. Furthermore, after the collapse, whole areas remained almost uninhabited for a thousand years - virtually until the 1970's.

Some of the environmental changes appear to have been as long-lasting as the loss of population. Lakes that were apparently centers of settlement in the Maya time have not even today recovered the state of productivity that made their shores good living places more than 1,000 years ago.

Such clues to the past as these were found, in eight years of research, by scientists at Florida State University and University of Chicago. Their research, still continuing, showed there was an exponential growth in Mayan population during at least 1,700 years in the tropical lowlands of what is now Guatemala.

According to the new estimates, human numbers doubled every 408 years. This trend may have caught the Maya in a strange trap. Their numbers grew at a steadily increasing pace, but for many centuries, the growth was too slow for any single generation to see what was happening.

Over the centuries, the increasing pressure on the environment may have become impossible to maintain. Yet the squeeze could have been imperceptible until the final population spurt at the end. Some specialists believe that in more northerly regions, the quality of Mayan civilization may have deteriorated without so great a population drop.

The new estimates for the southern lowlands are based largely on a detailed survey of traces of residential structures that were built, occupied and abandoned over the centuries.

The studies are focused on the region of two adjacent lakes, now called Yaxha and Sacnab, in the Peten lake district of northern Guatemala. The area was inhabited as early as 3,000 years ago and the first agricultural settlements appeared there about 1,000 B.C. The land was largely deforested by A.D. 250.

Gradually-intensified agriculture seems to have caused severe cumulative damage to an originally verdant environment. To this was added the impact of increases in human dwellings and other major architectural works on the land. Essential nutrients washed, slid and were moved downhill to be lost in the lakes, diminishing the fertility of agricultural land.

Increases in phosphorus in the lakes from agriculture and human wastes showed that pollution must have aggravated the environmental damage. Scientists even confirmed the population trend by estimates of the per capita increase in phosphorus going into the lakes as human numbers rose.

Authors of the research believe theirs to be the first published report giving documented estimates for the growth pattern of Mayan population and correlating this growth with the damage to the environment that went with it.

The authors pointedly omit any claim that they have solved the long-standing mystery of the Maya collapse.

They do say the kind of environmental pressure their studies suggest may have been one important factor among several. The scientists also believe the research may have some useful practical implications.

'We believe we are doing a kind of archaeology that has a tremendous amount of relevance' said Dr. Don S. Rice, assistant professor of archaeology at the University of Chicago and adjunct assistant curator of archaeology at the university museum.

Dr. Edward S. Deevey, leader of the research team, notes that the Mayan was one of the few high civilizations ever to flourish in lowland tropical forests and that the problems these ancient Central Americans faced may offer lessons for the present and the future. Dr. Deevely is graduate research curator of paleoecology at the Florida State Museum, Gainesville. The other authors, in addition to Prof. Rice are H.H. Vaughan, Mark Brenner and M.S. Flannery of the University of Florida and Prudence M.Rice assistant curator of archaeology and assistant professor at Florida University.

The research to date offers evidence that traditional perceptions of the civilization of the Maya may need to be revised, said Prof. Don Rice. 'Certainly, population was denser that the traditional models suggest' he said. So much population would require advanced agriculture and trade, and evidence is emerging that both existed.

'It adds support to a belief that the Maya were not a peaceful, theocratic, fun-loving, tropical group,' he said. Instead, the accumulating evidence suggests, they have become a complex society with advanced agriculture, a managerial elite class and all the stresses, strains and conflicts that go with such size and complexity. Then their way of life collapsed, leaving a severely damaged environment and a cultural vacuum that persisted for many centuries.

Study Depicts Mayan Decline

By HAROLD M. SCHMECK Jr.

ONE of the great mysteries of human history has been the sudden collapse of one of the main centers of Mayan civilization in Central America at a time when it was apparently at a peak of culture, architecture and population around A.D. 800.

No one knows exactly why this society of several million people collapsed, but new research shows a gradually tightening squeeze between population and environment that may have been crucial to the fall.

Tropical environments are notoriously fragile. By understanding what the Mayas did to theirs, modern humans may get some useful guidance on how to treat tropical environments today — a realm of knowledge that could be particularly useful to the so-called Third World.

Just before the final cataclysm, the new research suggests, the population in one area ranged from about 200 to 500 persons per square kilometer (about four tenths of square mile). This population-density almost certainly required advanced agriculture or large-scale trade.

Within two to four Mayan generations, which probably meant less than 100 years, the population dropped back to what it had been almost 2,000 years before — 20 or less per square kilometer and sometimes far below even that sparse population. Furthermore, after the collapse, whole areas remained almost uninhabited for a thousand years — virtually until the 1970's.

Study Depicts Mayan Decline - Harold M. Schmeck Jr. - New York Times. Oct 23 1979

Open a new model (File - New model) and define those settings:

INITIAL TIME= -1000 FINAL TIME= 1000 TIME STEP=1 Units for time=Year

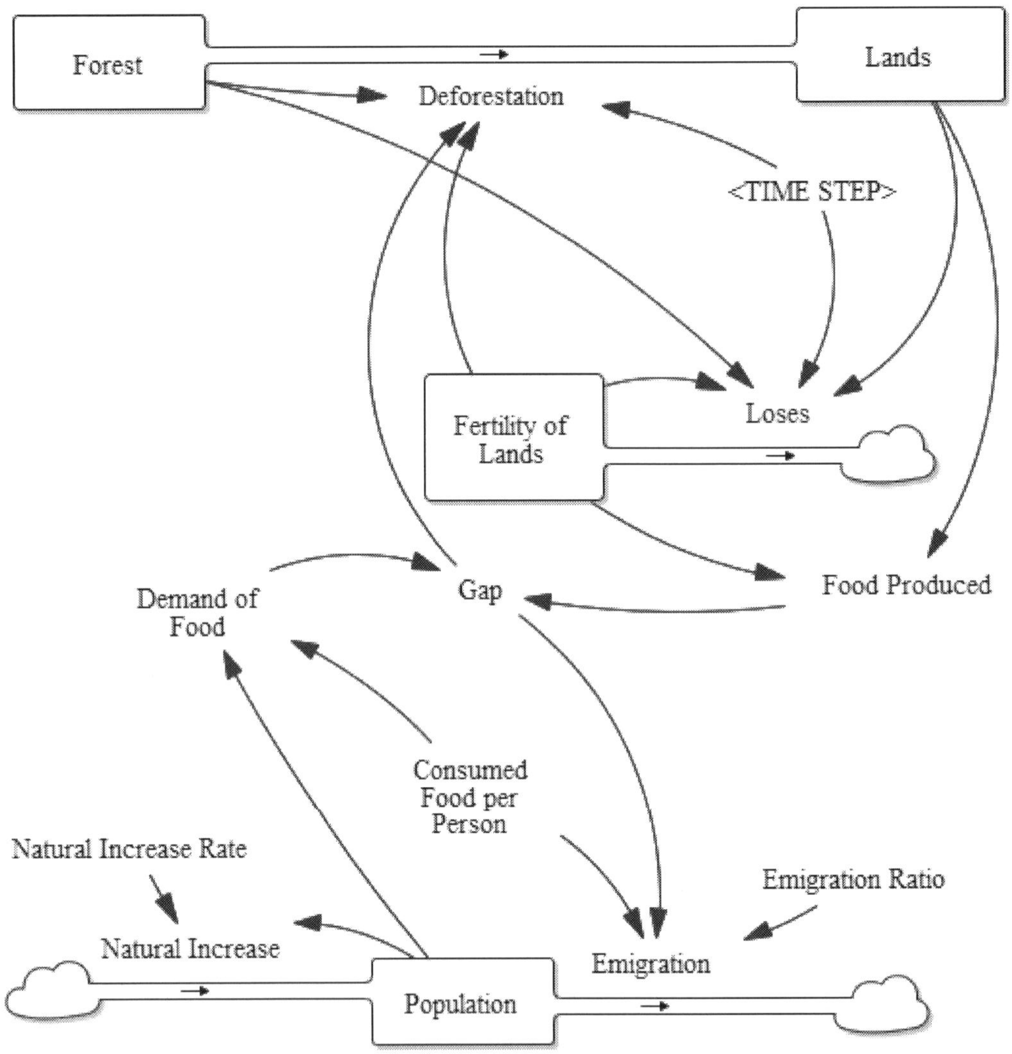

Start drawing the Stocks and then the Flows

Equations 

(01) Deforestation=MIN(Gap/MAX(Fertility of Lands,1),Forest/4)/Intensity
Units: km2/Year
Basically they deforest the amount of km2 that they need to close the Gap (kg) considering the Land Fertility (kg/km2). Function MAX is used to prevent error if model need to divide by 0 . Function MIN is used to atemperate the response of model. Intensity gives us a way to regulate the speed of the model. So, Intensity=1 means that every year they deforest all the forest they need.

(02) Consumed Food per Person= 400
Units: kg/(person*Year)
We assume a few more than 1 kilo per person and day.

(03) Demand of Food=Consumed Food per Person*Population
Units: kg/Year

(04) Emigration=(Gap/Consumed Food per Person)*Emigration Ratio
Units: person/Year
We assume that if know the amount of Gap of food (kg/year) and the consumption per person and year (kg/person/year), we can obtain the equivalent number of people without food (person/year). We correct the obtained value, applying the Emigration Ratio (5%) assuming that only a small part of this people really emigrates (they redistribute the existing food).

(05) Emigration Ratio= 0.05
Units: 1/Year

(06) Fertility of Lands= -Loses
initial value: 5000000
Units: kg/(km2*Year)
With the initial value of 5,000,000 kg/km2 we can obtain 40,000,000 kg if we have 8 km2 of Land. This is equal to the initial food demanded for 100,000 inhabitants with a Consumed food per person equal to 400 kg every year.

(07) Food Produced= Fertility of Lands*Lands
Units: kg/Year

(08) Forest= -Deforestation
Initial value: 5000
Units: km2
Initial value is obtained from an assumed density of 400 person/km2, and from the final inhabitants assumed: 2.000.000 person (2 000 000/400=5,000)

(09) Gap= Demand of Food-Food Produced
Units: kg/Year

(10) Intensity = 1
Units: Year
The same variable is used to regulate: 1.- The period of time that population decide to close the gap, and deforest all that they need. 2.- The speed in with the Fertility of lands are taken place.

(11) Natural Increase Rate=0.0017
Units: 1/Year
If the population doubles every 408 years, as text say, the percentage must be 0,170 % as we can see (1+0.0017)^408=2

(12) Lands= Deforestation
Initial value: 8
Units: km2
Initial value is obtained from the total food consumed 40,000,000 kg every year for the initial value of inhabitants: 100.000 persons, and an Initial fertility of Land of 5.000.000 kg/km2 and year. (40,000,000/5,000,000=8)

(13) Loses=Fertility of Lands*MIN(2,(Lands/Forest)^1.9)/Intensity
Units: kg/(km2*Year)/Year
We can accept that Losses of Fertility are proportional to the Fertility that exists, and of the difference from the Lands and Forests with a quadratic relation (we like to avoid the use of a Table). We need to use as exponent 1.9 to see the collapse in year 800, as text explains. If we use 2 as exponent, the collapse occurs in year 1000.

(14) Natural Increase=Population*Natural Increase Rate
Units: person/Year
The calculated value of 0.17% is the net value from birth and deceases. With this rate population goes from 100,000 to 2,000,000 inhabitants in 1,800 years.

(15) Population= Natural Increase-Emigration
Initial value: 100000
Units: person
If we assume there were 2,000,000 inhabitants at the moment of collapse (+800 A.D.), we can obtain the initial value in year 1000 B.D., with a rate of 0.17%. They were 100.000 inhabitants approximately.

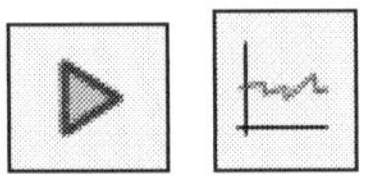
Simulate and display the results

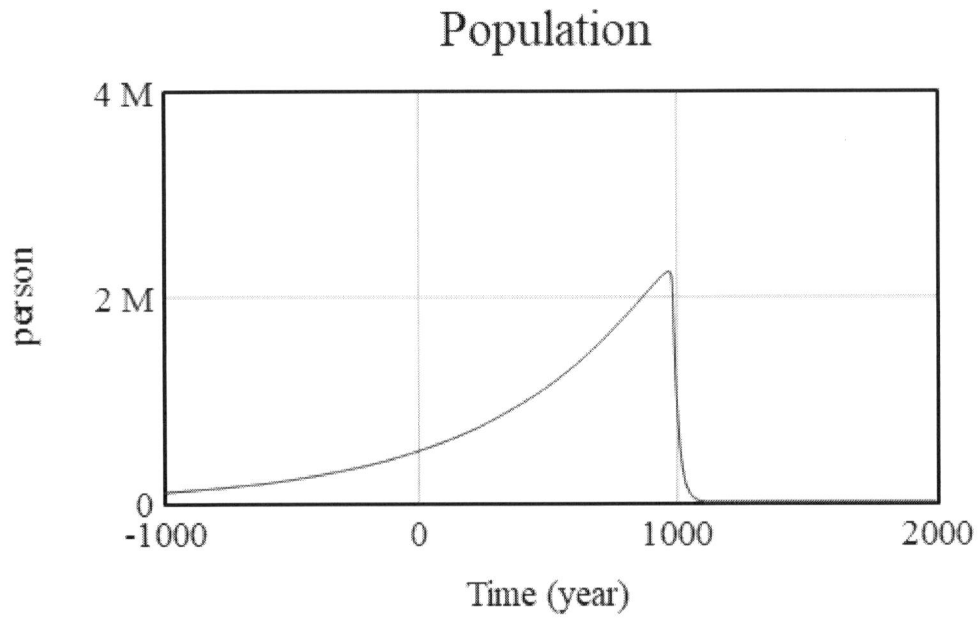

3.4. The fishery of shrimp

The shrimp fishery in Mexico is the third most important fishery after mojarra (any of several species of small, silvery, mainly tropical American marine fishes) and oysters on the coast of the Gulf of Mexico. On the Pacific coast, only tuna and sardine surpass it. The economic value of the shrimp catch makes it the most important fishery of the country, contributing 40% of the total fisheries income. The boats of shrimp fishers represent 66% of the present fishing fleet, while in 1970 they represented 82%.

In the Bay of Campeche, at the south of the Gulf of Mexico, the most important species is the pink shrimp. In the Nineteen-Seventies, more than 20,000 tons per annum were caught but the catch diminished constantly and is presently 4,000 tons per annum.

The fishermen accuse PEMEX, the state oil company, of causing the collapse by intruding on and contaminating their areas of fishing, although the fishing takes place more to the north and the sea currents concentrate the contamination to the south.

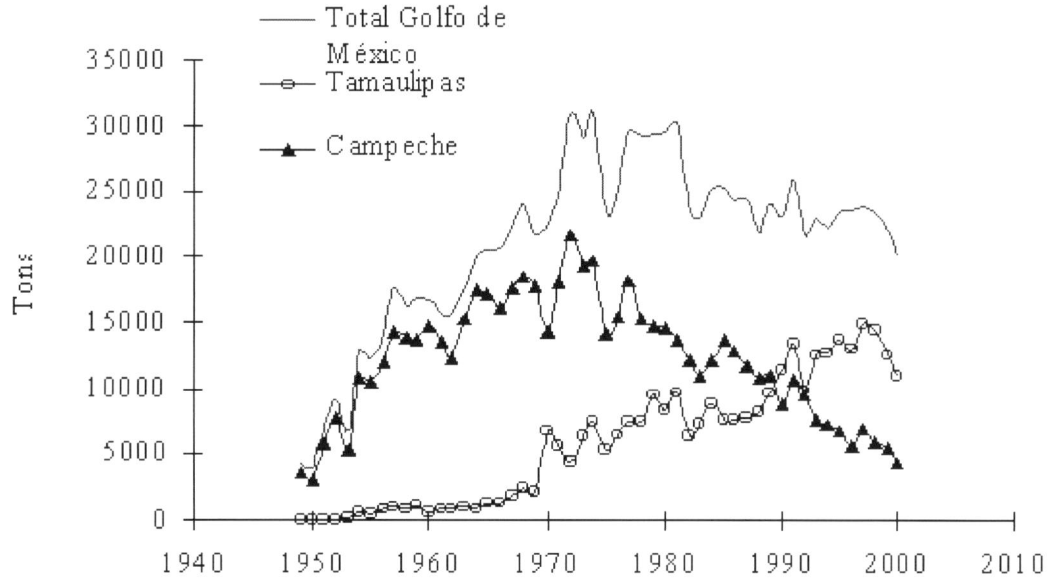

Although the estimations in the graph point clearly at over-exploitation, an interesting controversy has arisen between the investigators. One of the best investigators of the country suggested that if the number of boats had diminished, mortality by fishing also must be reduced. And, if the catch has not increased, the population is hidden somewhere, or if the population really has diminished, it must be by another cause.

He supported his argument with Schaeffer's model, (the same as Verhulst-Pearl's model but applying biomass instead of the number of individuals and including the fishing) assuming the fishery is in balance (the capture equals the population growth). In this model, if the catch activity is reduced, measured by the number of boats, the population of shrimp increases. (See Model 1 below)

Nevertheless, it is possible to argue that the boats currently used are more powerful than those used in the mid-Seventies, the fishing nets are larger, and the trips are longer (60 days now against 15 days then). The shrimp population is concentrated now in fewer places and is more vulnerable to the fleet. Only a quarter of the historical maximum number of boats now exist. In spite of that, Schaeffer's model is commonly used. More than the model itself, the problem could be its use in this case. This model has the formula:

$$\frac{dB}{dt} = rB(1 - \frac{B}{K}) - C$$

$$C = f * q * B$$

B = Biomass of the population; r= % of growth of the biomass; K= greatest Biomass of population; C = fishing (tons or tonnes); f = fishing Effort (number of boats); q = Coefficient, the fraction of the biomass captured by each unit of effort.

It is difficult to have confidence in this model, based on a balance of capture and population growth, when observed data shows strong variance. Using this model is equivalent to assuming that the observed capture has been equal to the population growth. That assumption originated in the days before computers - using it simplified the process of determining the parameters, thereby allowing one to complete a regression analysis. At the present time, however, many researchers have abandoned that assumption and recommend instead using methods not based on a balance (that is, to use methods that simply accept that the catch amount can be different from the population growth).

The second problem is that this model does not say anything to us of how the fleet reacts to the changes in the fishing catch (tonnes). Hilborn and the Walters (1992) proposed a change that includes the modification of the effort in correlation with the profit (net gain):

$$\frac{df}{dt} = k(Cp - cf)$$

Constant k = determines the speed of change of the effort (i.e., new boats); p = Price of the produced fish; c = Cost of operation by effort unit (boat).

When the total income is positive, new boats enter into the fishery. If it is negative, the boats leave this fishery, both at a speed determined by k. When the number of boats diminishes, the population of shrimp increases and vice versa. This could cause oscillations in the population of shrimp, in the fishery, and in the effort. See Model 2 and Model 3 below.

Hilborn, R y C.J. Walters 1992 Quantitative Fisheries Stock Assessment; Choice, Dynamics and Uncertainty. Chapman & Hall. New York. Hoppennsteadt, F.C. 1982 Mathematical methods of population biology. Cambridge University Press. London.

Model 1

This model is a simple population model, Schaeffer's, with increase coming from 'birth rate' represented as '% biomass increment' and with decrease represented by 'catch' (Captures). The equation for increase in population utilizes a multiplier with the ratio of current population to maximum attainable population serving as a means for establishing a linearly approached limit on something, a device often used in SD models. When the value of the variable quantity equals that of the limit, the multiplier becomes zero.

The population is in equilibrium in this model. We have been taught that establishing a system in equilibrium is a good place to start the iterative process of model building. In this model, as 'catch' coefficient k, representing the number of boats, approaches the biomass increment, the catch and population decline precipitously but follow each other in a more or less equilibrium manner through time. Conversely, if the number of boats is reduced (k<0.1), the population tends toward its maximum naturally sustainable 'highest population'.

Model - Settings
INITIAL TIME=1950   FINAL TIME= 2000   TIME STEP= 1   Units for time=Year

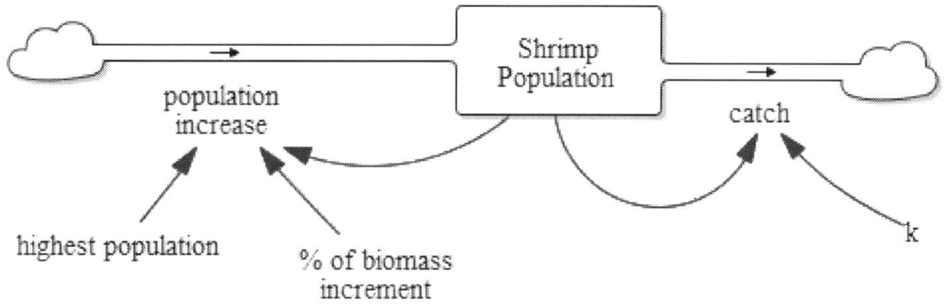

(01) Shrimp Population= +population increase-catch
Initial value: 30000
Units: Tons

(02) % of biomass increment = 0.5
Units: 1/Year

(03) catch = k*Shrimp Population
Units: Tons/Year

(04) highest population= 37500
Units: Tons

(05) population increase= % of biomass increment*Shrimp Population*(1-(Shrimp Population/highest population))
Units: Tons/Year

(06) k= 0.1
Units: 1/Year

Model 2

Model 2 attempts to represent the conceptual model that the fishing boats of today are more efficient in obtaining their catch than previously, extending the simplistic Schaeffer model. It also relates changes in fleet size to the profitability of fishing.

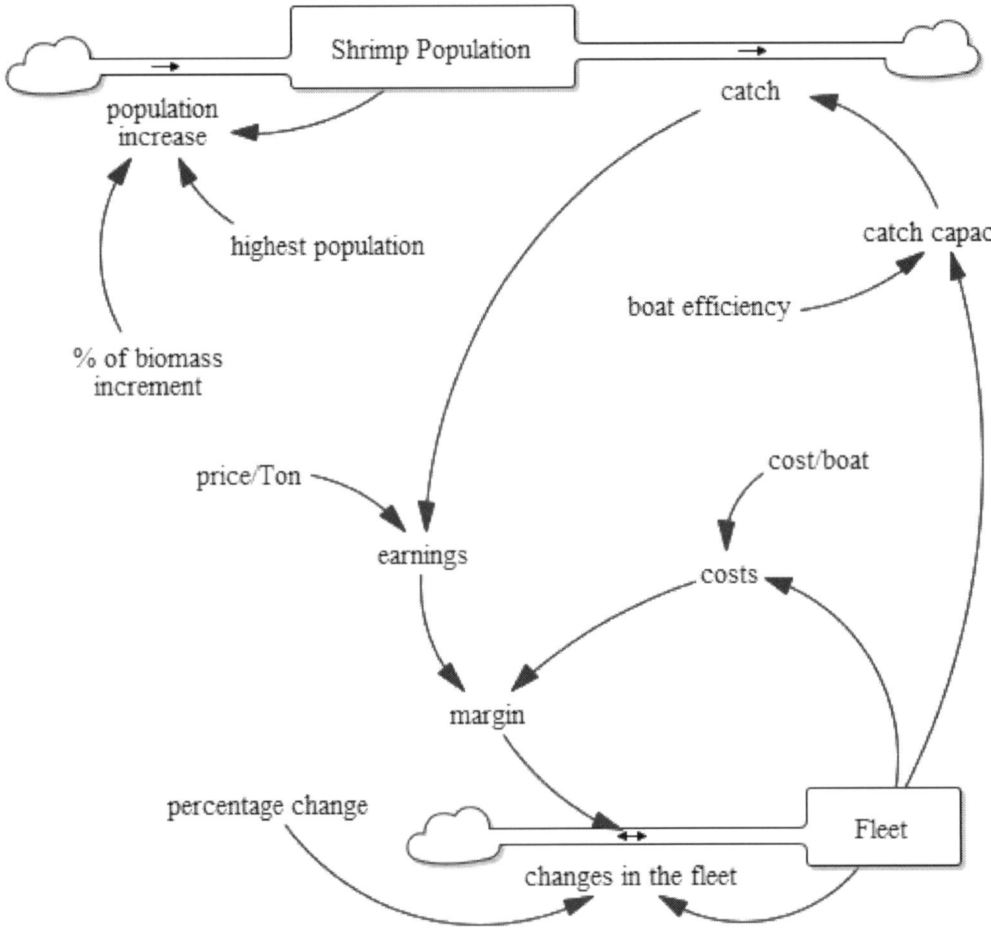

(01) % of biomass increment=0.5
Units: 1/Year

(02) captures= catch capacity
Units: Tons/Year

(03) catch capacity = Fleet* boat efficiency
Units: Tons/Year

(04) changes in the fleet=IFTHEN ELSE(margin>0.2, Fleet*percentage change, 0)
Units: boat/Year
When margin is more than 20% appears an increase in the number of boats.

(05) cost/boat= 1
Units: $/boat/Year
Milion $ per Year per boat. Costs of amortization and maintenance of the boat every year

(06) costs= Fleet*cost/boat
Units: $/Year

(07) earnings= catch*price/Ton
Units: $/Year

(08) boat efficiency = 60
Units: Tons/boat/Year
initial value 3,000 Tons captured / 50 boats = 60 Tons/boat

(09) Fleet= changes in the fleet
Initial value: 50
Units: boat
Number of boats

(10) highest population= 37500
Units: Tons

(11) population increase=% of biomass increment* Shrimp Population *
(1-(Shrimp Population / highest population))
Units: Tons/Year

(12) margin= (earnings-costs)/costs
Units: Dmnl

(13) percentage change= 0.05
Units: 1/Year
% of annual variation in the number of boats based on the margin.

(14) Shrimp Population = population increase - catch
Initial value: 30000
Units: Tons

(15) price/Ton= 0.02
Units: $/Tons
million of $ per Ton of shrimp

Comment 1. At the initial values of 50 boats and 3000 tons catch, the system is in equilibrium. Because the price of shrimp is fixed and the costs per boat are fixed, the profit 'margin' is zero so boats neither enter nor leave the fleet. Also in this model, the boats will catch as much as they are able ('catch capacity'). However, this catch appears to be sustainable as population is steady.

Comment 2 Just at a small increase in boat efficiency (say, 60.5 tons/boat instead of 60 tons/boat by staying out an extra day), the catch increases. Therefore, the earnings increase as would margin. But the fleet increases in response and so the number of boats and hence, fleet costs increase almost in parallel. Thus the margin actually remains the same. This is basic macro-economics although there is no realistic delay between change in margin and entry of new boats to the fleet.

However, as the fleet increases, behind the scenes, the shrimp population drastically declines, reaching zero in about 20 years. Thus the model accurately represents the real situation in Campeche.

This reflects the fact that there is no connection in this model between the catch and the population and between population and long-term fleet efficiency and costs. Fleet size and hence, catch are driven purely by short-term profit interests. In that sense, the model #2 is not suited for good management of the resource or the industry.

Note also that any decline in catch due, for example, to decreased boat efficiency (e.g., it takes more days of fishing with a diminished population) is not reflected in the 'Changes to fleet 'IF THEN ELSE'' formulation.

Model 3

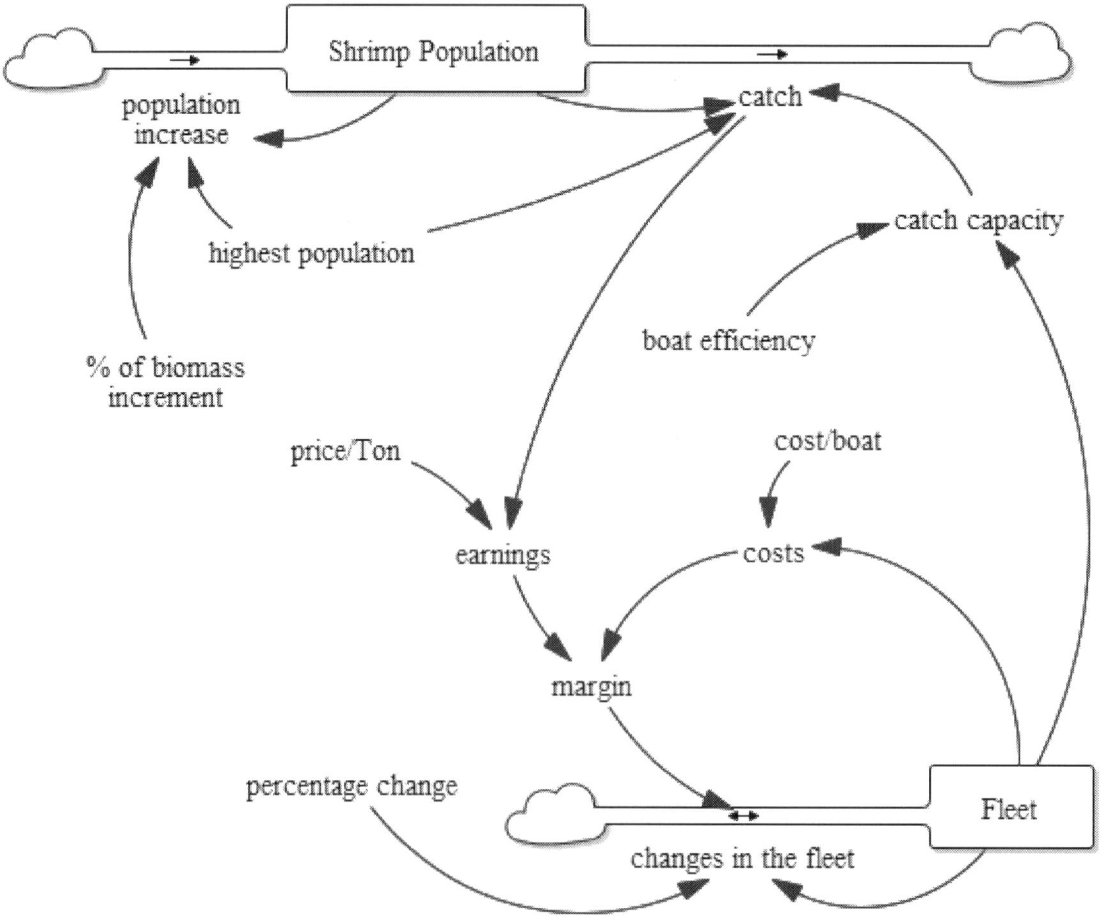

(01) % of biomass increment= 0.5
Units: 1/Year

(02) catch= catch capacity*(Shrimp Population / highest population) *RANDOM NORMAL(0.8,1.2,1,0.1,5)
Units: Tons/Year
Captures are based in the Captures Capacity, and also in the Population of shrimps related to the highest possible value. We add a random component to simulate occasional changes.

(03) catch capacity = Fleet*boat efficiency
Units: Tons/Year

(04) changes in the fleet= IF THEN ELSE(margin>0.2, Fleet *percentage change*margin, -Fleet*percentage change*ABS(margin))
Units: boat/Year
When margin is more than 20% appears an increase in the number of boats. The change of the fleet in proportional to the margin.

(05) cost/boat= 1
Units: $/boat/Year
milion $ per Year per boat. Amortization and maintenance of the boat every year

(06) costs= Fleet*cost/boat
Units: $/Year

(07) earnings=catch*price/Ton
Units: $/Year

(08) boat efficiency =60+RAMP(3,1950,2050)
Units: Tons/boat/Year
Initial value 3,000 Tons captured / 50 boats = 60 Tons/boat. We include the efficiency increase based on the aspects described in the text.

(09) Fleet= changes in the fleet
Initial value: 50
Units: boat
number of boats

(10) highest population= 37500
Units: Tons

(11) population increase=% of biomass increment*Shrimp Population *(1-(Shrimp Population/highest population))
Units: Tons/Year

(12) margin= (earnings-costs)/costs
Units: Dmnl

(13) percentage change= 0.05
Units: 1/Year
% of annual variation in the number of boats based on the margin.

(14) Shrimp Population = population increase-catch
Initial value: 30000
Units: Tons

(15) price/Ton= 0.02
Units: $/Tons
million of $ per Ton of shrimp

Simulate and display the results

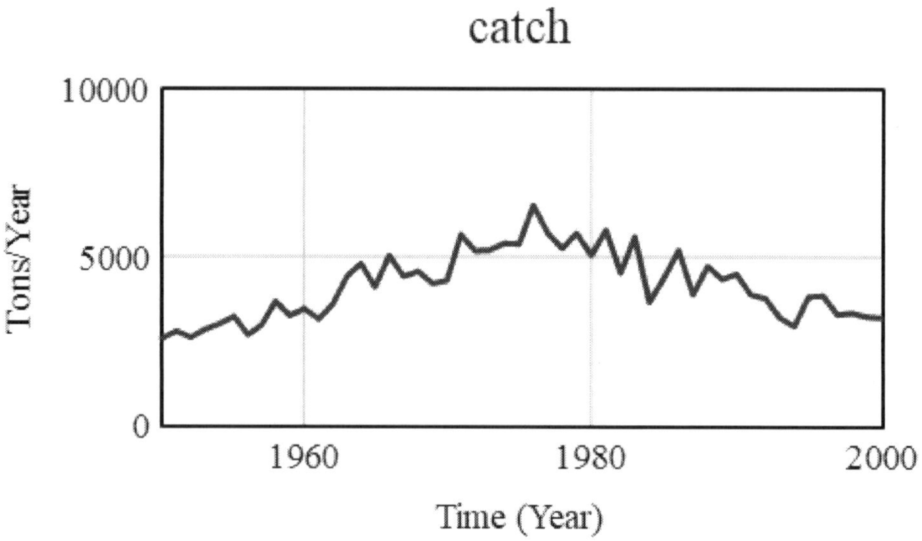

The output replicates even more dramatically the scenario 1 produced in Comment above. As the boat efficiency increases (RAMP function), the catch increases until the margin becomes great enough to attract more boats (~1960). Then the fleet rapidly increases in response to margin so that the catch begins to exceed the replacement.

Then, around 1975 the catch falls in response to the much lower ‘shrimp to highest population’ ratio, the sustainability factor.

The introduction of a random variability in the catch, perhaps realistically related to fleet capacity - affecting storms, the price of fuel (c.f. 1967 and 1973) and seasonal variations in highest population carrying capacity - make the results conform believably to the actual data giving us confidence in the model.

3.5. Rabbits and foxes

This model gathers the dynamics of two typical populations of prey and predator. We will use this example to see if it is possible to create a model without having drawn the causal diagram. For this we can start with a type of problem 'the population of rabbits has important oscillations that we want to reduce'. A subsequent investigation has demonstrated that foxes are the cause of this oscillation. Here is a list of elements that can be related to a problem of this type:

- population of rabbits
- populations of foxes
- birth of rabbits
- death rate of foxes
- birth rate of rabbits
- birth of foxes
- decrease in the number of foxes
- life expectancy of foxes
- birth rate of foxes
- sustainable population of rabbits

A good way of identifying the stocks of a system is to make a mental photograph of the system and assign the characteristic levels to those elements that appear in the image. In this case, the populations of foxes and rabbits would be the stocks. The flows are temporary variations in the stocks and here we will consider the births and decreases in the populations of foxes and rabbits. The rest of the elements are auxiliary variables.

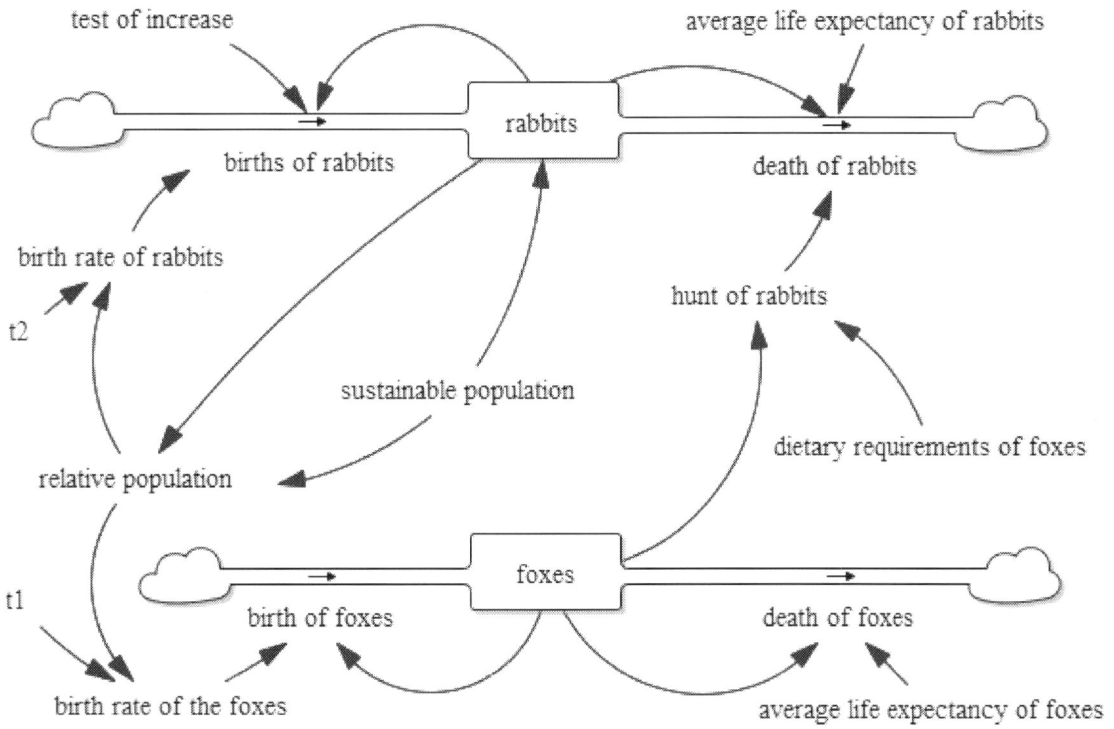

In our case, we are going to design the model so that it reproduces a stable situation of the populations of foxes and rabbits and then we will simulate the effect of a minor change in one of the variables, namely the birth of rabbits. In this way we can see with clarity if the structure of the system decreases or increases this small change or if oscillations are created.

Model - Settings

INITIAL TIME= 0 FINAL TIME= 100   TIME STEP = 1 Units for time=Month

Equations 

(01) average life expectancy of foxes = 4

(02) average life expectancy of rabbits= 2

(03) birth of foxes=foxes*birth rate of the foxes

(04) birth rate of rabbits= t2(relative population)

(05) birth rate of the foxes= t1(relative population)

(06) births of rabbits=(rabbits*birth rate of rabbits)+test of increase

(07) death of foxes= foxes/average life expectancy of foxes

(08) death of rabbits=MAX(rabbits/average life expectancy of rabbits, hunt of rabbits)

(09) dietary requirements of foxes = 25

(10) foxes= +birth of foxes-death of foxes
initial value = 40

(11) hunt of rabbits=foxes*dietary requirements of foxes

(12) rabbits= +births of rabbits-death of rabbits
initial value = sustainable population

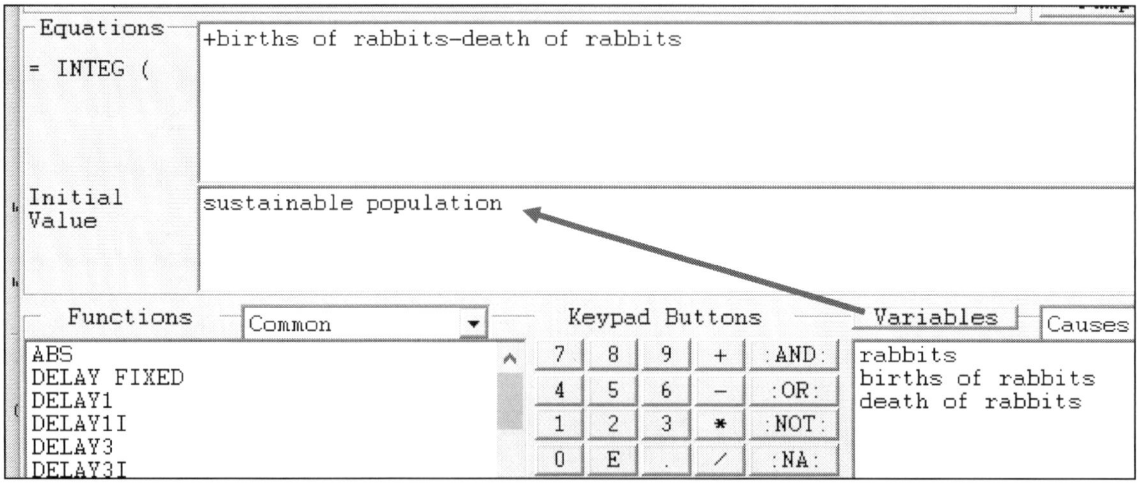

(13) relative population = rabbits/sustainable population

(14) sustainable population= 500

(15) t1 = (0,-0.12),(0.5,0.12),(1,0.25),(1.5,0.27),(2,0.3)
Note: the balanced value is 0.25 when the relative population = 1
Figures in brackets do not affect model calculations.

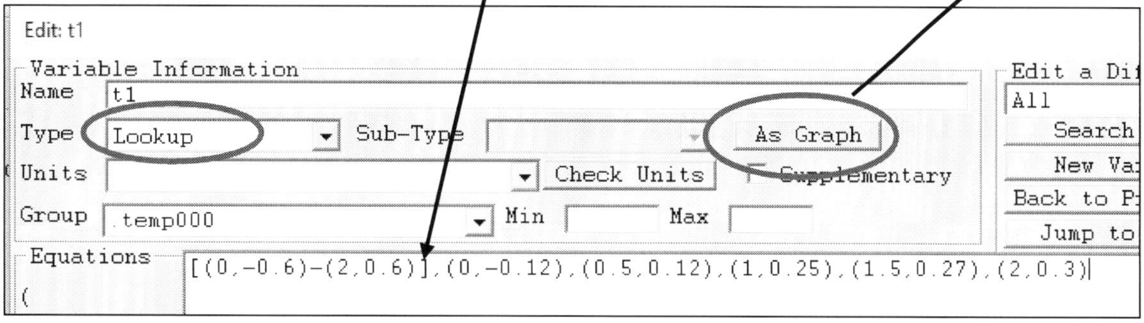

(16) t2 = (0,4),(0.5,3.5),(1,2),(1.5,1.25),(2,1)
Note - the value 2 is the point of equilibrium for the relative population = 1

(17) test of increase= PULSE(10,1)*100

Simulate and display the behaviour

It is observed that this system responds to a small disruption producing an oscillatory system of constant amplitude.

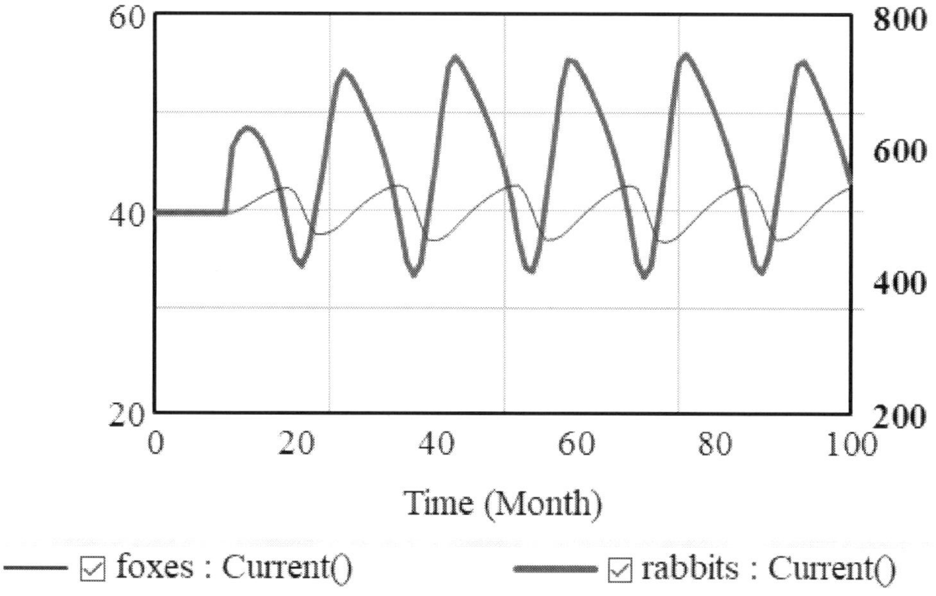

This graph has been created using Control Panel – Custom Graph – New.

3.6. A study of hogs

Many systems tend to oscillate in response to external disturbances. The national economy, for example, shows cycles of high and low Gross National Product and employment. Animal populations (such as rabbits and foxes) often exhibit oscillations.

In this exercise we will explore an example of cyclical behaviour: the production and consumption of pork products in Spain. While the pork example is interesting in itself, it is also important because it focuses on some general questions about the nature of oscillations and the stability of systems in response to disturbances.

The production and consumption of pork products involve three groups - farmers, butchers, and consumers. Butchers buy hogs from farmers, dress the hogs to obtain pork products, and then sell the pork products to consumers.

In order to understand the cycles that take place in the production and consumption of pork products, it will be helpful to build two models - one focusing on the process of breeding and fattening hogs on the farm, and the other focusing on the sale of pork products. The two models can then be combined to generate cyclical behaviour. It is easier to begin with a model of the sale of pork products, then turn to the breeding of hogs, and finally combine the two models into one larger model and analyze the behaviour the combined model generates.

Model 1: The Sale and Consumption of Pork

The main element in the model of the sale of pork is the inventory of pork maintained by butchers. When hogs are slaughtered on the farm, the pork products that are produced are added to the inventory kept by butchers. When pork is sold, it is taken from the inventory. The amount of pork people buy depends on the price of pork. The price of pork, in turn, depends on the size of the pork inventory. When the inventory is high, prices fall. When the inventory is low, prices rise.

We will assume that the number of hogs slaughtered each month is a constant - 750,000 per month. Hogs weigh about 100 kilos each, and the 'dressed yield' of pork is about 80 percent (this means that each hog produces 0.80*100= 80 kilos of pork). Thus a slaughter rate of 750,000 hogs per month corresponds to a pork production rate of 60 million kilos of pork each month.

On the average, each person in Spain normally consumes about 1.5 kilos of pork each month. With a population of 40 million people, the total amount of pork normally consumed each month is 1.5*40 million = 60 million kilos of pork per month. But when the price is high, relative to the normal price of pork, people usually consume somewhat less than 1.5 kilos each month. When the price is low, they consume somewhat more.

The price of pork depends on the price of hogs, and the price of hogs depends on supply and demand. Assume that butchers try to keep on hand about 2 weeks' worth of the amount of pork products they usually sell to avoid stocks-outs (we use the term 'coverage' for this). When the available inventory of pork runs low, relative to their normal inventory, butchers are willing to pay higher prices to get hogs. When the butchers' inventory of pork is high, butchers are less willing to buy and the price of hogs falls.

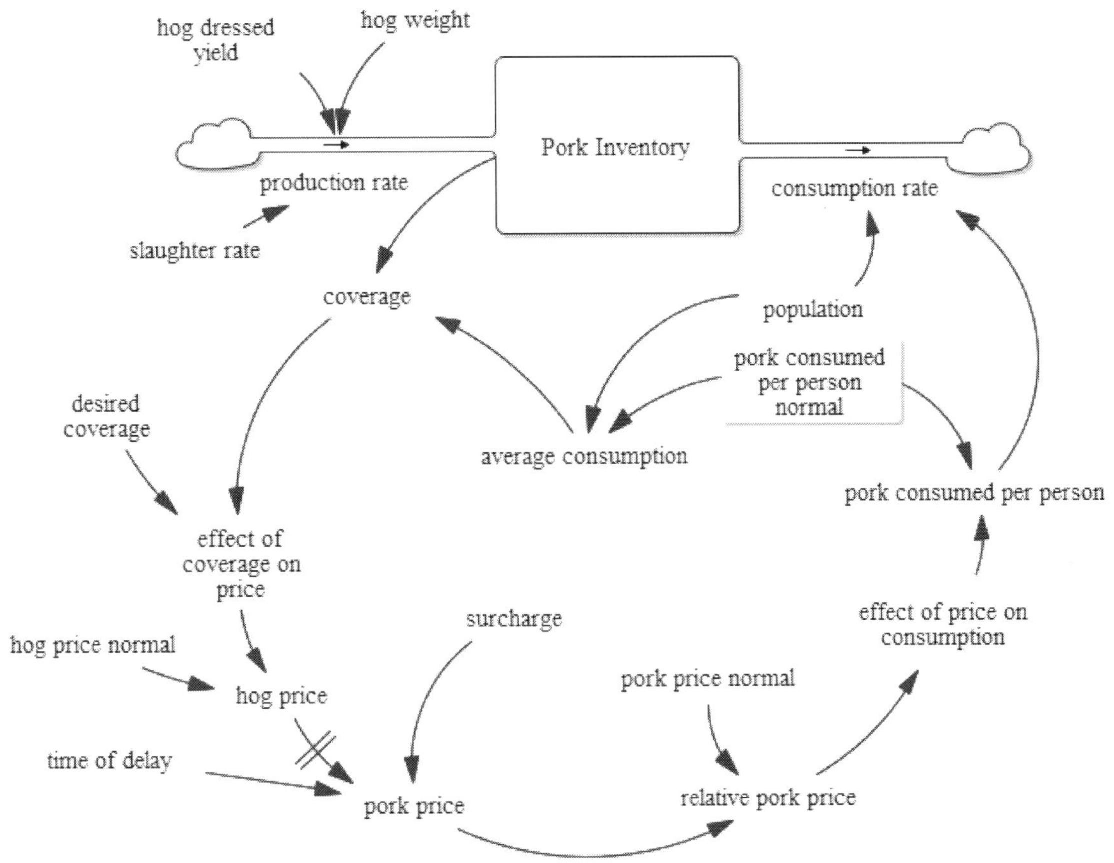

Assume that the normal price of hogs at slaughter is 3 euros per kilo and that butchers charge consumers a 7 euro/kg premium over their price of hogs (3 euros/kg), resulting in the price of pork to consumers of 10 euros/kg.

Now, draw the flow diagram of the sale and consumption of pork, and write equations for the model. Put the equation of slaughter rate= 750,000 and see if the model is in equilibrium. If it isn't, you have made a mistake in the equations :-)

Once you have obtained a model that runs in equilibrium, try testing its response to exogenous disturbances. For example, how does the system respond to a 6 month 10,000 increment in the hog slaughter rate, starting at month 6? This can be done using the adding the function PULSE(6,6)*10000 in the equation of slaughter rate.

Let's simulate 48 months. Click File – New Model and define as settings:
INITIAL TIME=0 FINAL TIME= 48 TIME STEP=1 Units for time=Month

Equations 

(01) average consumption= population*pork consumed per person normal
Units: kg/Month

(02) consumption rate= population*pork consumed per person
Units: kg/Month

(03) coverage= Pork Inventory/average consumption
Units: Month

(04) desired coverage= 0.5
Units: Month

(05) effect of coverage on price = coverage/desired coverage
Lookup: (0.4,1.5),(0.9,1.2),(1,1),(1.6,0.9),(1.8,0.8),(2,0.5)
Units: Dmnl

(06) effect of price on consumption = relative pork price
Lookup: (0.5,1.5),(1,1),(1.5,0.9),(2,0.75)
Units: Dmnl

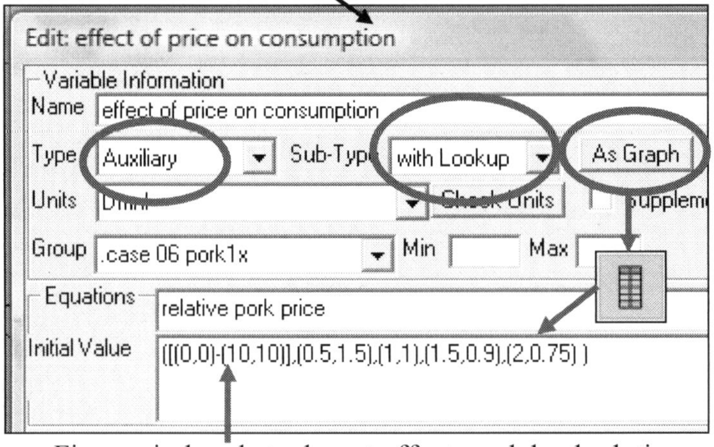

Figures in brackets do not affect model calculations

(07) hog dressed yield= 0.8
Units: Dmnl

(08) hog price= effect of coverage on price *hog price normal
Units: Euro/kg

(09) hog price normal= 3
Units: Euro/kg

(10) hog weight= 100
Units: kg/hog

(11) population= 40000000
Units: person

(12) pork consumed per person= pork consumed per person normal*effect of price on consumption
Units: kg/(person*Month)

(13) pork consumed per person normal= 1.5
Units: kg/person/Month

(14) pork price= SMOOTH(hog price+surcharge, time of delay)
Units: Euro/kg

(15) Pork Inventory= +production rate-consumption rate
Initial value: 30000000
Units: kg

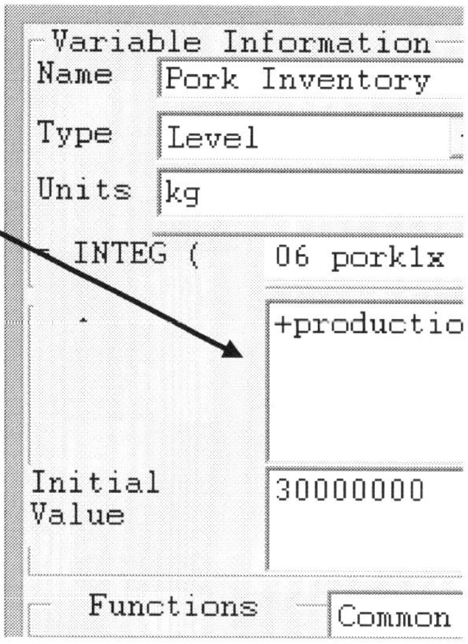

(16) pork price normal= 10
Units: Euro/kg

(17) relative pork price= pork price/pork price normal
Units: Dmnl

(18) slaughter rate= 750000+PULSE(6,6)*10000
Units: hog/Month

(19) surcharge= 7
Units: Euro/kg

(20) time of delay= 3
Units: Month

(21) production rate= slaughter rate*hog weight*hog dressed yield
Units: kg/Month

Take a look to the PULSE and SMOOTH functions in the Annex chapter.

Simulate and save the model as Pork1.

Model 2: The Breeding and Maturation of Hogs

In Model 1, we assumed the hog slaughter rate was an exogenous constant. Now let us develop a model of the breeding and maturation of hogs to simulate the slaughter rate. For this model, we will assume that the price of hogs is an exogenous constant. In Model 3, we will combine the two models to simulate both the hog price and the slaughter rate.

Farmers distinguish between two kinds of hogs - hogs for market and hogs for breeding stock. Hogs for market (which can be either male or female) are fattened for about 6 months after they are born, and then they are slaughtered. Females intended for market are not bred. Females to be bred (called sows) are raised separately as 'breeding stock,' and they are used entirely for breeding. We will assume that they are not slaughtered for pork.

Farmers change the size of their hog herd by adjusting the number of sows in their breeding stock. When the price of hogs is higher than normal, farmers increase the size of the breeding stock, and when the price of hogs is lower than normal, they decrease the size of the breeding stock.

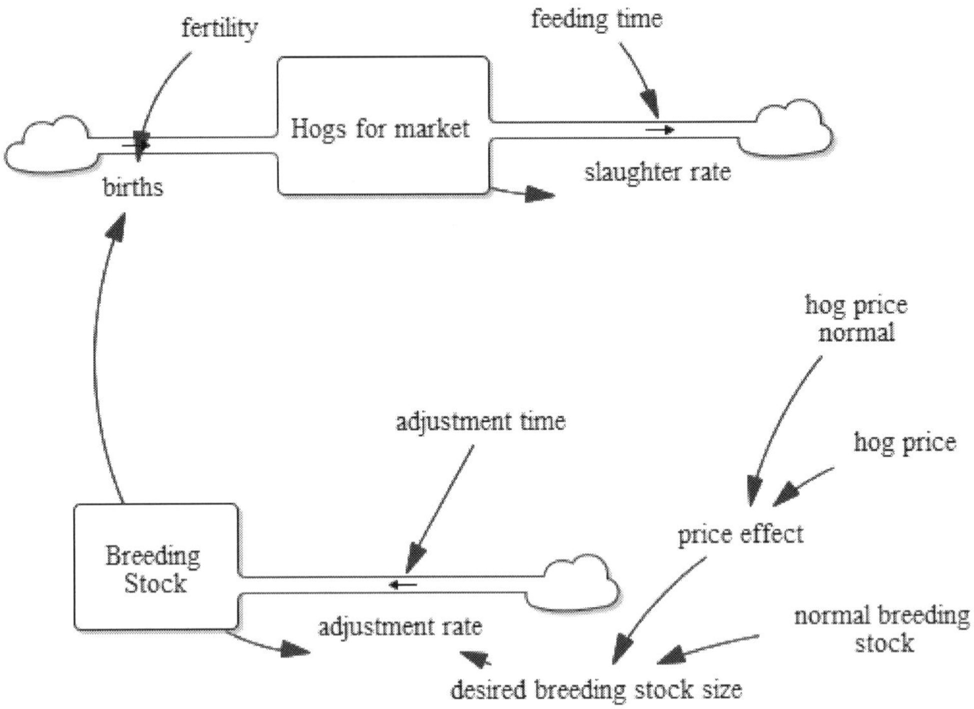

Draw the flow diagram and write the equations for the model of the breeding and maturation of hogs. Assume that each sow gives birth to 18 hogs per year (1.5 per month). Also, assume that hogs must be fattened for 6 months before they are ready for market. The most difficult part of the model is the 'breeding stock adjustment rate.' We will formulate this using a 'desired breeding stock size' as a function of hog price. This adjustment can be expressed most easily as a percentage of the actual breeding stock size. The breeding stock adjustment rate can then be formulated as the difference between the 'desired size' and the 'actual size,' divided by an adjustment time. Set the hog price at 3 euros/kilo. We will choose values for the breeding stock and for number of hogs that will produce an equilibrium slaughter rate of 750,000 hogs per month (the value we assumed in Model 1.)

Click File – New Model and define:
INITIAL TIME=0 FINAL TIME= 48 TIME STEP=1 Units for time=Month

(01) adjustment rate=(desired breeding stock size-Breeding Stock)/adjustment time
Units: hog/Month
We like to simulate that in every period the farmer compares how many breeding stock he has, and how many he wants, and he don't make immediately an adjustment equal to this difference. He makes a progressive adjustment. To simulate this adjustment we use the adjustment time.

(02) Hogs for market=births-slaughter rate
Initial value: 4500000
Units: hog

(03) births=Breeding Stock*fertility
Units: hog/Month

(04) Breeding Stock= adjustment rate
Initial value: 500000
Units: hog

(05) desired breeding stock size= normal breeding stock*price effect
Units: hog

(06) normal breeding stock= 500000
Units: hog

(07) feeding time= 6
Units: Month

(08) fertility= 1.5
Units: 1/Month

(09) hog price= 3
Units: Euro/kg

(10) hog price normal= 3
Units: Euro/kg

(11) adjustment time= 3
Units: Month

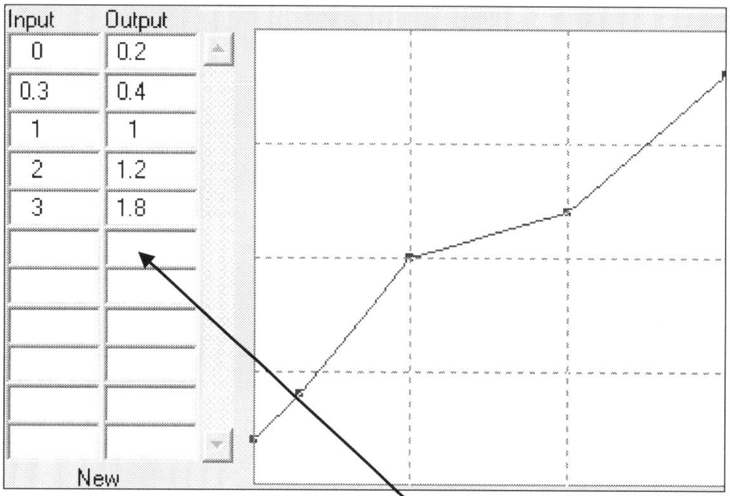

(12) price effect = hog price/hog price normal
Lookup: (0,0.2),(0.3,0.4),(1,1),(2,1.2),(3,1.8)
Units: Dmnl

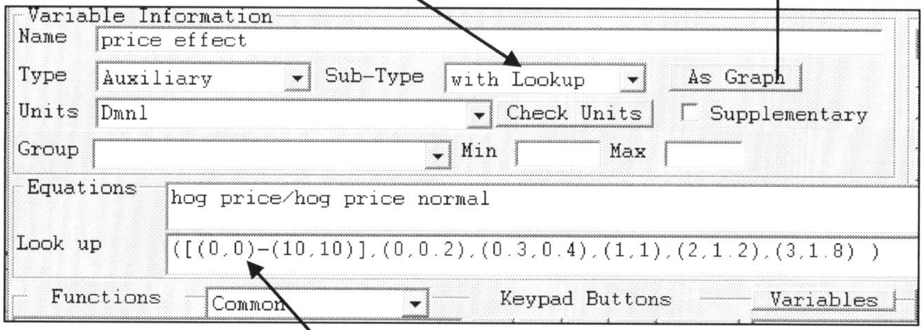

Figures in brackets do not affect model calculations

(13) slaughter rate=(Hogs for market/feeding time)
Units: hog/Month

Simulate and save the model as Pork2. Graphs shows constant values for all the variables.

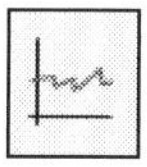

You can test if the units are correct clicking the icon: 

Model 3: Combining the Two Models

Now let us merge the models developed, please see the next page.

If you have the model Pork2 in the screen, save it and click Select All – Copy, later File – Open model, and choose the Pork 1 model. Click Edit –Paste and move to the right (with the paste icon activated) the added elements. Save the new model as Pork3.

Now you must use the hog price from model 1 in place of the hog price in model 2, and the slaughter rate from model 2 in place of the slaughter rate in model 1.

Modify the equations for "price effect" and "Production rate", using the variables shown in the list.

Simulate the model and examine the results. If you use the same initial values as those you chose in Models 1 and 2 (with slaughter rate= 750000), the model should be in equilibrium.

Once you have obtained an equilibrium run, you can test the model's response to external disturbances. For example, activate the test as you did in model 1. What happens? Can you explain why the cycles occur?

Two features of cyclical behaviour are particularly important - period and damping. The period of a cycle is the time that elapses from one peak to the next. Damping refers to the degree to which oscillations fade away over time. For the parameter values we have chosen, the hog model produces stable oscillations.

Based on the book of Dennis L. Meadows, Dynamics of Commodity Production Cycles (Cambridge, Mass.: MIT Press).

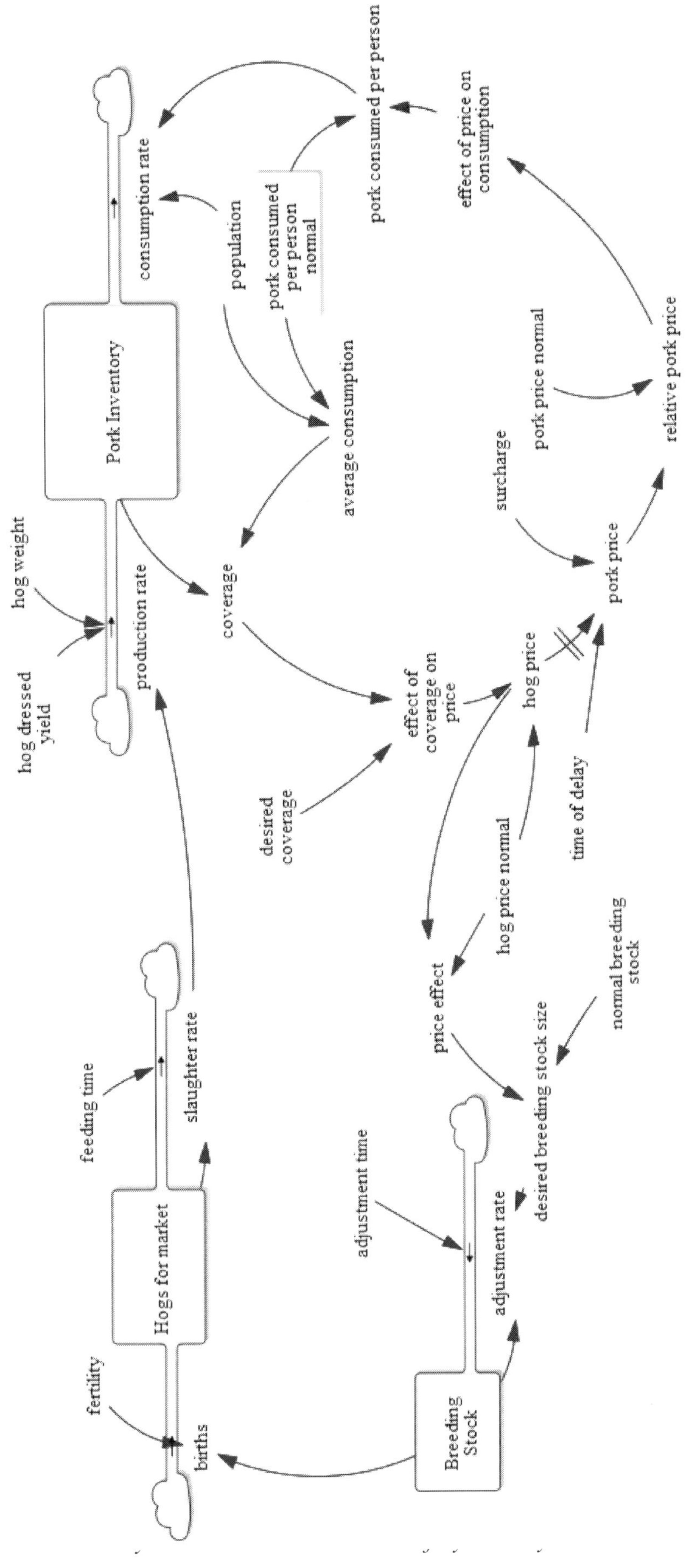
Pork Inventory
consumption rate
production rate
hog weight
hog dressed yield
population
pork consumed per person normal
pork consumed per person
effect of price on consumption
average consumption
coverage
desired coverage
effect of coverage on price
hog price
hog price normal
time of delay
pork price
surcharge
pork price normal
relative pork price
price effect
normal breeding stock
desired breeding stock size
Hogs for market
slaughter rate
feeding time
fertility
births
Breeding Stock
adjustment rate
adjustment time

(01) adjustment rate=(desired breeding stock size-Breeding Stock)/adjustment time
Units: hog/Month

(02) adjustment time=3
Units: Month

(03) average consumption=population*pork consumed per person normal
Units: kg/Month

(04) births=Breeding Stock*fertility
Units: hog/Month

(05) Breeding Stock= adjustment rate
Initial value: 500000
Units: hog

(06) consumption rate= population*pork consumed per person
Units: kg/Month

(07) coverage= Pork Inventory/average consumption
Units: Month

(08) desired breeding stock size= normal breeding stock*price effect
Units: hog

(09) desired coverage= 0.5
Units: Month

(10) effect of coverage on price = coverage/desired coverage
Lookup: (0.4,1.5),(0.9,1.2),(1,1),(1.6,0.9),(1.8,0.8),(2,0.5)
Units: Dmnl

(11) effect of price on consumption = relative pork price
Lookup: (0.5,1.5),(1,1),(1.5,0.9),(2,0.75)
Units: Dmnl

(12) feeding time= 6
Units: Month

(13) fertility= 1.5
Units: 1/Month

(14) hog dressed yield= 0.8
Units: Dmnl

(15) hog price=effect of coverage on price*hog price normal
Units: Euro/kg

(16) hog price normal=3
Units: Euro/kg

(17) hog weight=100
Units: kg/hog

(18) Hogs for market= births-slaughter rate
Initial value: 4500000
Units: hog

(19) normal breeding stock= 500000
Units: hog

(20) population= 40000000
Units: person

(21) pork consumed per person= pork consumed per person normal*effect of price on consumption
Units: kg/(person*Month)

(22) pork consumed per person normal= 1.5
Units: kg/person/Month

(23) Pork Inventory= +production rate-consumption rate
Initial value : 30000000
Units: kg

(24) pork price=SMOOTH(hog price+surcharge,time of delay)
Units: Euro/kg

(25) pork price normal= 10
Units: Euro/kg

(26) price effect = hog price/hog price normal
Lookup: (0,0.2),(0.3,0.4),(1,1),(2,1.2),(3,1.8)
Units: Dmnl

(27) production rate= slaughter rate*hog weight*hog dressed yield
Units: kg/Month

(28) relative pork price= pork price/pork price normal
Units: Dmnl

(29) slaughter rate= (Hogs for market/feeding time)+ PULSE(6,6)*10000
Units: hog/Month

(30) surcharge= 7
Units: Euro/kg

(31) time of delay= 3
Units: Month

To display the equations of a model click the icon Document or Document All.

Simulate and display the results

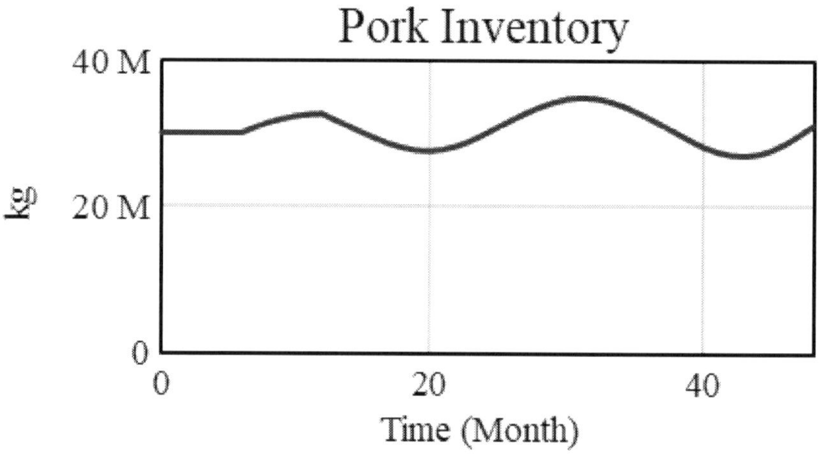

X-Y graphs

We can better see the trend of a system with oscillations using the X-Y graph. Go to the Control Panel - Custom Charts - New and define the two variables that we want to see together. In this case we will use the Pork Inventory and Hogs for market.

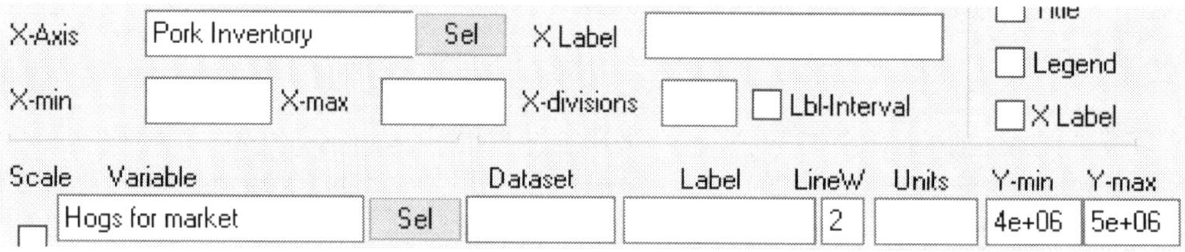

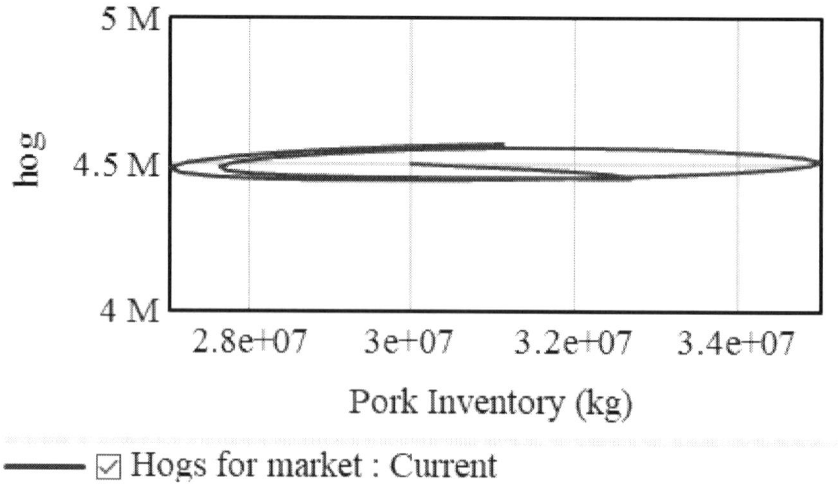

The graph shows the evolution of the values in each period of these variables and how a spiral is generated, which is the other way to visualize increasing oscillations. If we look closely at the center of the spiral it corresponds to the initial values of these variables, and thus Pork Inventory has a value of 30,000,000 and the Hogs for market are 4,500,000

3.7. Ingestion of toxins

The human being has recognised the venomous character of certain plants and animals from the first stages of history. Nevertheless, the enormous development of industrial activity, particularly the chemical industry, has caused the creation of a large number of artificial chemical substances, both in the quality of primary materials and in the residues of certain processes. Many of these chemicals have adverse effects on nature and human beings.

The study of these substances and their effects on humans has introduced a relatively new science, toxicology, which in its development has incorporated elements from other sciences such as physiology, pharmacology, biochemistry and epidemiology. What arises from this is that the study of the problematic derivative from the exposure to toxic substances requires the collaboration of various disciplines. Therefore it is interesting to view these problems from the System Dynamics point of view that incorporates the different elements that compose them in a model of analysis and measure that operates systematically and consequently allows us to observe the mutual influences that take place.

When an organism is exposed to a toxic substance, a series of complex processes of absorption, distribution, metabolism and elimination (ADME) take place whose speeds are difficult to calculate.

The simplest outline enables us to imagine an organism receiving a determined quantity of the toxin and at the same time eliminating a proportion of the toxin. If the elimination is very rapid, it is probable that the organism can tolerate new quantities in divided doses. However, if the elimination is slow, exposure during a long period of time can cause an accumulation of the toxin that goes over the limit resulting in anything from allergic reactions to death.

Nevertheless, the danger of a substance is not enough to characterise the risk of intoxication. All the factors which contribute to the organism incorporation of toxins must be considered, such as the exposure scenario, the concentration in the contaminated environment, the channel, the exposure's frequency and duration and the characteristics of the individual.

On the other hand the analysis of risk is also a tool which allows us to implement mitigation measures.

The following model simulates the emission of a gaseous toxin and its inhalation by an individual exposed to the influence. Various aspects of the process are organized in the following way:

1) The emission is toxic from a specific source, calculated in molecules/hour in a given temperature expressed in K° from an evaporation surface measured in m2. The pressure of the vapour is the substance in consideration, at the temperature in the environment. The system of transference to the environment is characterised by a coefficient of the transference of the mass.

2) The scenario is constituted by the location where the source of emission is contained. Its volume allows us to calculate the effective concentration that is expressed in mg/m3 after multiplying the emission in mol/hour by the molecular weight of the substance. This concentration is compared with the permitted limit (in this case the dose of reference for inhalation) and through the relationship between both elements, the rate of air renewal is regulated as a mitigation measure.
3) Finally, the stock of toxin accumulated represents the quantity in the organism exposed. In the organism exposed, a quantity known as Incorporation is measured in mg/hour. This is determined by the variable concentration, the rate of inhalation, the amount of air breathed by a healthy individual every hour, and the effective exposure that allows us to simulate a working week with a set shift pattern, daily breaks and weekends.

The toxin accumulated is lowered by elimination, which responds to a kinetic of the first order. In other words, the quantity accumulated is proportional to the quantity present depending on the Elimination Coefficient. As additional data the Internal Concentration is calculated dividing the toxin accumulated by the body weight.

The model

When a toxic substance enters in a living organism, an extremely complex series of reactions take place. The process involves absorption, metabolism and elimination mechanisms, the velocity of which is very difficult to calculate.

The simplest scheme is to imagine a known quantity of toxin entering the organism, which is capable of eliminating a part of the entering toxin. If the elimination rate is very high, it is probably because the organisms tolerate new ingestions of toxic. Adversely, the organisms could reach or exceed the toxicity threshold, undergoing adverse consequences - thus allergenic reaction until death.

Consequently, the knowledge about the toxic characteristics of a substance is necessary to take into account the duration, concentration and frequency during which the organism is exposed to ingestion.

1) **Accumulated Toxic** Stock that simulates the exposed organism.

2) The organism incorporates a toxic quantity called **Incorporation,** in mg/h, which depends on **Concentration** (environmental toxic concentration) and **Exposure Hours** (the time in hours during which the organism is exposed to toxin)**.** In addition, **Exposure Hours** allows the simulation of a work week with certain **shift**, daily and week-end rest.

3) Some toxic quantity is released by the organism, through **Elimination** flow, according a first order kinetics. In other words, the eliminated quantity is proportional to accumulated quantity. The equation for this phenomena is:

$$dC/dt = k\,C$$

where ***k*****, the elimination coefficient**, is a first order constant, measured in fractions per unit time, and involve all the phenomena complexity.

This kind of reaction allows an easy estimation of 'half time', the time necessary to reach a concentration equal to a half of the initial concentration. The equation is:

$$t_{1/2} = -(\ln C/Co)/k \qquad \text{where} \quad C/Co = 0.5$$

Replacing C/Co in the equation, $t_{1/2}$ depends only on k.

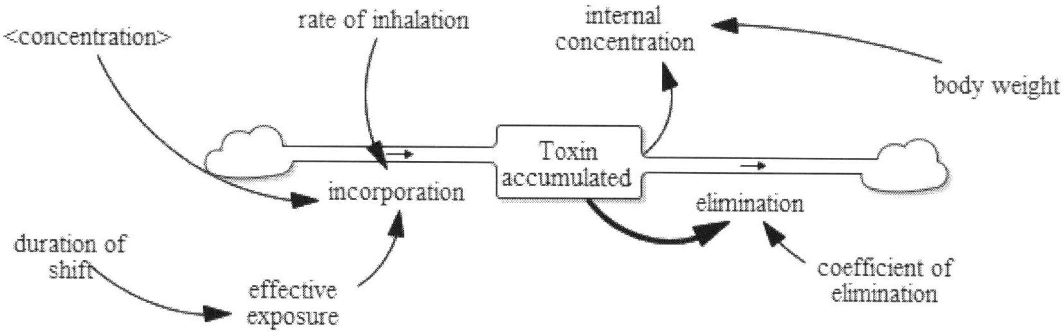

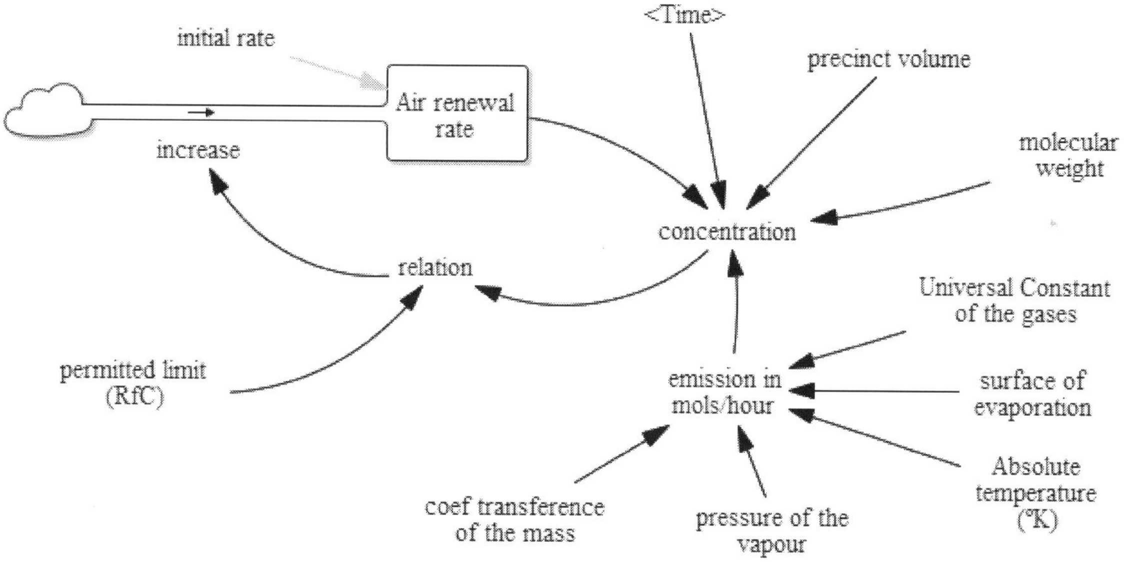

Note: Use the Shadow Variable tool to write the <Time> variable 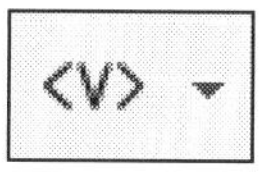

Model - Settings

INITIAL TIME= 0 FINAL TIME= 168 TIME STEP=1 Units for time =h (hours in a week)

Equations 

(01) coef transference of the mass = 8.7
Units: m/h
It measures the velocity of the transference from the liquid phase to the vapour phase.

(02) coefficient of elimination= 0.035
Units: 1/h
It contains all the complexity of the organic process for the elimination of the toxin. The greater the coefficient, the greater the elimination.

(03) concentration = emission in mols/hours *molecular weight* (1-EXP(-Air renewal rate * time))/(Air renewal rate* precinct volume)
Units: mg/m3
Concentration of toxin in the air.
USE THE SHADOW VARIABLE ICON TO ADD THIS VARIABLE IN THE SECOND PART OF THE DIAGRAM 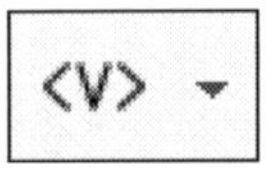

(04) internal concentration = toxin accumulated / body weight
Units: mg/kg
Internal concentration of the toxin.

(05) Universal Constant of the gases = 8.314
Units: Jule/mol ºK

(06) duration of shift = 8
Units: h
Duration of the shift at work

(07) elimination = coefficient of elimination * toxin accumulated
Units: mg/h
Simulate a kinetic of elimination of the first order, proportional to the concentration of toxins present in the organism.

(08) emission in mols/hour = surface of evaporation * pressure of the vapour* coefficient transference of the mass/(universal constant of the gases * Temperature absolute (ºK)).
Units: mols/h
Mols in balance in the gas phase.

(09) effective exposure = PULSE TRAIN (0, shift duration, 24, 119)
Units: h
This represents the number of hours during which there is toxic exposure during the working shift, in which case the response of PULSE TRAIN is 1. The subject is not exposed during breaks given that the PULSE TRAIN is 0.

(10) incorporation = effective exposure * concentration * rate of inhalation
Units: mg/h

(11) increase=IF THEN ELSE(relation>=1.1,1, IF THEN ELSE(relation<= 0.9,-0.5 ,0))
This allows an increase or decrease in the rate of air renewal according to the relation value.
Units: 1/h/h

(12) permitted limit (RfC) = 0.0003
Units: mg/m3
This is the maximum environmental concentration allowed that assures the preservation of health.

(13) body weight =64
Units: kg
Weight of the individual exposed to the toxin.

(14) molecular weight = 200600
Unit: mg
Molecular weight of the toxin expressed en mg

(15) pressure of the vapour = 0.301
Units: Pascal
Vapour pressure of the substance that it is evaporating (Pascal = N/m2)

(16) relation = concentration / permitted limit (RfC)
Units: Dmnl (Dimensionless)
Relations between the concentration of the environment and the concentration of permitted exposure.

(17) surface of evaporation = 0.02
Units: m2
Free surface through which toxic evaporation takes place

(18) rate of inhalation = 0.8
Units: m3/h
The normal inhalation of air of an adult in m3/h

(19) initial rate = 1
Units: 1/h
The initial number of times per hour that the air recycles itself

(20) Air renewal rate = increase
Initial value: Initial rate
Units: 1/ h
The number of times the air renews itself per hour, variable in order to meet the permitted limit.

(21) Absolute temperature (ºK) = 298
Units: ºK
Absolute temperature of the studied environment.

(22) Toxin accumulated = + incorporation – elimination
Initial Value: 0
Units: mg
It is the quantity of the toxin that accumulates during the working week

(23) precinct volume = 1000
Units: m3
The volume of the location where the toxin is produced and where the exposure takes place.

Conclusions

1) The model allows to analyze basic interrelationships between toxic concentration, frequency and duration of the organism exposure, and its elimination capability.

2) Environmental toxic concentration impacts in a direct manner on the maximum value reached in the accumulation stock. Under certain circumstances, it is possible to exceed the elimination organism's capability.

3) For a defined k, the previous estimation of '**t** $_{½}$' allows to diminish the number of necessary simulations, because its value combined with the frequency and duration of the exposition gives a first approximation to organisms elimination possibilities.

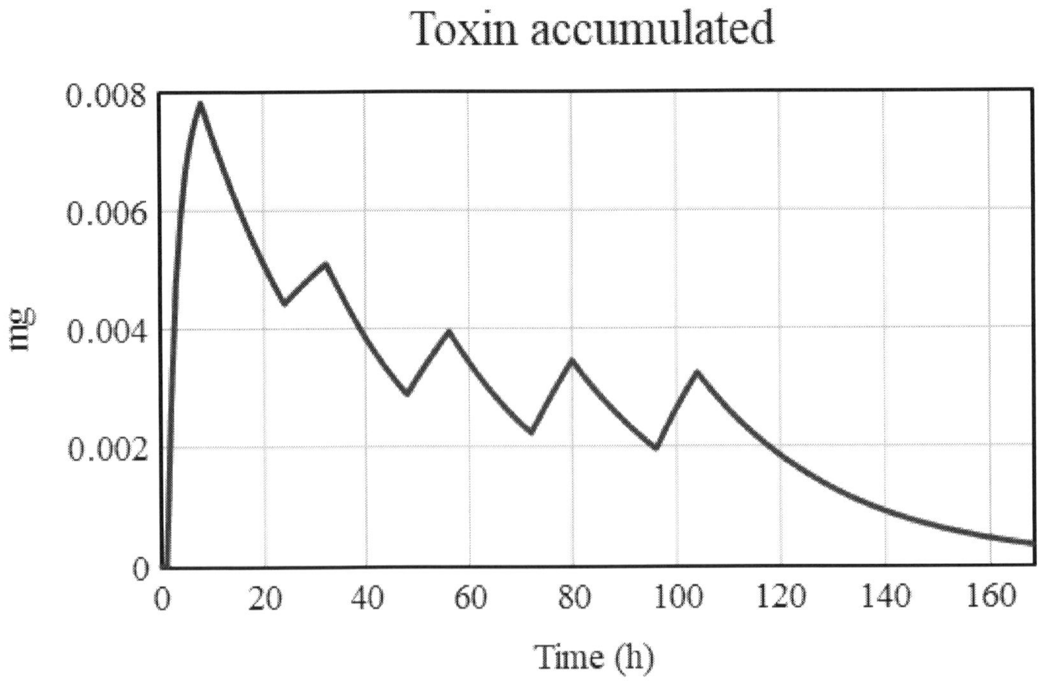

Comments

a) It is possible to define an auxiliary as the initial value of a stock. In this case we indicate that the stock of the rate of renovation of the air takes the value of the variable Initial Rate as the initial value.

b) When we have a very complex Stock and Flow Diagram, we can use the shadow variables that are no more than a copy of the original variable. Its use avoids the existence of an excessive number of arrows that cross in the diagram.

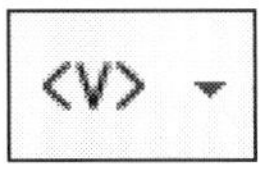

c) Check the use of the red triangle icon. For this you will need to move the cursors shown around the screen and you will be able to see the sensibility of the system to changes in the concentration and other constants.

We can observe the result of the variables of the model.

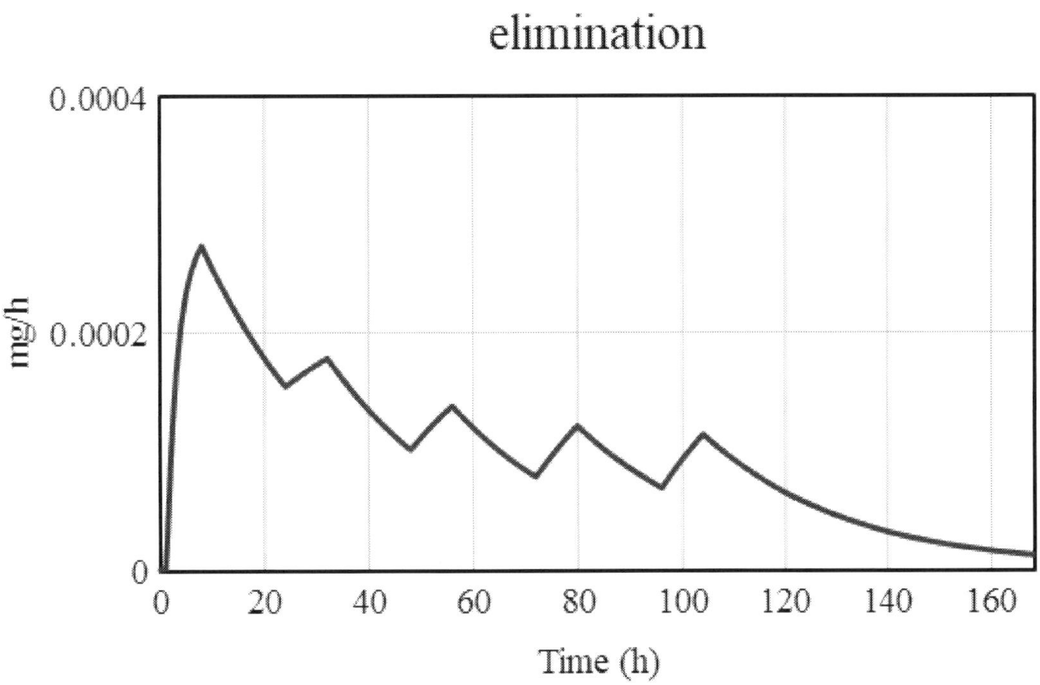

3.8. The barays of Angkor

The city of Angkor was the capital of the Kmer Empire from the 8th to the 14th century. It was situated where today we find Cambodia, together with the delta of the river Mekong and some 400 km to the north of the capital Phnom Penh. The name of the city came from the Sanskrit 'Nagara', which means Capital city.

The emplacement of the area was of great importance for the trade between Iran, India, China and the former Mediterranean provinces of the Roman Empire. Thanks to its good commercial relations, it quickly came to be a city of great economic, cultural and technological prosperity; and as a consequence of the great prosperity, the population increased.

Until the 9th century it was a weak empire formed by many independent powers that were known as Andripura, but at the beginning of the 9th century, the King Joyavarman II conquered the area called Aninditapura and founded the Kmer Empire. During the reign of Indravaraman 1, in the year 880, the capital moved to Harihalaya 15 km from where Angkor can be found today. Its population was formed by 100,000 people who obtained 4,00 Tm of rice per year. In this period, the temple Preah-Ko was built, the centre of Bakong with its five-level pyramid and the first Baray. The Baray were rectangular pools of variable dimensions which collected water for the irrigation of rice. There were also regular annual inundations caused by the monsoon climate.

This first Baray had dimensions of 3.8 km by 1 km yielding a capacity for 10 million cubic meters. These dimensions allowed for enough water to fill 40 km2 of the rice cultivation. With its system of canals fed from the river Stung Rolous, the Kmer Empire managed to multiply its food generation capacity and became capable of obtaining 130 Tm/Km2 per year in 3 harvests.

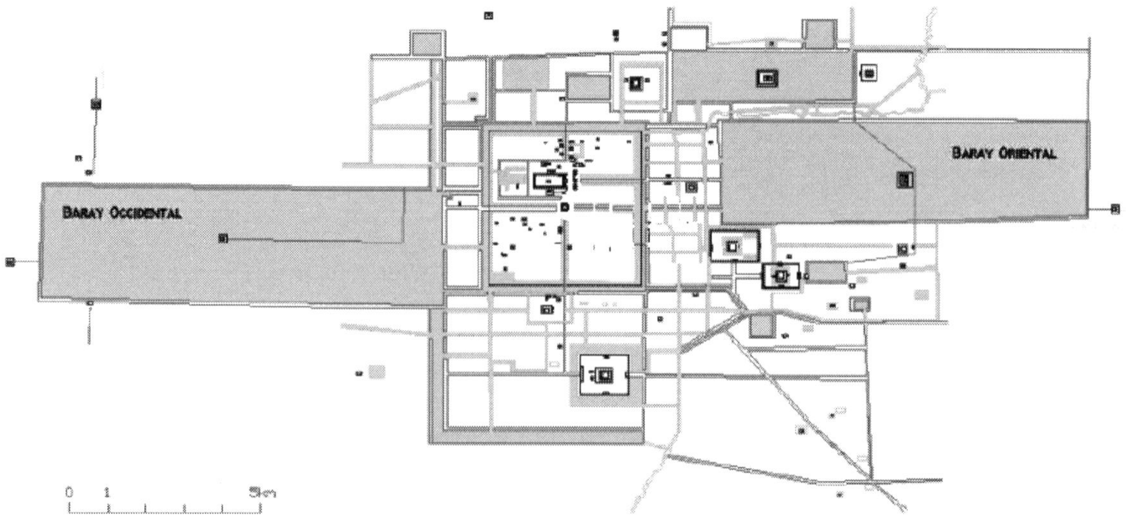

In the year 897, Yasovarman I started the construction of a second Baray called 'oriental'. In the middle of it, an Island, Loley was built. On the island, an Asharama, dedicated to the ancestors of the king, was constructed.

This Baray was five times bigger that the first, 7 km by 2.5 km with a capacity of 50 million m3 of water that was fed from the river Stung Siem Reap that made possible the farming of 200 more km2 of land. This King also moved the capital to what is now considered the first Angkor. The city was created and protected by a wall 4 km by 4 km. Finally, in the year 1010, Sûryavarman I built a third Baray called 'occidental', bigger than the first two, 8 km by 2 km, fed by the river Stung Puok with 57 million m3 of water. The farmable area was increased by 100 km2.

Succeeding ruling kings reorganised the Baray but without increasing its capacity. Gradually, at a rate of 0.1% a year of its capacity, the Baray stopped being feasible owing to an increase of land and a decrease in its capacity to store water. With this model we will study the evolution of the principle parameters of the system and its variables.

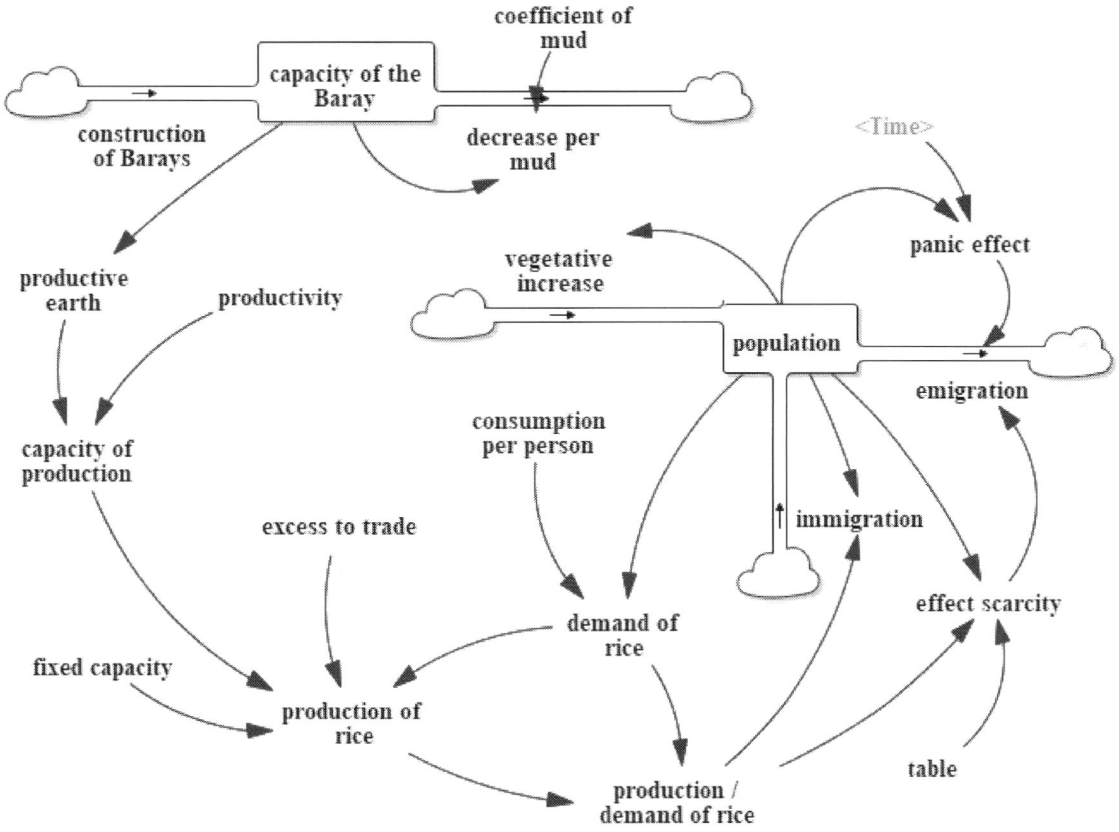

Key data of the model:

3 Baray with a total of 117 million m3 in total.
Area of irrigation 340 km2
Three harvests implied 130 Tm of rice per km2
Consumption of rice per person is 44kg of rice/year.

If there is over-production, immigration is produced. If there is under-production, emigration is produced.
Catastrophe when the soil covers 75 % of the capacity of the Baray.
Coefficient per mud is 0.1% per year of the capacity of the Baray
Initial population: 100,000 people.
Horizon of the simulation: between the years 800 to 1500.

Construction of the model. Aspects to highlight:

Use of the Time element

On occasions, we want to define the evolution of an element of the system to determine periods of time, in other words, we want to introduce Time as a variable of the system. In this case, we click the same icon that we have used to create the Shadow Variables and there we will find the variable Time. We select it and we can use it as any other element of the system.

If we have defined the temporal horizon of the period from 0 to 100, the variable Time will take the values 0, 1, 2, 3, etc. On the other hand, the model works from the period 800 to 1500, the variable Time will be worth 800, 801, 802, 803, etc.

Tables

To demonstrate the relationship that exists between two variables, one independent and the other dependent, we have recourse to the equations. On occasions, we will not find an arithmetical equation that represents this relationship and we will use a table. The tables are an assembly of points that show the value that a dependent variable takes when the independent has different values. In this way, the point (0.6, 0.04) represents that we consider that when the independent variable takes a value of 0.06, the dependent variable takes the value of 0.04.

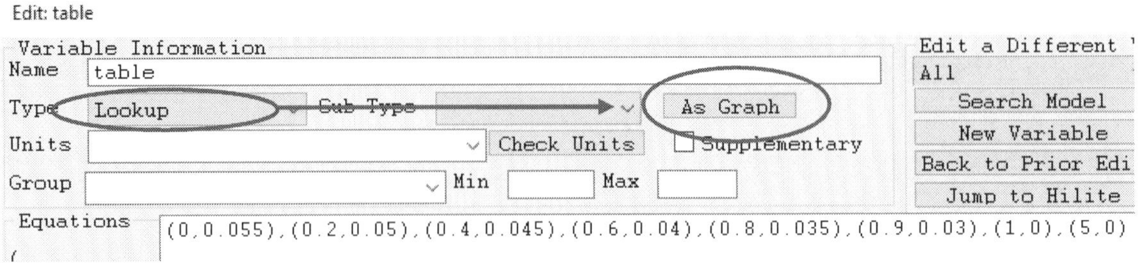

In this case we consider that the effect 'scarcity' is the dependent variable that depends on the independent variable 'production/demand of rice'. To represent this in a model, we need to select the Type = Lookup when writing the equation on the table.

If we click the button As Graph, we can enter the data more comfortably. We also see its graphic representation. In this case we see that when the quotient between the production and the demand is equal to or larger than 1, scarcity has a value of 0.

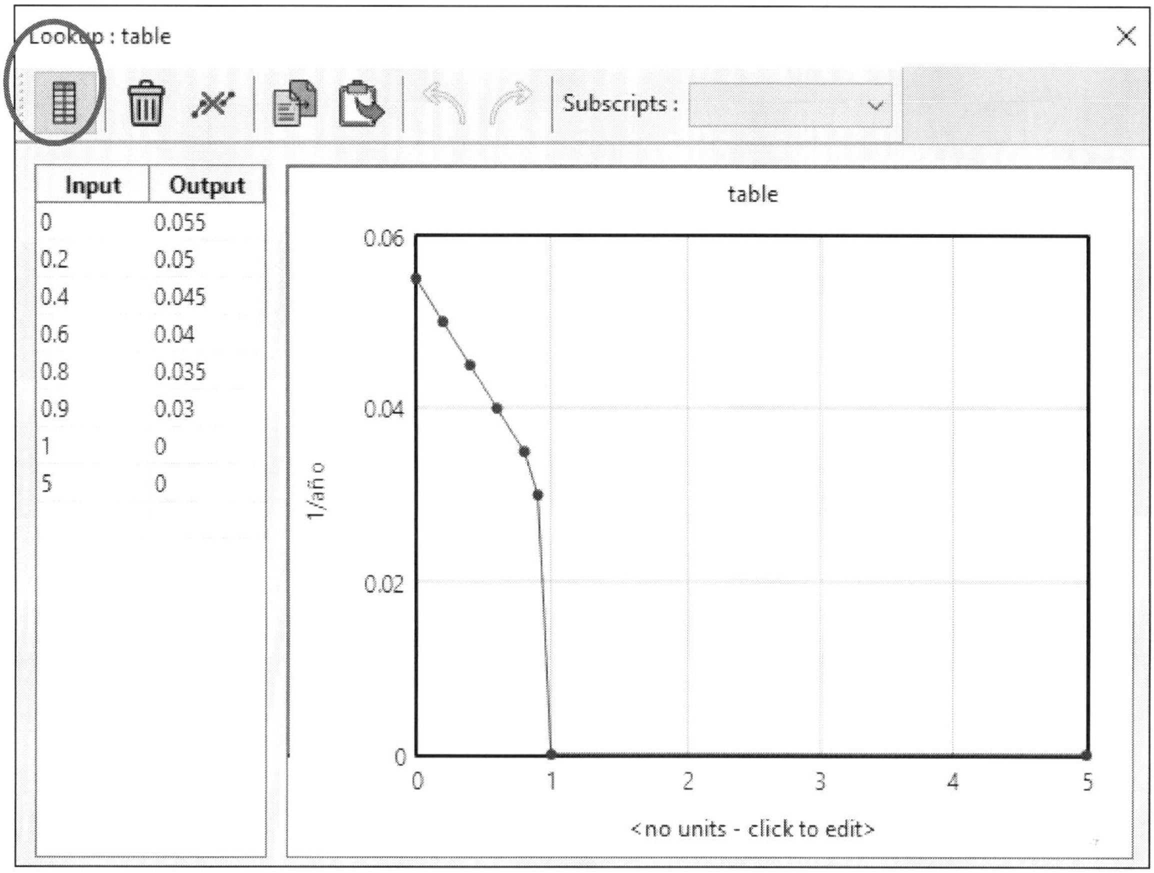

Model - Settings

INITIAL TIME=800 FINAL TIME=1500 TIME STEP=1 Units for time= Year

Equations 

(01) capacity of the Baray = construction of Baray - decrease per mud
Initial value: 0
Units: Million m3
Millions of cubic meters of water

(02) capacity of production = productive earth * productivity
Units: Tm/Year

(03) fixed capacity = 4400
Units: Tm/Year

(04) coefficient of mud = 0.001
Units: 1/Year
Coefficient that collects the lost capacity per accumulation of mud.

(05) construction of Barays =
10*PULSE (880,1)+50*PULSE(897,1)+57*PULSE(1010,1)
Units: Mm3/Year
Construction of the first Baray in the year 880, the second in 897 and the third in 1010, with their respective capacities, in total 117 million m3 of water.

(06) consumption per person = 0.044
Units: Tm/person/Year
Tons of rice per person and year.

(07) demand of rice = population * consumption per person
Units: Tm/ Year

(08) decrease per mud = capacity of the Baray * coefficient of mud
Units: Mm3/Year

(09) effect scarcity = population* table (production / demand of rice)
Units: person/year
Emigration is produced owing to the scarcity of aliments, and a non-lineal form as it is represented in the table.

(10) panic effect= IF THEN ELSE (Time>1200:AND:population<750000, population*0.1,0)
Units: person/Year
A panic is produced starting from the year 1200 when the population falls below 750,000 which affects 10% of the population annually.

(11) emigration = MAX(SMOOTH(scarcity effect, 10), panic effect)
Units: person/Year
The population emigrates if there is scarcity or, even if there is no scarcity, panic is produced because the expectations are not good.

(12) excess to trade = 0.2
Units: Dmnl
% that wants to produce rice in excess to trade with other products.

(13) vegetative increase = 0.002* population
Units: person/Year
Net vegetative rate of increase is taken as (births – deaths) 0.2 %

(14) immigration = SMOOTH (IF THEN ELSE (production / demand of rice > 1.2, 0.05* population, 0),20)
Immigration is produced when the production increases more that 20 % to the demand of rice. These immigrations are equivalent to 5% of the population, and they have a certain time delay.

(15) population = + immigration + vegetative increase – emigration
Initial Value: 100000
Units: person

(16) production / demand of rice = production of rice / demand of rice
Units: Dmnl

(17) production of rice = MIN (capacity of production, demand of rice* (1+ excedent to trade) + fixed capacity
Units: Tm/Year
Tons of rice per year. The minimum between the capacity of production and the demand is produced considering the excess to trade and the fixed quantity.

(18) productivity = 130
Units: Tm/Km2/Year
Tm per Km2 obtained in three annual harvests.

(19) table
(0,0.055),(0.2,0.05),(0.4,0.045),(0.6,0.04),(0.8,0.035),(0.9,0.03),(1,0),(5,0)
Units: 1/Year

(20) productive earth = capacity of the baray * 340/117
Units: Km2
340 Km2 with 117Mm3 are cultivated

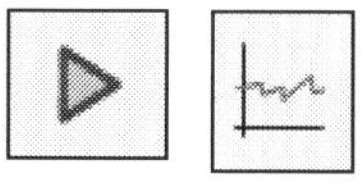

After simulate the model, if we click the icon Graph we can see the behaviour of the variables of the system. Clicking twice on the name of the variable that we want to see and afterwards choosing between the icons to the left of the screen in the format visualisation (graphic of table) we will obtain the following graphics:

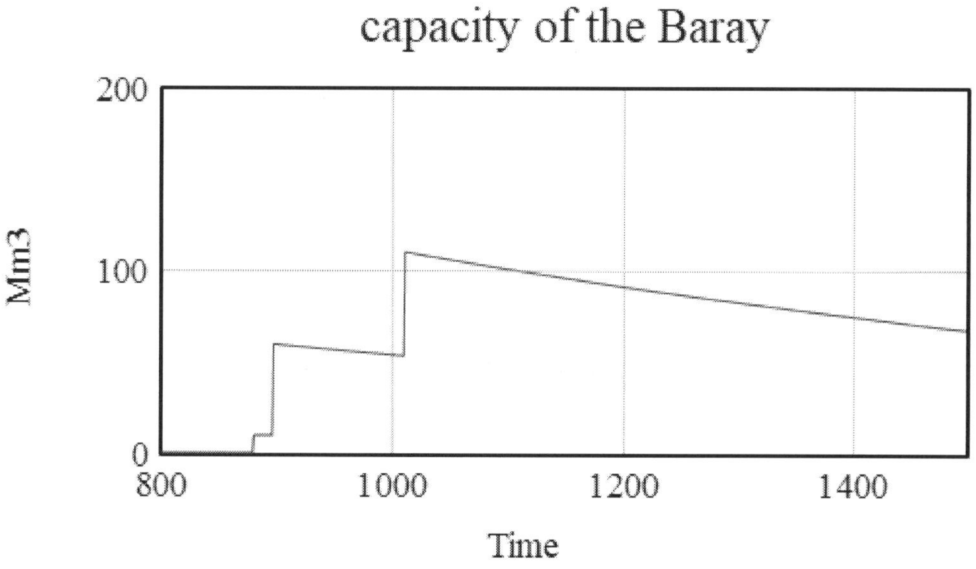

We can see the effect of simulating different alternatives and see the results on the same graphic representations. Let us suppose that we want to simulate the effect of a consumption per person of 0.033 instead of 0.044 Tm from the original model.

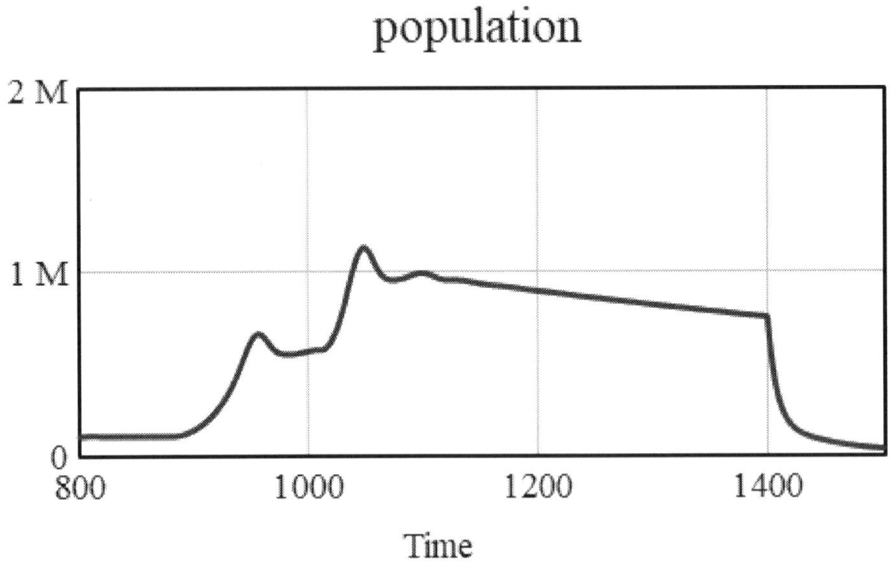

Edit the variable 'consumption per person' and change the value to 0.033.

Next click the icon Simulate and the message 'Data set Current already exists. Do you want to override it?' appears, click 'No' and give a new name to the simulation (as current2). As we display the results we obtain the graphics for both simulations.

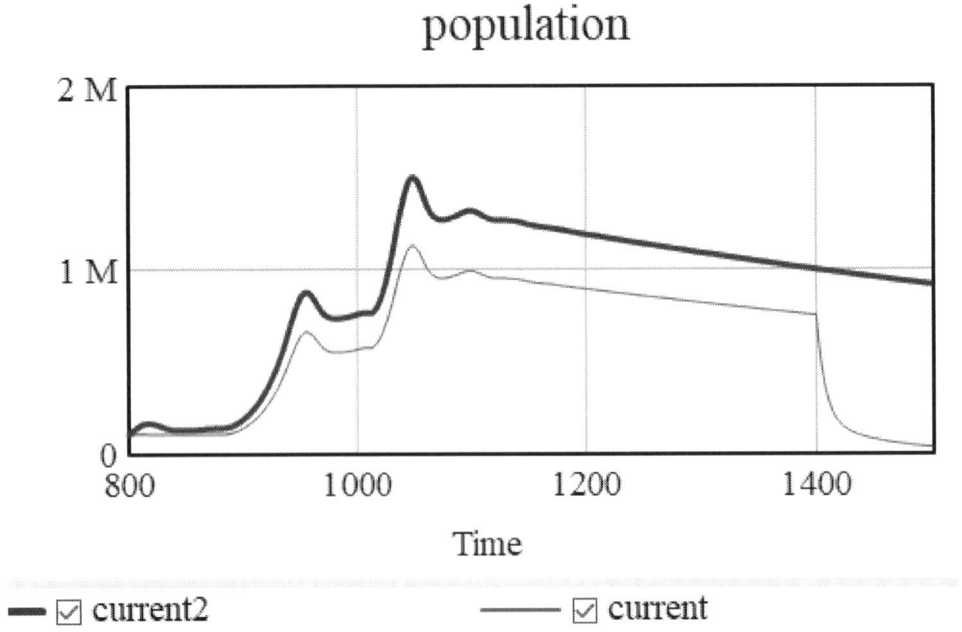

If we want to manipulate different simulations, selecting some and hiding others, we have to go to the superior bar through the option Control Panel – Dataset Manager and leave in the right window only the simulations that we want to see, and passing to the left part of the screen those that we do not want to see.

Finally, if we desire to see the behaviour of various elements in the graphic, we can by using a very similar option, go to Control Panel – Custom Graph – New and select the desired variables.

mark to show both variables with the same scale

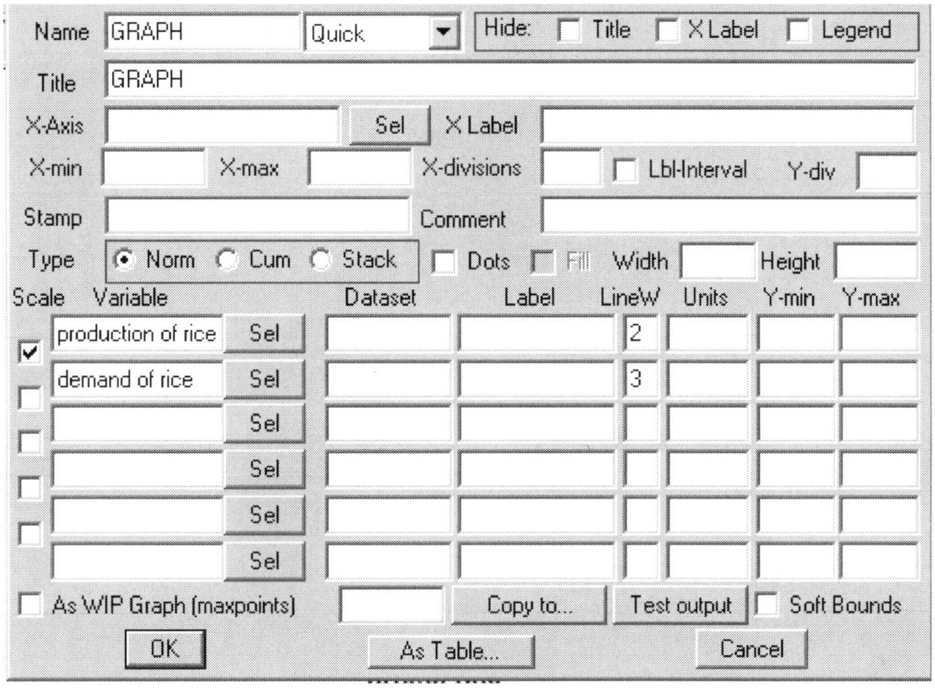

The following is the result:

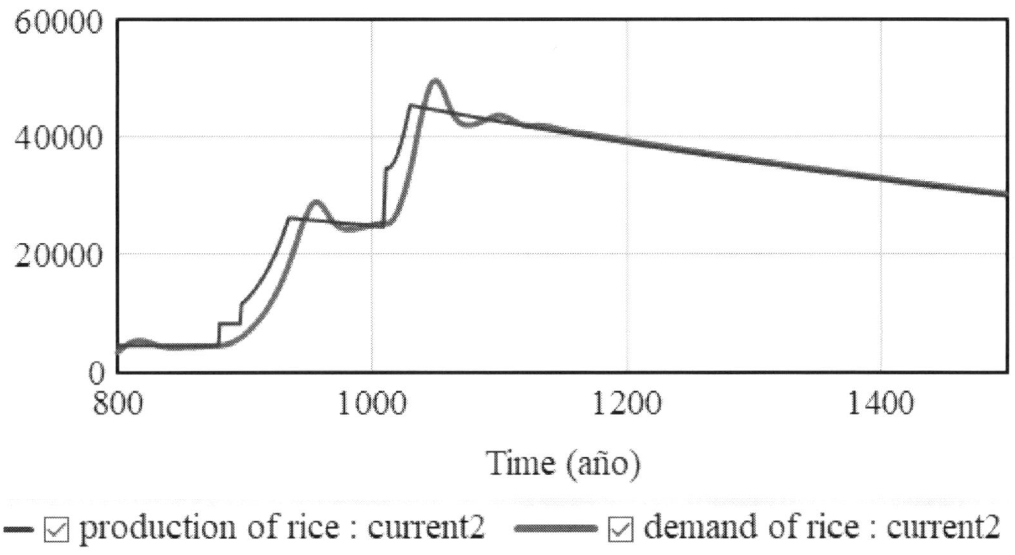

3.9. The Golden Number

Fibonacci (Leonardo Pisano) was a mathematician who lived in Pisa (Italy) between 1170 and 1250. He had a reputation for being someone who solved problems that were of no practical value. On one occasion, he was given the following problem:

> *'A pair of rabbits is placed in a closed yard to see how many descendants they have in the course of a year. It is assumed that each month, after the second month of their life, each pair of rabbits has a new one. Calculate the amount of rabbit pairs at the end of the year.'*

The solution was a sequence of numbers that became famous and is known as the Fibonacci sequence. In it, each term is the sum of the previous two. Therefore, it would be: 1,1,2,3,5,8..... These numbers give the amount of rabbit pairs as months goes by.

The curiosity of knowing if it was possible to repeat this situation by means of a model led to this chapter, which probably has no practical value. The model is really simple and consists of two flows, Flow 1 and Flow 2, two stocks, Box 1 and Box 2 and two auxiliary variables, N1 and N2, which makes it possible to change the initial values of Box 1 and Box 2. Let us see how the Fibonacci sequence is produced.

An analysis of how the software works will make it understandable. Simplifying the nomenclature and making Flow 1 = F1, Flow 2 = F2, Box 1 = B1 and Box 2 = B2, and using these variables with subscripts that refer to time, we will have the initial values:

t= 0 F1(0) = 1 B1(0) = 1 F2(0) = 1 B2(0) = 1

In t = 1 and t= 2:

t= 1 F1(1) = 2 B1(1) = 1 F2(1) = 1 B2(1) = 2
t= 2 F1(2) = 3 B1(2) = 2 F2(2) = 2 B2(2) = 3

In the model diagram, we can see that F1 = B2 and that F2 = B1. In addition, that at any time, t, the flow value, and the stock that defines that value, are the same, so for example F1(1)=B2(1) = 2 and F2(1)=B1(1)=1. But we must also remember that the value of a specific stock at a specific moment in time is obtained by summing the value it had at the previous period in time, plus the input minus the output at the previous period in time.

Thus, for any period in time 'n' we have

t=n F1(n) = B2 (n)
C1(n) = B1(n-1) + F1(n-1) - F2(n-1)
F2(n) = B1 (n)
B2(n) = B2(n-1) + F2(n-1)

Substituting F1 and F2 in terms of B1 and B2 and simplifying them we have:

(1) $B1(n) = B2(n-1)$

(2) $B2(n) = B2(n-1) + B1(n-1)$

Working on equation (2):

(3) $B2(n-1) = B2(n-2) + B1(n-2)$

But $B2(n-2) = B1(n-1)$, so replacing everything in (1)

$B1(n) = B1(n-1) + B1(n-2)$

Which is precisely what Fibonacci's sequence does, in other words, each new term is the sum of the previous two. That is how the sequence appears in Box 1 (and actually in all the variables).

But let us go on. If you use the quotient between each term and the previous one, you get the Golden Ratio, which converges to **The Golden Number**, in other words, 1.618034.

Now, the question arises, is there a relation in this sequence that explains the population growth dynamic of a theoretical case about rabbits?

The second part of this simple model explains it. If we use the Golden Ratio just as it is calculated in the first block, as the rate of vegetative growth, we obtain the solution to the rabbit problem they asked Fibonacci.

The model's behaviour can be tested giving whole values to N1 and N2 (preferably $N1 < N2$).

The model

We are going to indicate the values in the Model – Settings

INITIAL TIME = 0
FINAL TIME = 12
TIME STEP = 1

We are going to work with two screens. On one screen, we will calculate the Golden Number as has been described and on the other screen, we will apply it to calculating the rabbit population. To do so, we will create a second window (View 2). To do so, we will first click View 1 located in the lower left hand portion of the screen, and we will select 'New'.

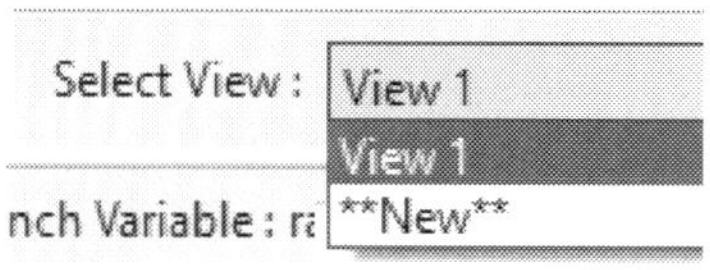

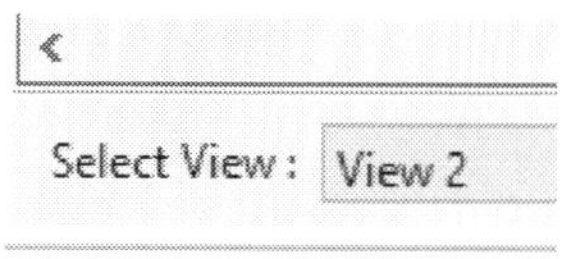

Enter the following variables on the first screen (View 1) to calculate the Golden Number:

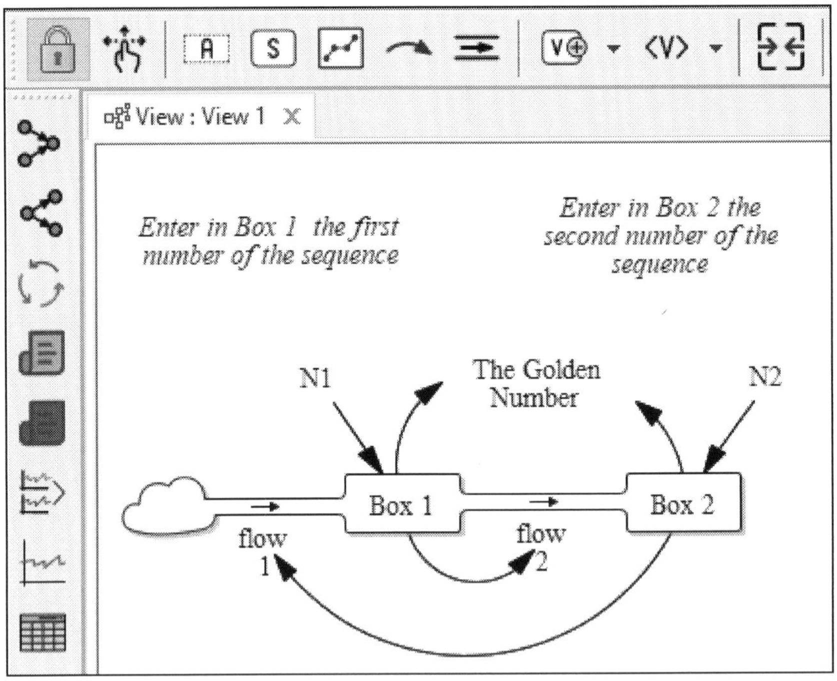

The explanatory text, '*Enter in Box...*' is added to the screen by clicking the icon: 

In the second view, enter the variables to calculate how the rabbit population evolves.

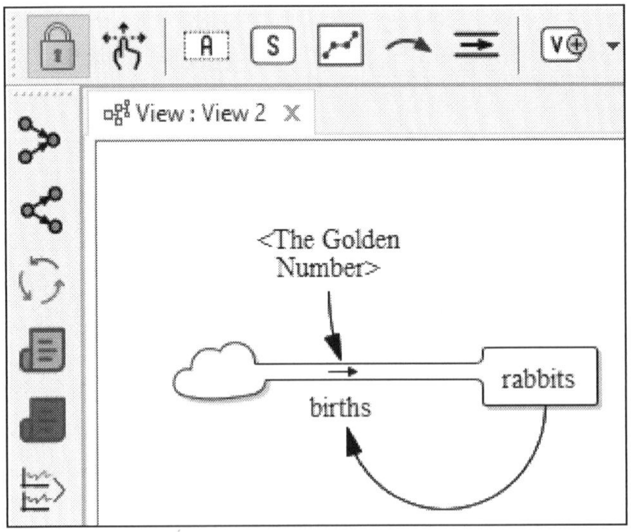

On this view, we use the 'Golden Number' variable that is calculated on the first view. To do so, create it as a Shadow Variable by clicking the icon: 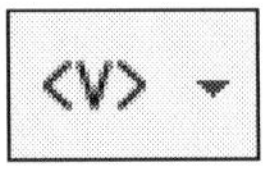

Equations

(01) Box 1 = flow 1-flow 2
Initial Value: N1

(02) Box 2 = flow 2
Initial Value: N2

(03) rabbits = births
Initial Value: 1

(04) flow 1= Box 2

(05) flow 2= Box 1

(06) N1 = 1

(07) N2 = 1

(08) births = rabbits*(Golden Number-1)

(09) Golden Number= Box 2 / Box 1

We can see the results by using one of the icons on the left side of the screen that show us the results in a numerical table format.

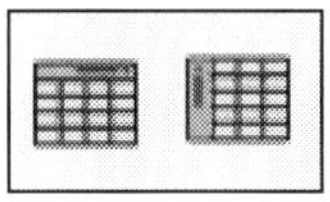

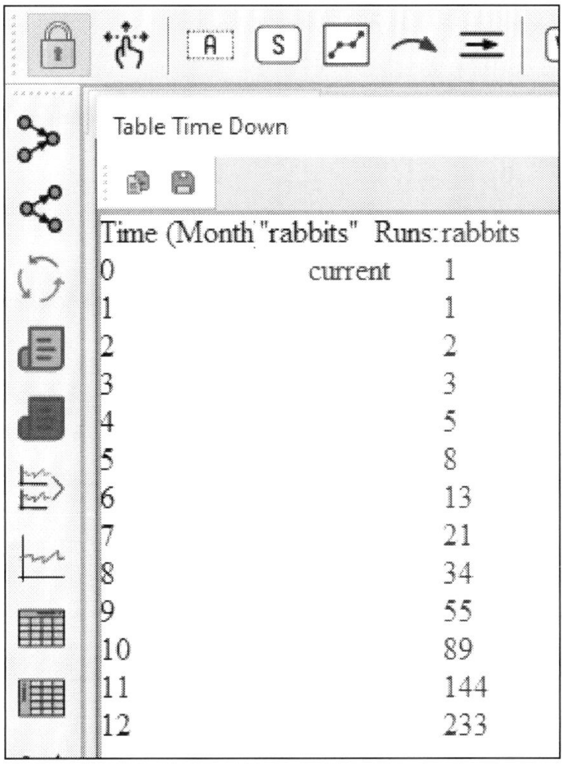

Time (Month	"rabbits" Runs:	rabbits
0	current	1
1		1
2		2
3		3
4		5
5		8
6		13
7		21
8		34
9		55
10		89
11		144
12		233

Management Area

3.10. Production and inventory

A manufacturing firm has experienced chronic instability in its inventory stock and production rate. Discussions with management and workers on the factory floor have revealed the following information:

1. The firm supplies its customers from a stock of finished inventory, which is generally sufficient to meet orders as they are received.
2. Desired production is determined by anticipated (forecasted) shipments, modified by a correction to maintain inventory at the desired stock.
3. The firm forecasts shipments by averaging past orders over an eight-week period as a way of smoothing out any noise or lumpiness in demand.
4. The firm tries to correct discrepancies between desired and actual inventory in eight weeks.
5. Desired inventory is four weeks' worth of anticipated shipments.

The firm's actual production rate is assumed to equal desired production. The initial value for inventory is assumed to be equal to desired inventory, and the initial value of the average order rate is assumed to be equal to the shipment rate. As a result, the model will begin in an initial equilibrium regardless of the initial customer order rate.

In this model customer orders will equal a constant of 1,000 units/week until the period 10, when orders step up by 10%, and then remain at the higher rate.

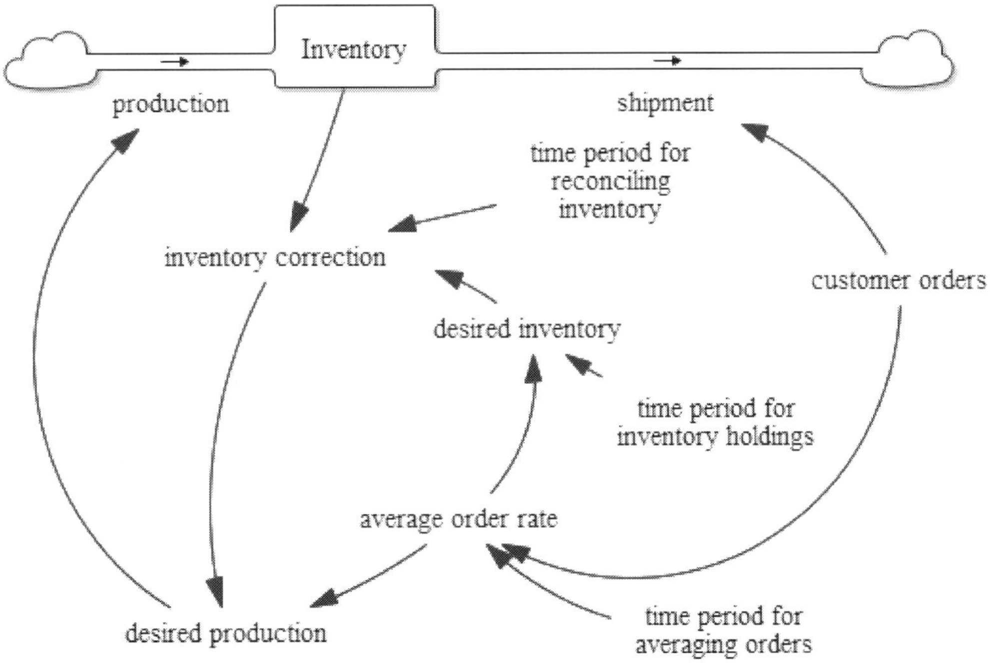

Preliminary question

Before making the model, draw a model diagram of the behaviour you expect the model will have when tested with the 10% step increase in orders (starting from equilibrium). Include the behaviour of orders, desired production, production, inventory, and desired inventory. Explain in one paragraph the reasons for the behaviour you expect.

Model 1

Click File-New model, and define: 

INITIAL TIME = 0   FINAL TIME = 100   TIME STEP =1   Units for time = Week

(01) average order rate=SMOOTH(customer orders, time period for averaging orders)
Units: units/Week
The firm forecast shipments by averaging past orders over an eight-week period as a way of smoothing out any noise or lumpiness in demand.

(02) customer orders=1000*(1+STEP(0.1,10))
Units: units/Week
In this model customer orders will equal a constant of 1,000 units/week until period 10, when orders step up by 10%, and then remain at the higher rate.

(03) desired inventory=average order rate*time period for inventory holdings
Units: units
Desired inventory is four weeks' worth of anticipated shipments.

(04) desired production=average order rate+ inventory correction
Units: units/Week
Desired production is determined by anticipated shipments, modified by a correction to maintain inventory at the desired level.

(05) Inventory= production-shipment
Initial value : 4000
Units: units
The initial inventory is four weeks' worth of shipments (1,000 x 4 = 4,000)

(06) inventory correction=(desired inventory-Inventory)/time period for reconciling inventory
Units: units/Week
The firm tries to correct discrepancies between desired and actual inventory in eight weeks.

(07) production=desired production
Units: units/Week

(08) shipment= customer orders
Units: units/Week

(09) time period for averaging orders= 8
Units: Week
Time to average order rate

(10) time period for inventory holdings= 4
Units: Week
Desired inventory coverage

(11) time period for reconciling inventory= 8
Units: Week
Time to correct Inventory

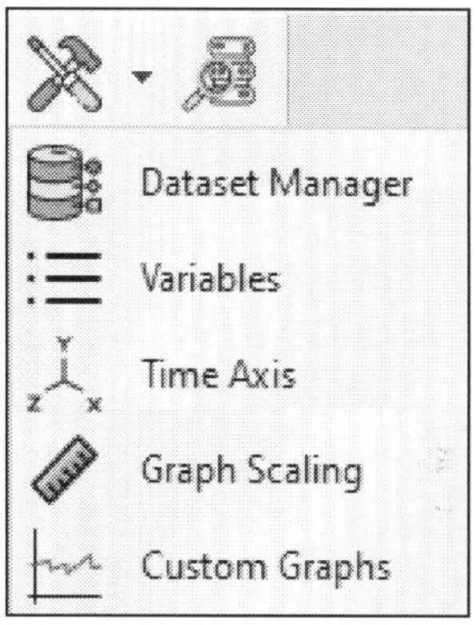

Simulate and show the results

To see more than one variable in the same graph go to Control panel – Custom Graph – New, New and select the variables, share the scale, and define the Y-min and Y-max values. OK. Then select and click Display

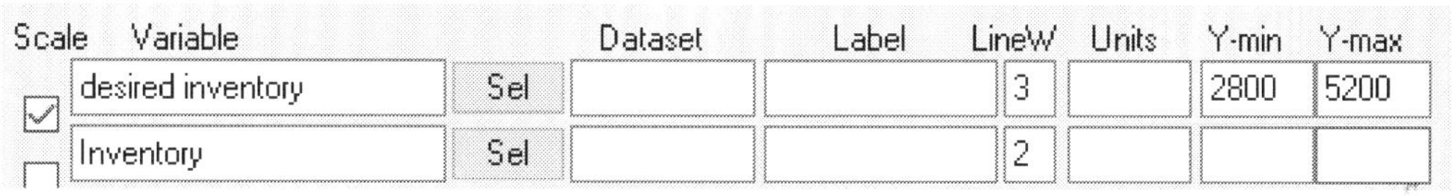

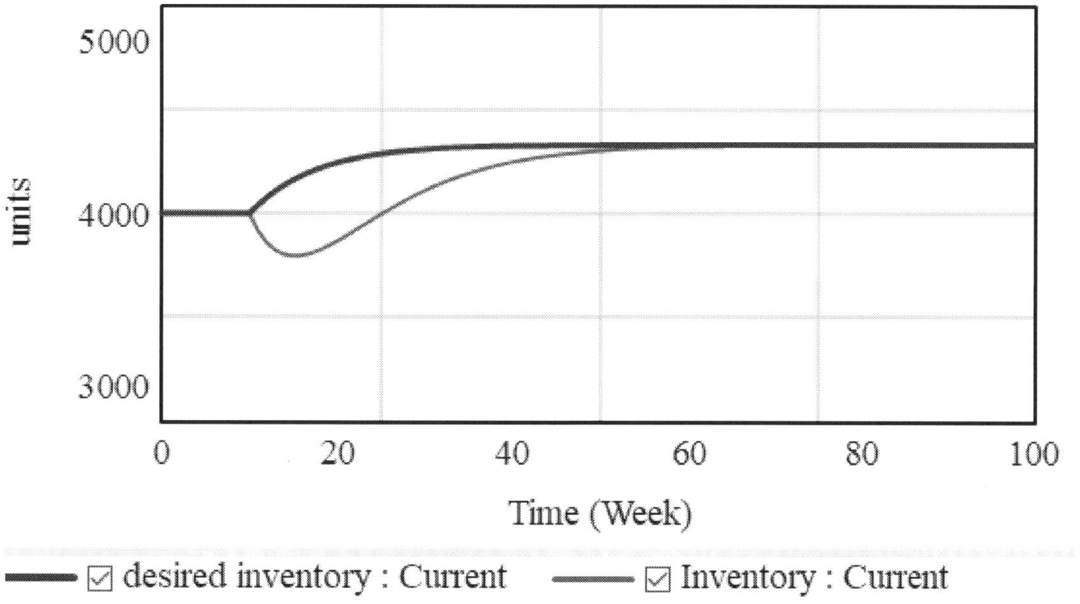

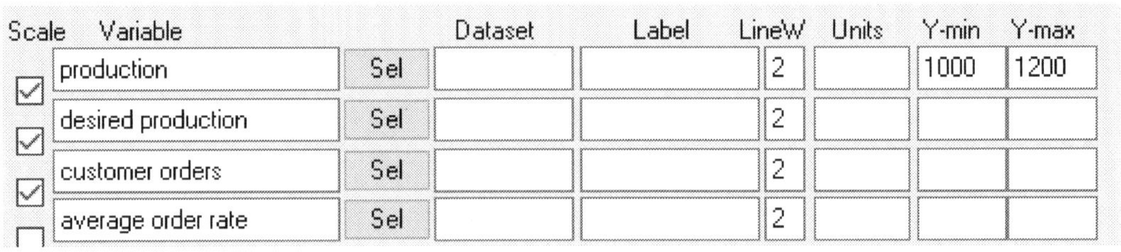

Scale	Variable	Dataset	Label	LineW	Units	Y-min	Y-max
☑	production	Sel		2		1000	1200
☑	desired production	Sel		2			
☑	customer orders	Sel		2			
☐	average order rate	Sel		2			

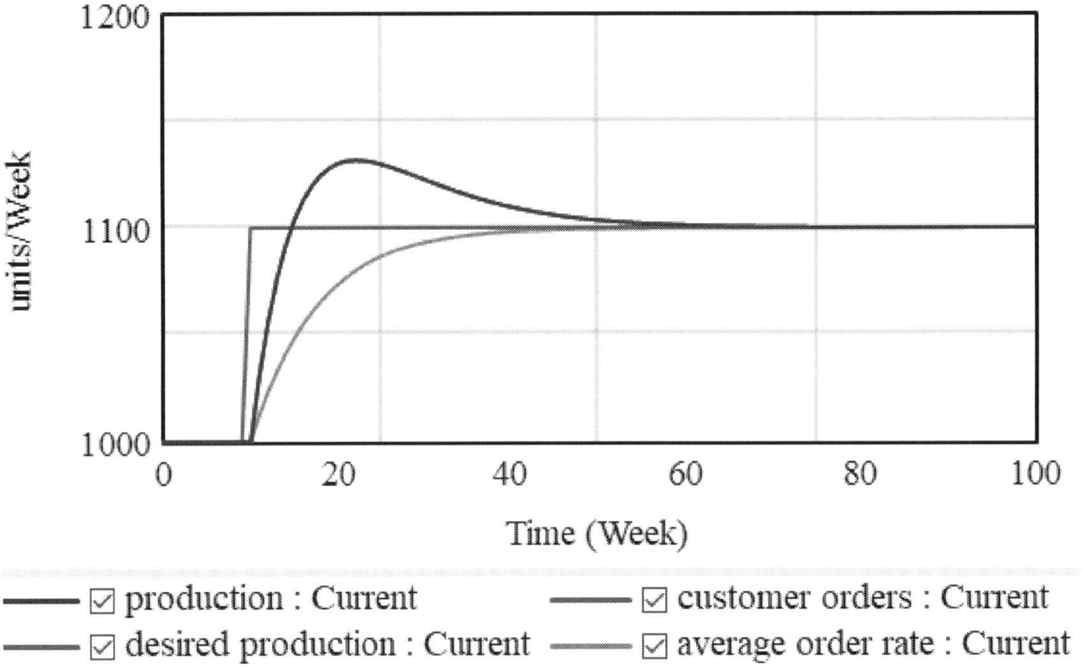

——— ☑ production : Current   ——— ☑ customer orders : Current
——— ☑ desired production : Current ——— ☑ average order rate : Current

Questions

a. After running the model compare the actual behaviour with your expected behavior for each of the variables, explaining the reasons for the actual behaviour and why it differs from your expectations (if it does).

b. Shorten 'time period for reconciling inventory' to four weeks.
　　　Does the behaviour mode change?
　　　Does the timing of the behaviour change?
　　　It is more or less stable?
　　　Explain why or why not.

Model 2

Noting that the current model assumes actual production equals desired production, you return to the company to check on the validity of that assumption. Further discussions reveal that while the firm has ample physical plant and equipment, labour cannot be hired and trained instantaneously. In fact, it takes approximately 24 weeks to advertise for, hire, and train new workers. The firm has a no-layoff policy, and workers stay with the firm an average of 50 weeks (one year).

The firm's hiring policy is to replace those workers who quit, modified by a correction to bring the actual workforce into balance with the desired workforce. Because workers must give two weeks' notice, there is no significant delay between quits and replacement hiring.

Because of the delays in hiring new people, it takes 24 weeks to make corrections to the workforce. The desired workforce is determined by desired production and average productivity, which is equal to 20 units per worker per week and is quite constant over time. Finally, union rules prevent the use of overtime or undertime.

Note in this model that the recruiting, hiring, and training delays have been aggregated together into the 'time period to hire new workers'

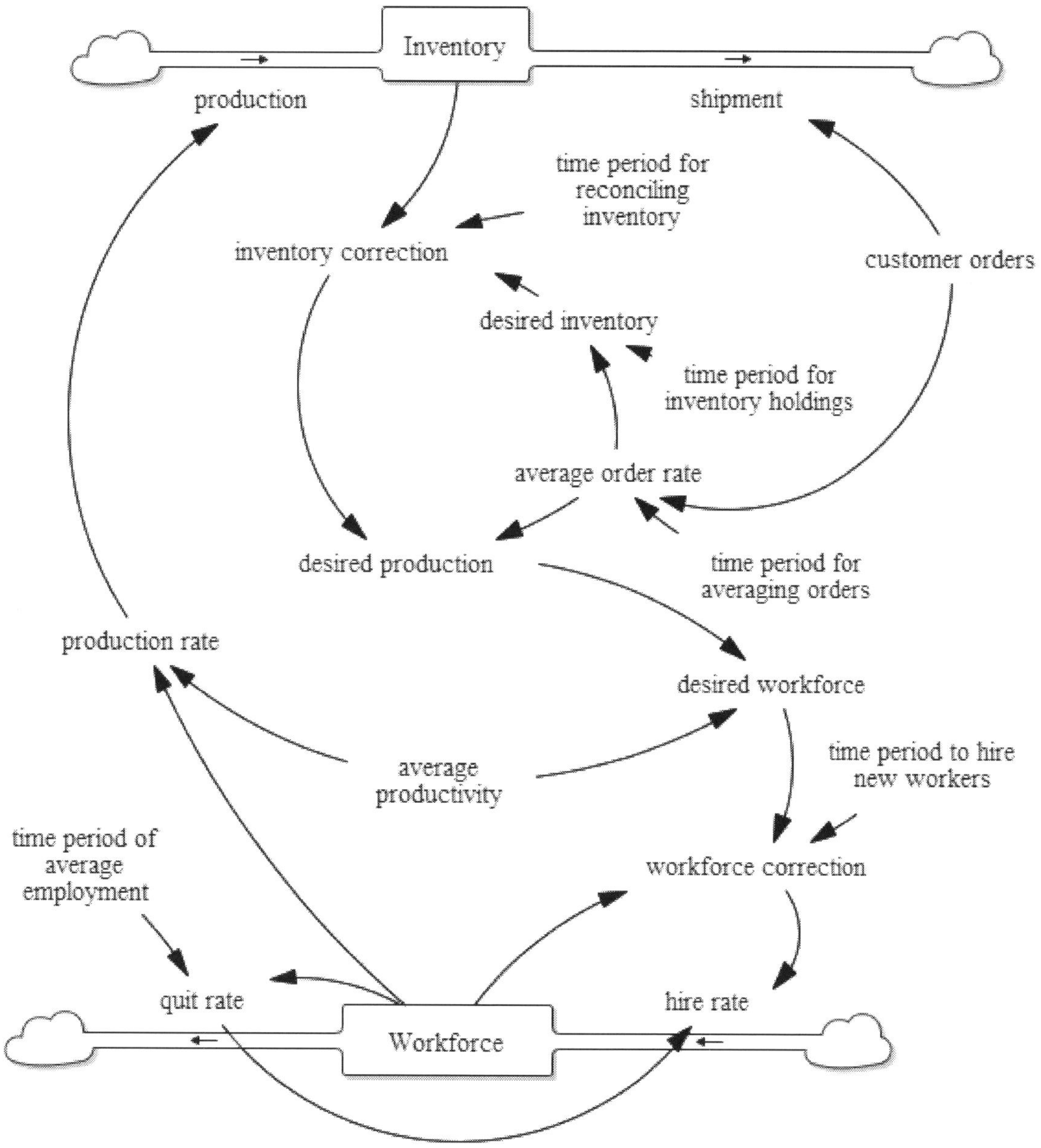

(01) average order rate=SMOOTH(customer orders, time period for averaging orders)
Units: units/Week
The firm forecast shipments by averaging past orders over an eight-week period as a way of smoothing out any noise or lumpiness in demand.

(02) average productivity= 20
Units: units/person/Week

(03) customer orders= 1000*(1+STEP(0.1,10))
Units: units/Week
In this model customer orders will equal a constant of 1,000 units/week until period 10, when orders step up by 10%, and then remain at the higher rate.

(04) desired inventory=average order rate*time period for inventory holdings
Units: units
Desired inventory is four weeks' worth of anticipated shipments.

(05) desired production=average order rate+ inventory correction
Units: units/Week
Desired production is determined by anticipated shipments, modified by a correction to maintain inventory at the desired level.

(06) desired workforce=desired production/average productivity
Units: person

(07) hire rate=quit rate+ workforce correction
Units: person/Week

(08) Inventory= production-shipment
Initial value: 4000
Units: units
The initial inventory is four weeks' worth of shipments (1,000 x 4 = 4,000)

(09) inventory correction=(desired inventory-Inventory)/time period for reconciling inventory
Units: units/Week
The firm tries to correct discrepancies between desired and actual inventory in eight weeks.

(10) production= production rate
Units: units/Week

(11) production rate= Workforce*average productivity
Units: units/Week

(12) quit rate= Workforce/time period of average employment
Units: person/Week

(13) shipment= customer orders
Units: units/Week

(14) time period for averaging orders= 8
Units: Week
Time to average order rate

(15) time period for inventory holdings= 4
Units: Week
Desired inventory coverage

(16) time period for reconciling inventory= 8
Units: Week
Time to correct Inventory

(17) time period of average employment= 50
Units: Week
Average duration of employment

(18) time period to hire new workers=24
Units: Week
Time to correct workforce

(19) Workforce= hire rate-quit rate
Initial value: 50
Units: person

(20) workforce correction=(desired workforce-Workforce)/time period to hire new workers
Units: person/Week

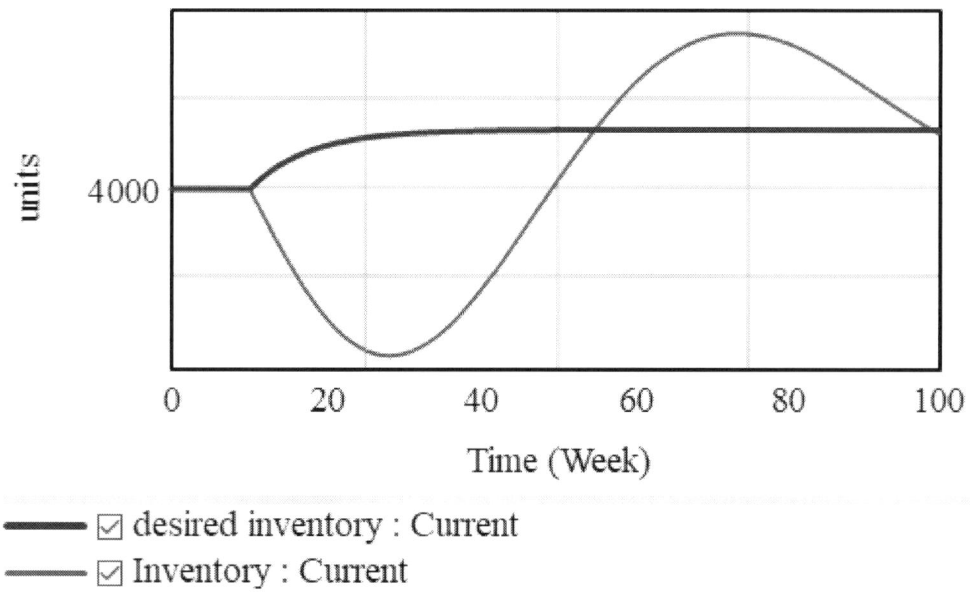

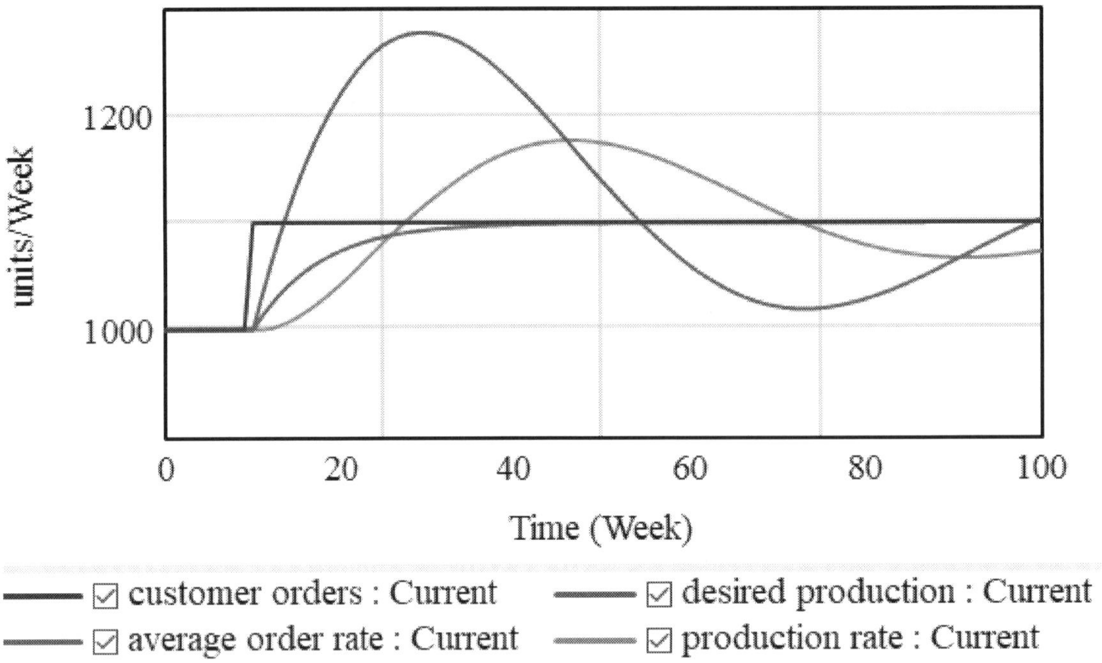

Questions

Managers control the 'time period for reconciling inventory', which represents how aggressively the firm tries to correct inventory discrepancies. A long 'time period for reconciling inventory' implies management is willing to allow a large inventory gap, while a short 'time period for reconciling inventory' implies swift corrections to eliminate a gap and bring inventory in balance with desired inventory rapidly.

a. To stabilize the system, should 'time period for reconciling inventory' be longer or shorter? Why? What do you mean by stability?

b. Pick a new value for 'time period for reconciling inventory' that you feel will increase the stability of the system, and run the model. Do the results match your expectations?

c. Explain your results. Should management be more or less aggressive in managing inventories?

Model 3

The firm wants to know whether they should click for the right to schedule overtime/undertime in upcoming contract negotiations with the union.

The normal workweek is 40 hours. Preliminary negotiations with the union reveal that they would be willing to allow a minimum of 35 hours per week and a maximum of 50 hours, at time-and-a-half, in return for increase in employee benefits and pension fund contributions. You agree to evaluate the benefits of overtime/undertime with the model so management knows what overtime is worth in improved stability.

Schedule Pressure is the ratio of desired to real production rate and measures the need for overtime. The 'Effect of Schedule Pressure on Production' relates the overtime/undertime actually scheduled to the pressure for overtime/undertime.

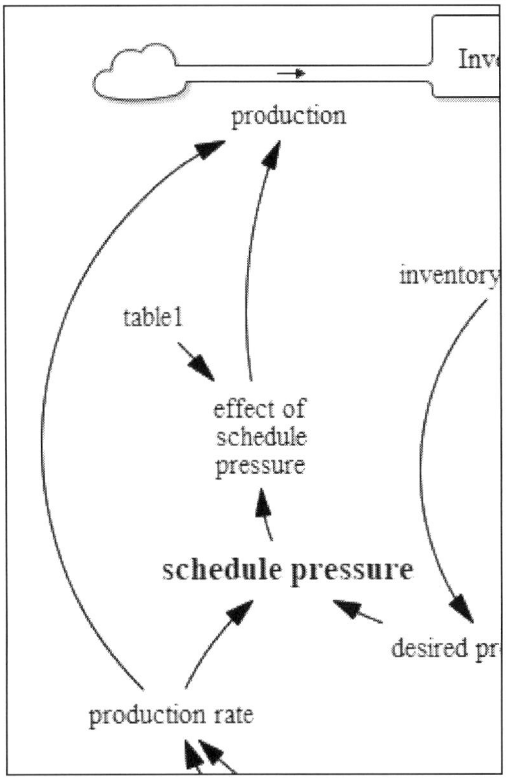

(01) average productivity=20
Units: units/person/Week

(02) average order rate=SMOOTH(customer orders, time period for averaging orders)
Units: units/Week

(03) customer orders=1000*(1+STEP(0.1,10))
Units: units/Week

(04) desired inventory=average order rate*time period for inventory holdings
Units: units

(05) desired production=average order rate+ inventory correction
Units: units/Week

(06) desired workforce=desired production / average productivity
Units: person

(07) effect of schedule pressure=table1(schedule pressure)
Units: Dmnl
The Effect of Schedule Pressure on Production relates the over / undertime actually scheduled to the pressure for over / undertime.

(08) hire rate=quit rate+workforce correction
Units: person/Week

(09) Inventory= production-shipment
initial value: 4000
Units: units

(10) inventory correction= (desired inventory-Inventory)/time period for reconciling inventory
Units: units/Week

(11) production=production rate*effect of schedule pressure
Units: units/Week

(12) production rate=Workforce*average productivity
Units: units/Week

(13) quit rate=Workforce/time period of average employment
Units: person/Week

(14) schedule pressure=desired production/production rate
Units: Dmnl
Schedule pressure is the ratio of desired to real production rate and measures the need for overtime.

(15) shipment=customer orders
Units: units/Week

(16) table1 = (0.8,0.875),(0.9,0.875),(1,1),(1.1,1.25),(1.2,1.25)
Units: Dmnl

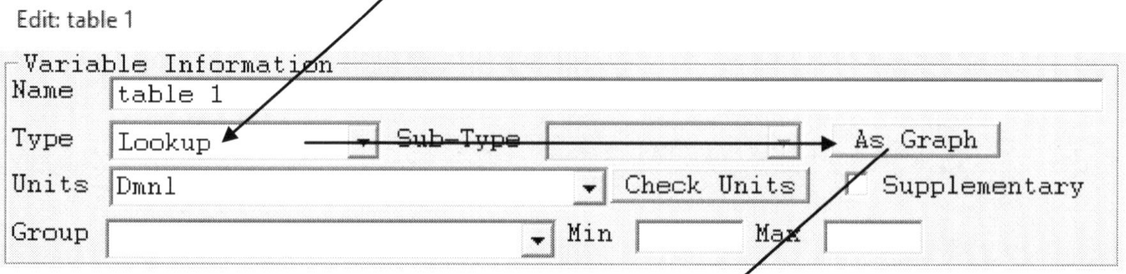

In a Lookup each row represents a point, this is a value for the "Output" variable that we know it takes for a value of the "Input" variable.
For example if we know that when Price=3 the Sales are 120, we define the point (3,120)

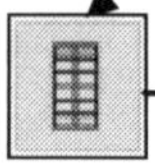

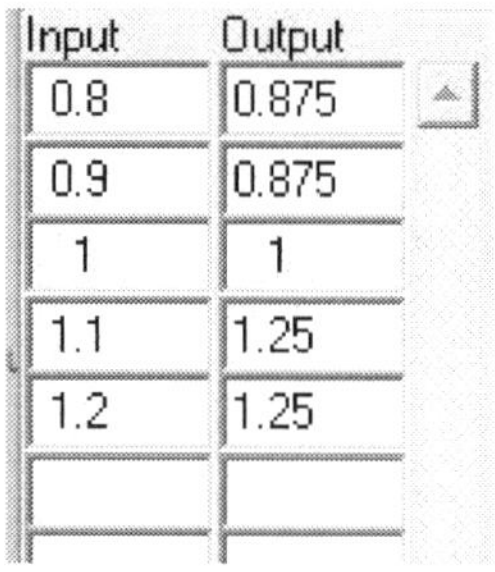

Input	Output
0.8	0.875
0.9	0.875
1	1
1.1	1.25
1.2	1.25

(17) time period for averaging orders= 8
Units: Week
Time to average order rate

(18) time period for inventory holdings= 4
Units: Week
Desired inventory coverage

(19) time period for reconciling inventory= 8
Units: Week
Time to correct Inventory

(20) time period of average employment= 50
Units: Week
Average duration of employment

(21) time period to hire new workers=24
Units: Week
Time to correct labor

(22) Workforce= hire rate-quit rate
Initial value: 50
Units: person

(23) workforce correction=(desired workforce-Workforce)/time period to hire new workers
Units: person/Week

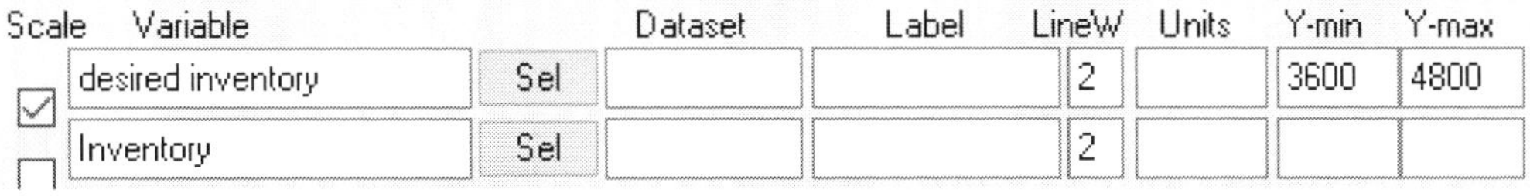

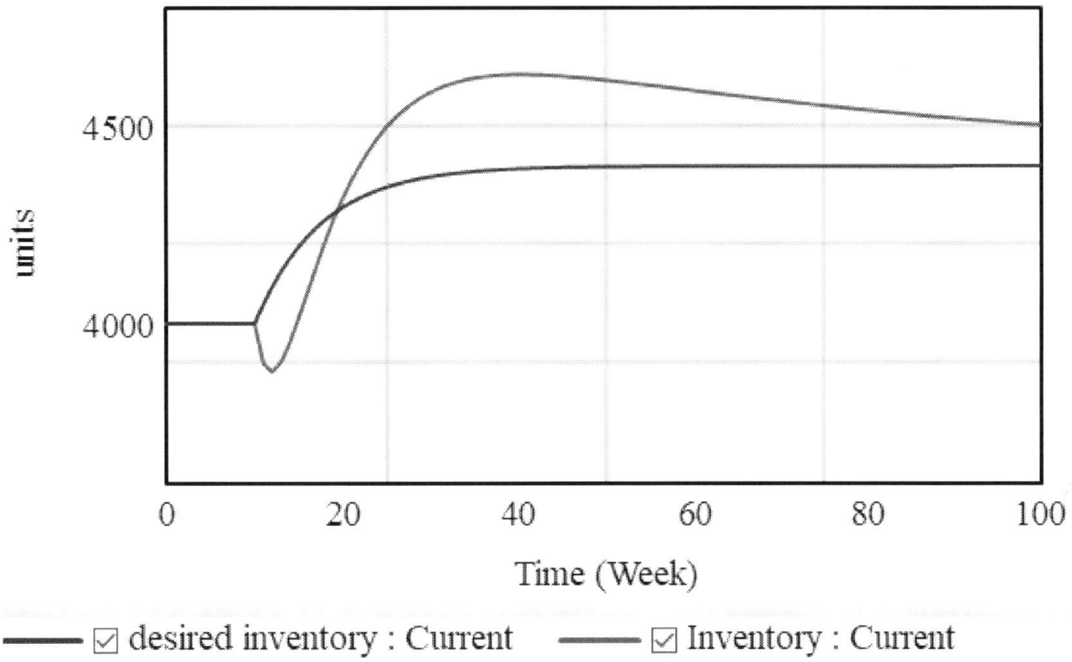

Click the icon Units Check, this message appears:

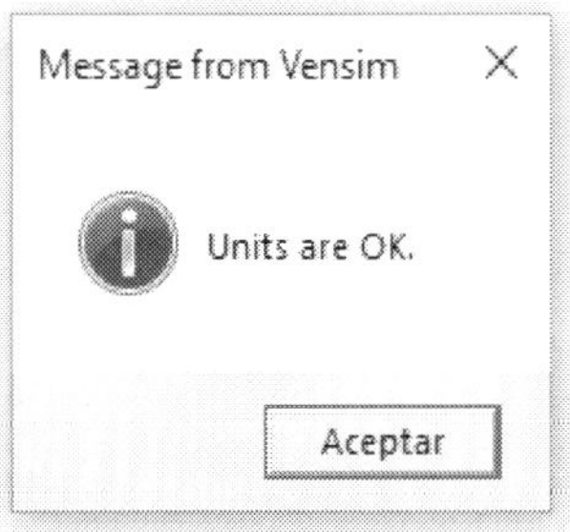

Questions

a. What does it mean when 'effect of schedule pressure' =1?
b. What is the effect of over / undertime on stability?
c. Generalize the results of the overtime policy. How can oscillatory systems be stabilized? (Don't forget about 'time period for reconciling inventory' and 'time period to hire new workers'.)

3.11. Global CO_2 emissions

We are asked to make a simple model of the emissions of CO_2 in the atmosphere. The model needs to be generic and adaptable afterwards to a determined period and geographical system. We are asked to base our model on the 'Identity of Kaya'

The 'Identity of Kaya' says that the emissions of CO_2 are equal to the Gross Domestic Product or GDP (€) by the Intensity of Energy (Kw/€) and by the Vector of Energy Production (CO_2/Kw).

CO_2 (tm) = GDP (€) x Intensity of energy (Kw/€) x Vector of Energy Production (CO_2/Kw)

The intensity of the energy is the quantity of energy demanded by each unit of annual GDP. We can calculate the total energy demanded based on the GDP.

Vector of Energy Production includes the emissions of CO_2 for every Kw of energy produced, and the distribution in percentages of the different sources of energy: coal, fuel, gas nuclear and renewable (hydroelectric, eolic and others). In this way we can calculate the emissions of CO_2 based on the energy needed and the Vector of Energy Production.

With the intention of fixing some initial parameters, it is indicated that the emission of CO_2 by the Kw produced is high with coal, medium with fuel and low with gas. There are no significant emissions with the other sources of energy. On the other hand, the Vector of Energy Production is formed by 50% fuel, 30% by nuclear and 10% by gas and the rest in equal measure between coal and renewable energy.

It is requested that the model allows, as well as introducing different scenarios of the GDP, the possibility of introducing improvements both in the efficiency of the Energy Intensity and in the emissions of CO_2 per unit Kw produced from the different sources of energy.

We want to create a model that gathers all these aspects and that ensures that the level of CO_2 in the atmosphere will be constant. The following scheme is proposed:

Emission = GDP x Energy Intensity x Vector of Energy Production

Being the respective measure units:

CO_2 = € x Kw/€ x CO_2/Kw

The model allows us to gather the identity of Kaya, in order to calculate the emissions of C02 based on the energy needed and on the Vector of Energy Production. The model also allows us to introduce the generic parameters that are indicated to us.

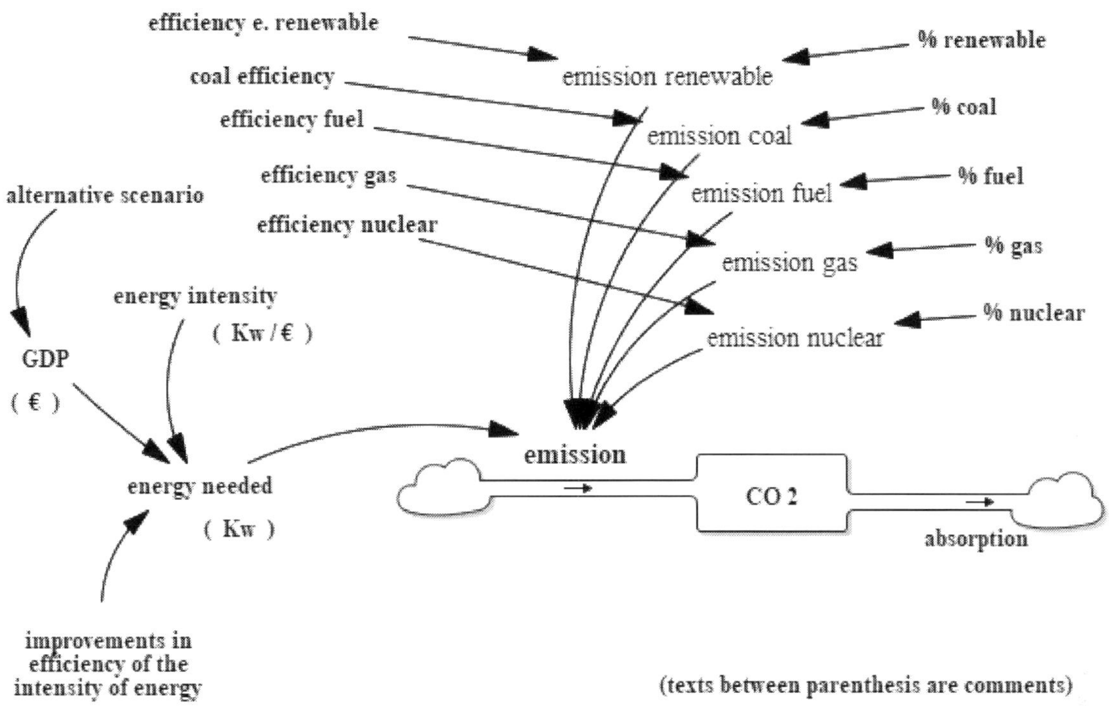

This is a model about emissions which are a Flow. A Stock has been added corresponding to the total quantity of CO2 in the atmosphere, although this is not strictly necessary. We can observe with the parameters indicated to us that the quantities of emissions are maintained constant, the same as the parameters used.

Model – Settings
INITIAL TIME= 1900 FINAL TIME= 2100 TIME STEP=1 Units for time=Year

(01) % coal = 0.05

(02) % fuel = 0.5

(03) % gas = 0.1

(04) % nuclear = 0.3

(05) % renewable = 0.05

(06) absorption= 12500
Units: CO2/Year

(07) alternative scenario = 1

(08) CO 2= +emission-absorption,
Initial value: 10000000
Units: CO2

(09) coal efficiency = 3
Units: CO2/Kw

(10) efficiency e. renewable = 0
Units: CO2/Kw

(11) efficiency fuel = 2
Units: CO2/Kw

(12) efficiency gas = 1
Units: CO2/Kw

(13) emission=energy needed * (emission coal+emission fuel+emission gas+emission nuclear+emission renewable)
Units: CO2/Year

(14) emission coal =% coal * coal efficiency
Units: CO2/Kw

(15) emission fuel = % fuel * efficiency fuel
Units: CO2/Kw

(16) emission gas= % gas * efficiency gas
Units: CO2/Kw

(17) emission nuclear= % nuclear * efficiency nuclear
Units: CO2/Kw

(18) emission renewable= % renewable *efficiency e. renewable
Units: CO2/Kw

(19) efficiency nuclear = 0
Units: CO2/Kw

(20) energy intensity = 10
Units: Kw/€

(21) energy needed = GDP*energy intensity*improvements in efficiency of the intensity of energy
Units: Kw

(22) GDP= 1000*alternative scenario
Units: €
Business as Usual.

(23) improvements in efficiency of the intensity of energy = 1

3.12. How to work more and better

The conflict that exists between the long-term and short-term solutions is something common in most systems, in the way that effective short-term solutions are not effective in the long-term and vice versa. For instance, in the business world, the limitations of preventative maintenance policies are well known. Although they are vital in the medium term, in the short term they cause breaks in the production that nobody wants to assume.

Similarly, in the quality environment, the obedience of strict quality control rules is the basis in order to achieve a prestigious reputation with clients. This can be understood in the medium term as achieving the trust of the clients and a bigger commercial margin. Nevertheless, a strict policy of quality control in the short term causes a rise in costs if merchandise that doesn't follow the norms is disposed of.

In the personnel training environment, an increase of hours dedicated to this activity increases productivity, quality and results in the company in the medium term. However, in the short-term, this implies higher costs and a reduction in the availability of staff available to work in the production.

Inside the company, we can also observe this situation in the need to increase the strength of the improvements in the production processes, and with this increase, the future productivity in the medium-term, or in other words, dedicate all the available efforts to dealing with the immediate needs of production. We can see an excellent approximation to the theme in the article 'Nobody Ever Gets Credit for Fixing a Problem that Never Happened' by Nelson P. Repenning and John D. Sterman (2001).

The basic cause of this conflict between short-term and long-term can be found in the fact that certain actions give immediate results while in others, there is a significant delay before the results can be seen.

This practical exercise attempts to illustrate the reality. It also attempts to show the usefulness of the causal diagram as an instrument to analyse a situation characterised by the existence of loops of feedback and delay.

The group of the Project Columbus 07 is owned by a company that is dedicated to the development of aerospace software and consists of IT technicians. Each employee has a great deal of autonomy in relation to the quantity of hours worked each week. The employees also have freedom to distribute these hours between three basic tasks. On one hand, the production of IT programmes to deal with current projects and on the other hand the analysis of IT tools – training in order to improve **productivity**. As this sector is very dynamic, it is important to dedicate a lot of hours to the **improvements** of productivity because technology can quickly become **obsolete.**

Every week, the employees receive comparative information between the real production and the desired production they develop. Based on the **difference of**

production perceived, **pressure is applied to vary the production**. This pressure can be understood in terms of an immediate adjustment to the **working hours in production**. **Pressure is also applied to improve the productivity** as a key aspect. In this way it is intended that more **hours are planned for productivity**.

The causal diagram formed by elements signalled in bold in the previous paragraphs allows us to illustrate this situation. We can identify two loops of feedback with the names 'work more' and 'work better'.

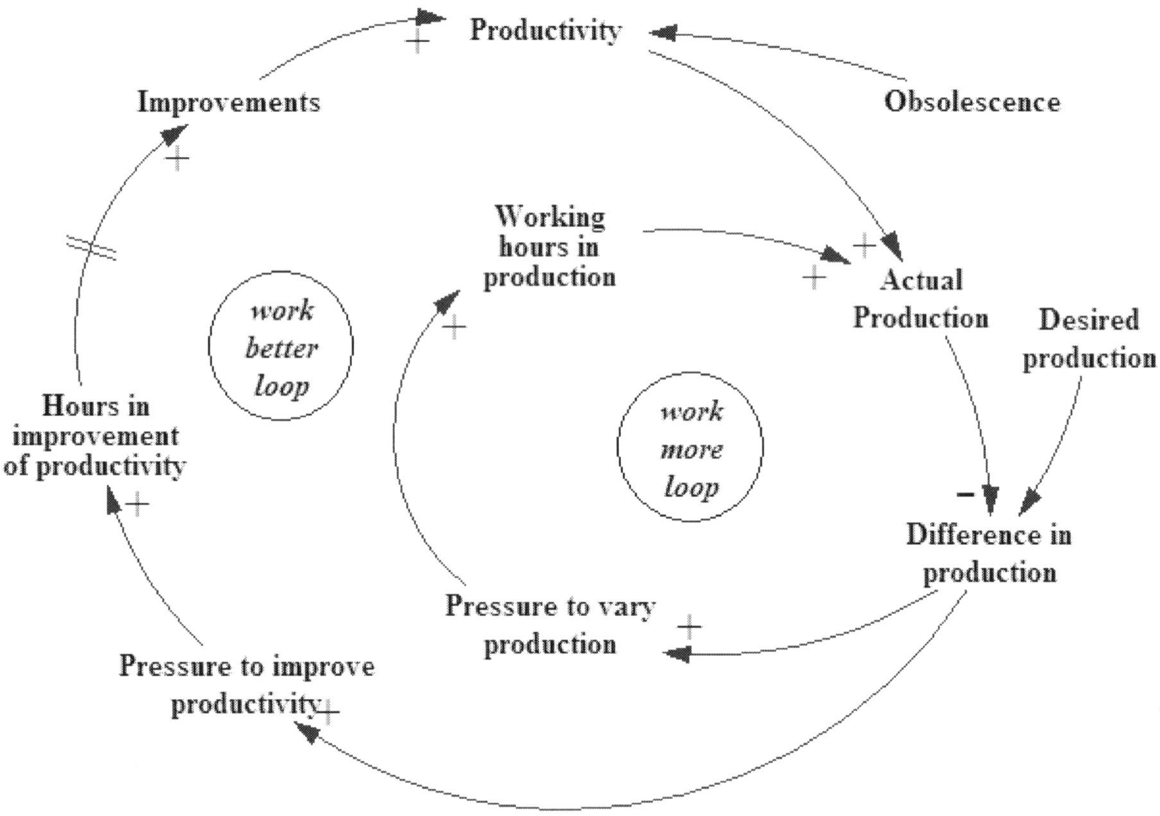

We can observe that they are two negative loops as they only have a negative relationship. They compare the state of the real production with the state of the desired production. The real state is adjusted using two different processes - the variation of the working hours – that allow immediate increases in production – and the improvement in productivity – which requires more time to show its effects.

The existence of a delay in the curve 'work better' will cause the other curve 'work more' to be the most active and the curve which offers results the quickest. It is for this reason that in companies that have high production targets, more emphasis is put on increasing the working hours rather than trying to increase the productivity

We are going to transform the causal diagram into a Stock and Flow Diagram SFD in order to be able work with the model in the computer. The necessary transformations will be minimal. The only change will be to define the levels of productivity and the working hours in production or in improvement of productivity. We will consider that the last two variables will change according to the pressure that each one receives.

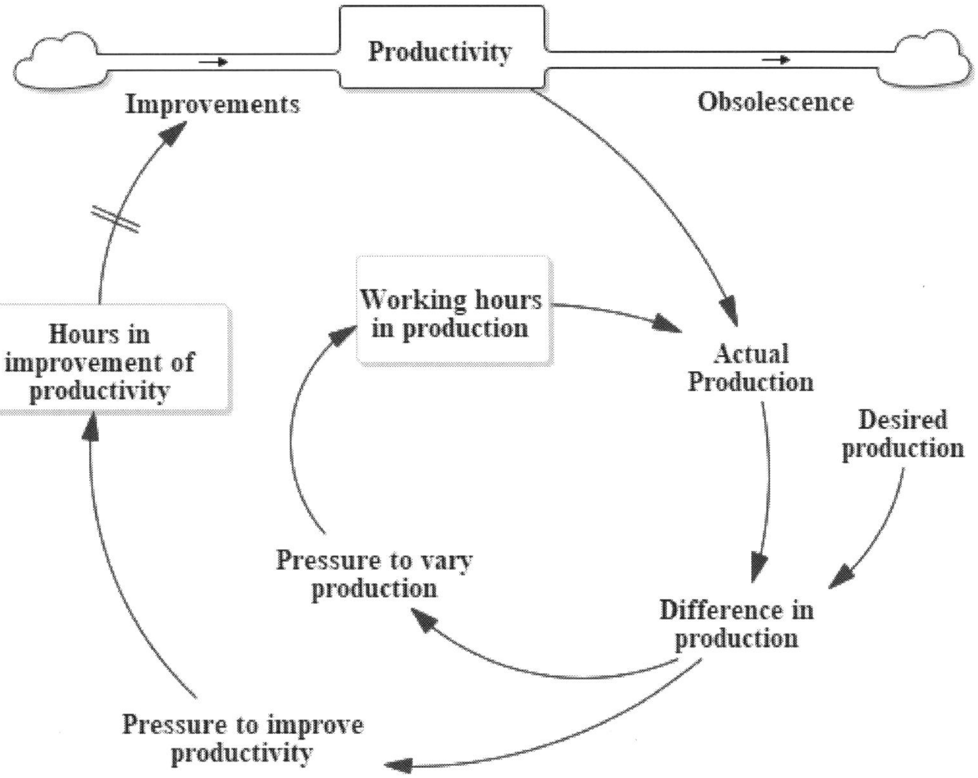

Now we need to have some additional information available to create the equations of the model. We will consider that the employee usually dedicates 35 hours a week to production tasks and 5 hours to tasks that improve productivity. The desired production is 3500 units. The productivity per working hour is 100 units/week. It is estimated that 20 weeks of delay exists between the hours dedicated to improve the productivity and in the time when this is manifested.

With the previous distribution of the hours it can be observed that the productivity is stable, with losses due to obsolescence and balanced improvements in the equivalent of 5 lines/week. With these indications we can formulate the equations of the model. Let us take a time scale of 100 weeks:

Model - Settings

INITIAL TIME= 0 FINAL TIME= 100 TIME STEP=1 Units for time = Week

Equations

(01) Desired production: = 3500
(02) Actual Production = Productivity*Working hours in production
(03) Productivity = +Improvements-Obsoleteness
Initial value= 100
(04) Difference in production = Desired production – Actual Production
(05) Working hours in production = Pressure to vary the production
Initial value: 35
(06) Hours in improvement of productivity =Pressure to improve the productivity
Initial value: 5
(07) Pressure to improve productivity = Difference of production / 500
(08) Pressure to vary production = Difference of Production / 100
(09) Improvements = DELAY 3I (Hours in improvements of productivity, 20, 5)
(10) Obsolescence = 5

We take as pressure to improve the production the units that exist as 'difference in production' divided by the 'productivity' initial 100. The 'pressure to improve the productivity' has to be much lower, as immediate results are not expected and for this reason we take a factor of 5 times the value of 'productivity'.

With these parameters we observe that all the elements of the system, especially the productivity, remain constants. It is important to have a model that represents the system in balance in order to appreciate better the effect of the changes in its structure or conditions.

To better observe the dynamic of the system we will see the effect of an increase of 1000 units of the desired weekly production. We will modify the equation in the following way:

Desired production = 3500+STEP(1000,10)

We can observe the following behaviour, and in summary, an increase in the desired production becomes firstly an increase in the working hours in production. Also, pressure to increase the productivity is increased. As this improves, the number of working hours can be reduced.

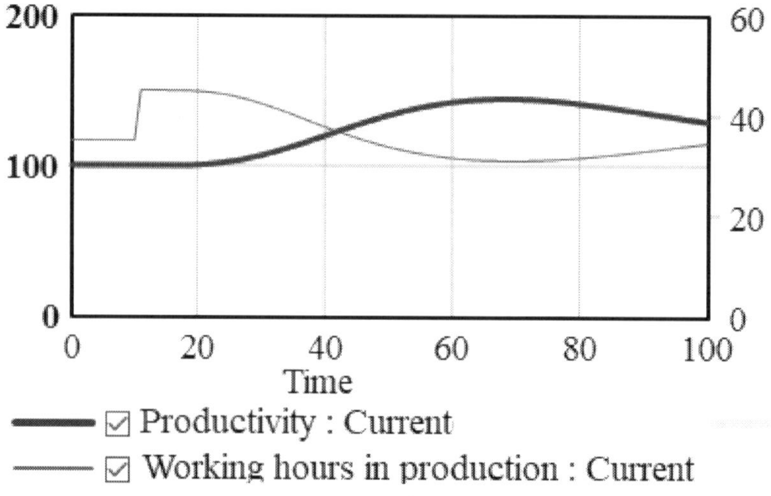

This model allows us to explain the reality of many companies, where although occasional efforts of production are needed, the progressive improvements of productivity allow the return to the initial situation after a certain time. On the contrary, **on other occasions we find that the pressure to increase the production influences the hours dedicated to the improvement of productivity.** This causes a progressive decrease in the level of productivity which in turn means that more hours of work are necessary. A curve appears every time employees feel obliged to work more as the productivity continues to decline.

Furthermore, the employee observes that if it dedicates **less hours to improving productivity, it has more hours available to work on production.** With this the difference between the real production and the desired production, there is less pressure of all kinds. We can see these behaviours represented by the darkest arrows in the diagram.

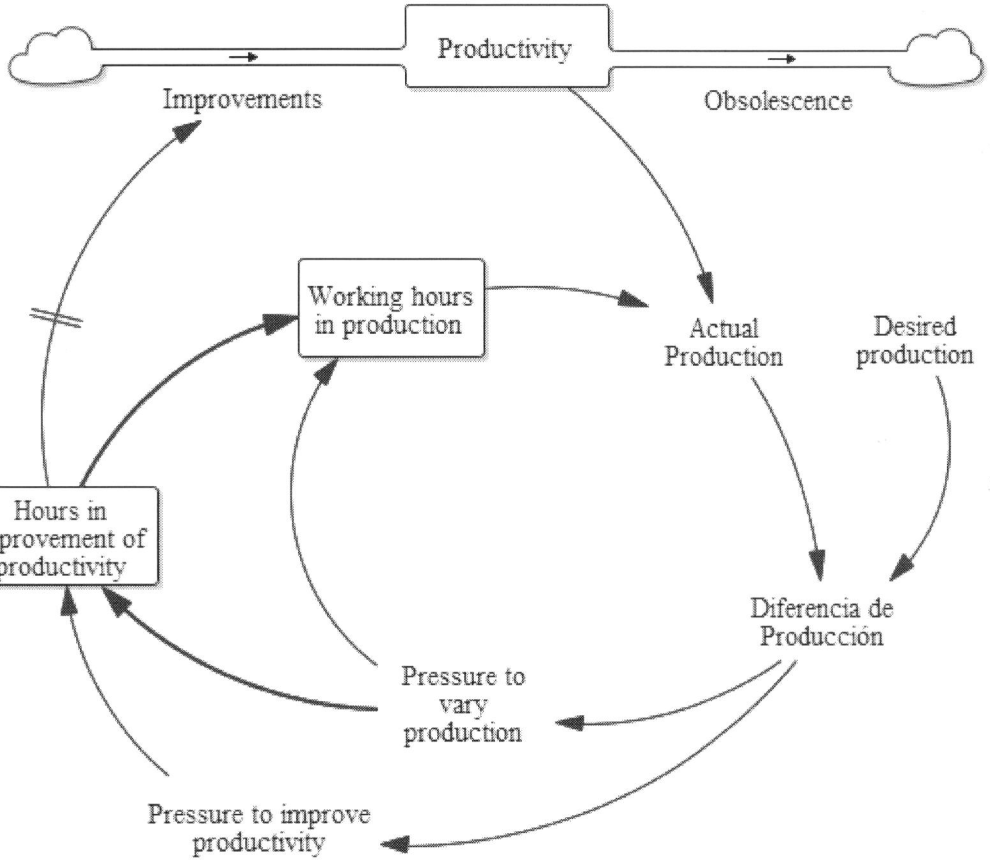

To represent this new situation we have to modify two equations in the previous model:

Hours in improvements of productivity = Pressure to improve productivity – pressure to vary the production/4

Working hours in production = pressure to vary the production – (Hours in improvement of productivity – 5)/5

In the first equation, we need to take into consideration the pressure to vary the production, reducing its proportion in relation to the pressure to improve productivity. So although it acts, it doesn't manage to annul. In the second equation we need to include the effect of the hours spent in improvement of productivity. Initially, there were 5 hours.

The dynamics that these new relations generate are especially attractive in the short term as the real production can be adjusted to the desired production very quickly. But in the medium term, it would be counterproductive because the obsolescence of the productivity isn't compensated by new improvements.

In this way we can observe how the productivity shows a progressive decrease as an increase in the hours dedicated to production

In conclusion we can observe how the natural tendency of the system could be to correct the differences between the desired production and the real production based on successive recourse to 'work more'.

This situation is logical as it achieves the desired effect quickly. Nevertheless, if the method of 'working more' impedes the necessary process to work better, it is inevitable that more and more hours will be needed owing to the drop in productivity.

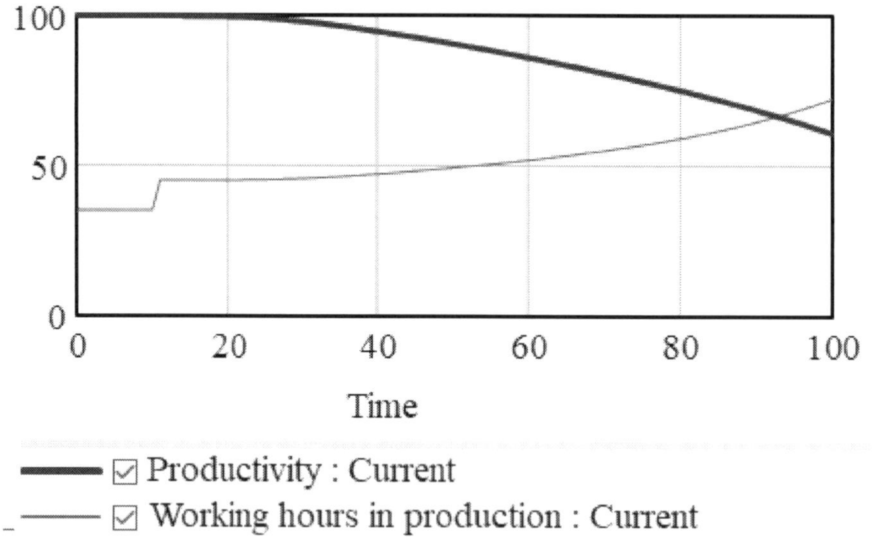

3.13. Management of faults

We want to simulate the effects that will provoke a decrease in the number of maintenance workers to the number of damaged pieces and the number of lost pieces. Create a new model with INITIAL TIME = 0 and FINAL TIME = 24 months

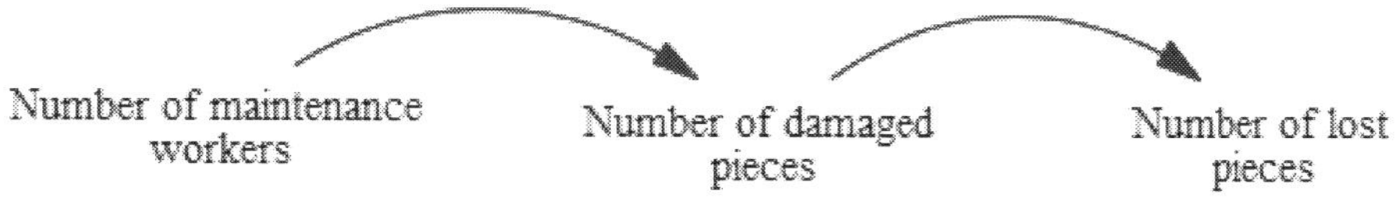

Let us suppose that the number of workers drops from 10 to 5 in the 6th month.

(1) Number of maintenance workers = 10-STEP(5,6)
(2) Number of damaged pieces = WITH LOOKUP (Number of maintenance workers (0,800),(2,600),(4,400),(6,200),(8,100),(10,0)
(3) Number of lost pieces= Number of damaged pieces+RAMP(1 , 0 , 24)

Now, we will define that the number of damaged pieces depends on the number of workers in that when there are 10 workers the damaged pieces are 0. If there are 8 workers we would have 100 pieces, if there are 6, we would have 200 pieces if there are 4 we would have 400 pieces, if there are 2 we would have 600 and if there are 0 we would have 800 pieces.

We will use an internal table for this. We will select Auxiliary Lookup as the type of variable number of pieces. The AsGraph button must be pressed and we will need to introduce the defined points as pairs of values of the independent variable (Input) and the dependent (Output). Once the information is entered, click OK.

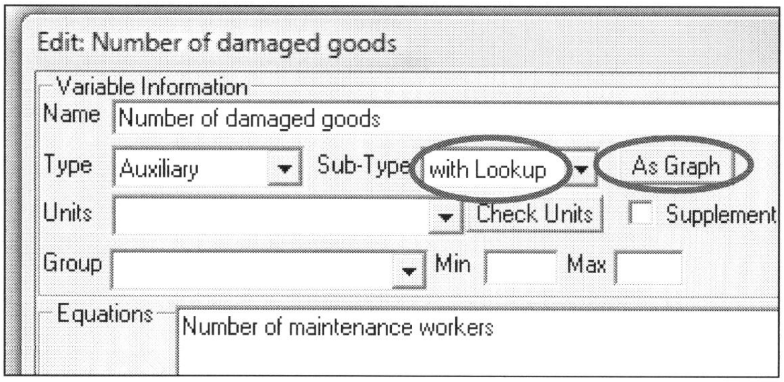

Also, we want to know when the number of maintenance workers is less than 10 and the number of lost pieces is more than 315. To do this we will use the Reality Check.

To do this add the shadow variables of “Number of maintenance workers” and “Number of lost pieces”, this avoid confusion with the real variables. Add the variable “RC of lost pieces” and the arrows.

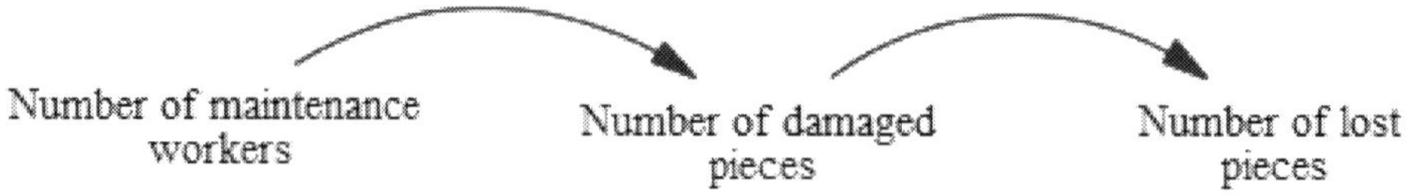

Define this new variable as Type: Reality Check and Sub-Type: Constraint, and enter the equation:

RC of lost pieces: THE CONDITION:
Number of maintenance workers<10:IMPLIES:Number of lost pieces>315

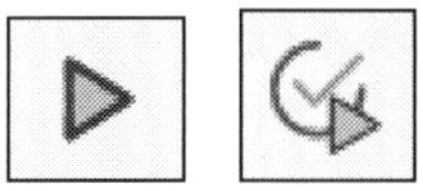

Simulate the model and then click the Reality Check icon. From the menu that appears, select "RC of lost pieces" and click Highlighted. A message appears showing that after month 16 this rule is not met.

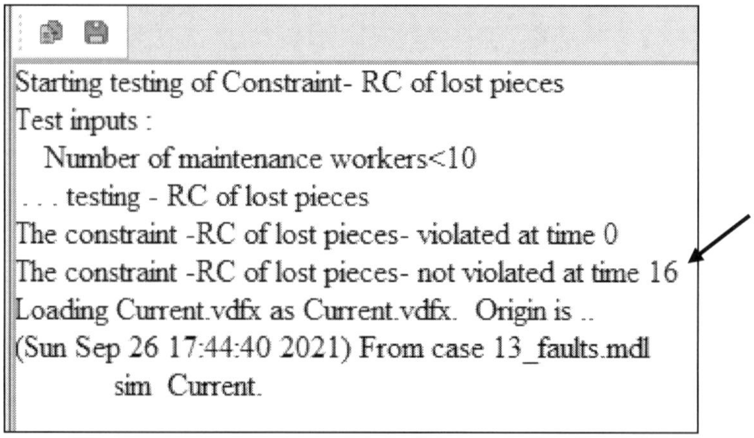

3.14. Project dynamics

There are a variety of activities, from writing term-papers to building nuclear power plants that have very similar dynamic characteristics. There is a goal and an expectation of what it takes to achieve the goal, followed by lots of work, and after some period of time, by a more or less successful achievement of the goal. Projects that are relative to the initial goals and expectations, take too long, cost too much, have poor quality and complete too little, are common.

In this exercise we will develop a model to help understand the processes involved in getting a project completed. For concreteness, we will think of this project as the design of a new building, though the model is directly applicable to other activities such as developing software, designing a new product, preparing a presentation and writing an exercise. The model would need to be extended to investigate activities such as construction in which material availability becomes very important, or scientific research in which experimentation plays a key role.

In conceptualizing and creating this model we will use an iterative approach. We will start with the simplest structure that is relevant to this problem, and continually refine it. This is a useful technique because it prevents a situation in which you feel you have completed a model but the situation results simply do not make sense. You will be simulating every step and seeing the effects of new model structure as it is added.

In developing this model we will be depending on the computer to continually give us feedback on the consequences of changing structure. While the computer is very good at this, it is also important to think about what we are doing. Before any simulation experiment is run you should ask yourself what you expect the results to be. If you are surprised, find out why. If you are right, make sure it is for the right reasons.

Model 1. Task accomplishment.

The most fundamental characteristic of any project is that there is something to do and it gets done.

Model 2. Stopping Work

There are some ways to shut down the project model. One is to simply stop simulating when the project has been completed. The other is to create a variable 'project is done' to control this. This last way will provide us with a mechanism for descoping the project if there are schedule or budget problems.

Model 3. Errors and Rework

So far, we have assumed that the work is being done without error. In general, this is not true. There are a number of places where errors can occur, including miscommunication among personnel, technical oversights and just plan mistakes. When errors occur they are not, however, immediately discovered. Errors remain undetected until there is a review or integration activity that brings them to light.

Model 4. Rework Discovery

The final stages of a project tend to see a big increase in re-work discovery. It is very much like finally putting the pieces of a puzzle together. In the end, it becomes quite obvious which pieces are missing or are the wrong shape. Example of problems that can occur are plumbing systems that depend on nonexistent access corridors and ventilation shafts that are the wrong size for the planned equipment.

Model 5. Schedule

The purpose of project management is to keep projects on schedule. To do this, it is necessary to know what the schedule is and adjust resources to meet that schedule.

Model - Settings

INITIAL TIME= 0   FINAL TIME=24   Units for time=Month   TIME STEP= 0.0625

If TIME STEP=1 the program will do 24 steps in the simulation process (FINAL TIME – INITIAL TIME). We would like to shorten this Time Step to 2 days. So we will use a TIME STEP equal to (12 months/year) x (2/365) = 0.065. We will choose the closest option in the Model – Settings menu: 0.0625

Model 1

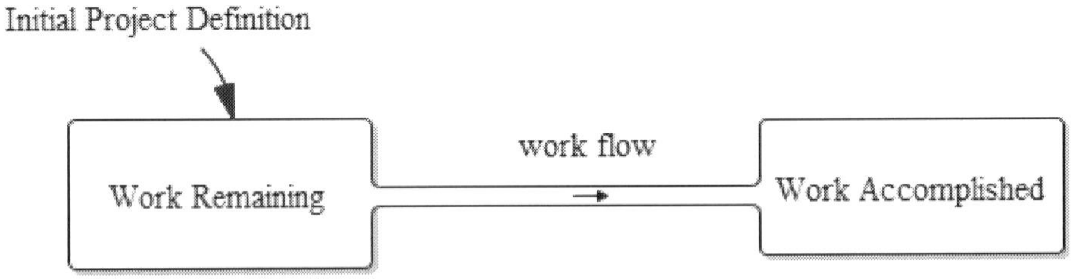

(01) initial project definition= 1000
Units: Tasks
We set 'Initial Project Definition' equal to 1000 tasks, and work flow 100. We are going to consider all the tasks are equal, because we will like to centre the objective of the model in the study of the errors, not in the tasks. We are dealing with an aggregate representation of a project.

(02) Work Accomplished= work flow
Initial value: 0
Units: Tasks

(03) work flow=100
Units: Tasks/Month
We are doing 100 tasks each month.

(04) Work Remaining=
-work flow
Initial value: Initial Project Definition
Units: Tasks

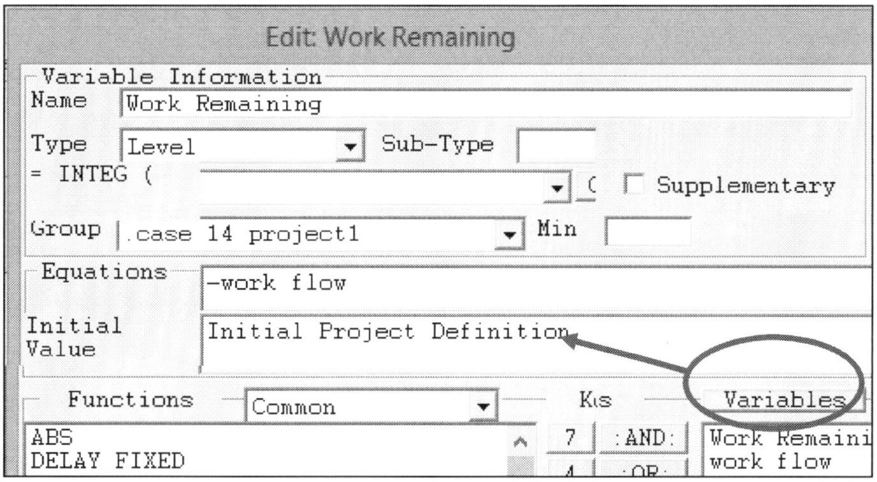

Note: In the equation of Work Accomplished' to prevent a warning message click 'Supplementary', because this variable has no influence on any other.

Simulate and results

To show two variables in the same graph, click the Control Panel icon, options: Custom Graph - New, and select the variables you want to see together.

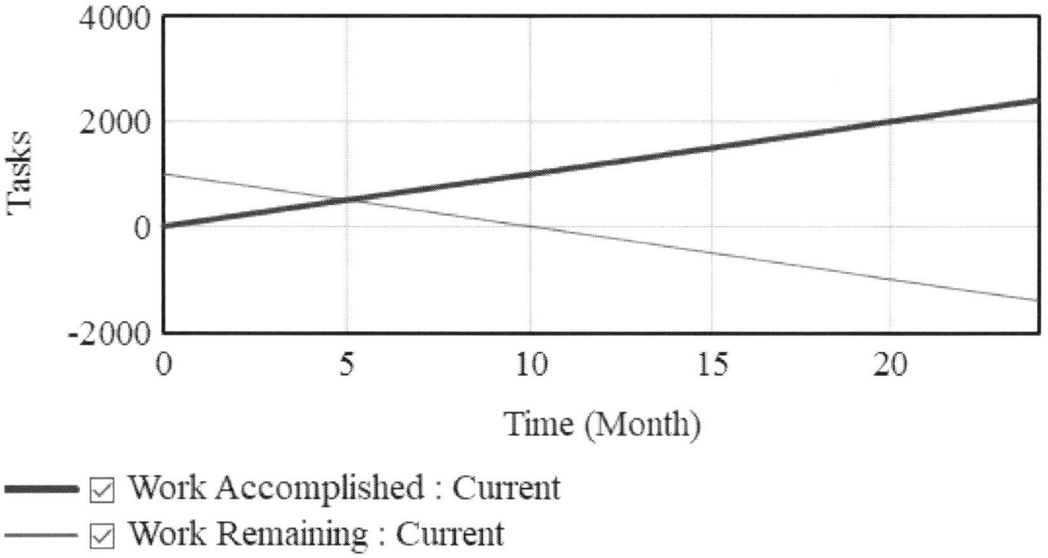

Model 2

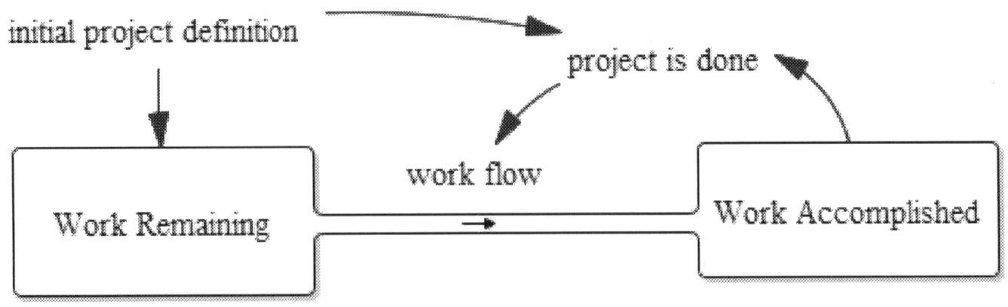

(01) initial project definition= 1000
Units: Tasks
We set 'Initial Project Definition' equal to 1000 tasks, and work flow 100. We are going to consider all the tasks are equal, because we prefer to centre the objective of the model in the study of the errors, not in the tasks. We are dealing with an aggregate representation of a project.

(02) project is done= IF THEN ELSE(Work Accomplished>=Initial Project Definition, 1,0)
Units: Tasks

(03) Work Accomplished= work flow
Initial value: 0
Units: Tasks

(04) work flow= IF THEN ELSE(Project is done=1, 0 ,100)
Units: Tasks/Month
We are doing 100 tasks each month until the project is done.

(05) Work Remaining= -work flow
Initial value: Initial Project Definition
Units: Tasks

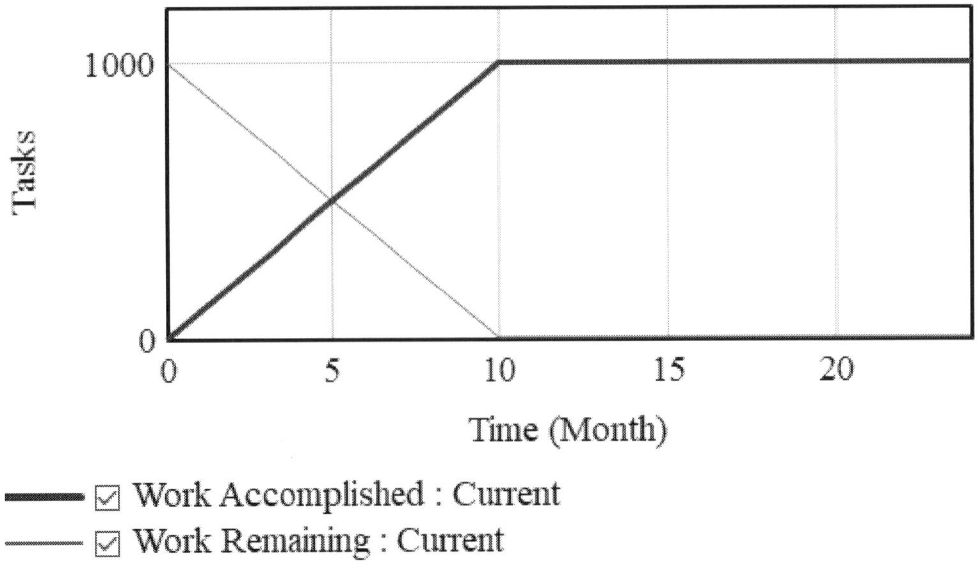

Model 3

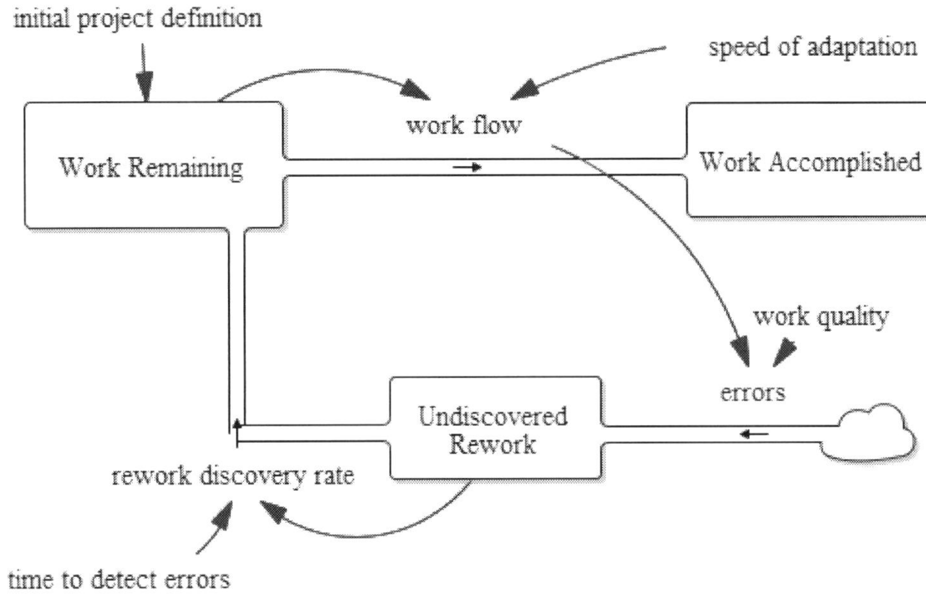

(01) errors=work flow*(1-work quality)
Units: Tasks/Month
In this model we will not assume that the tasks are done without errors. Errors are function of the tasks done (the work flow) and the percentage of tasks that do not accomplished the desired quality.

(02) initial project definition= 1000
Units: Tasks
We set 'Initial Project Definition' equal to 1000 tasks, and work flow 100. We are going to consider all the tasks are equal, because we will like to centre the objective of the model in the study of the errors, not in the tasks. We are dealing with an aggregate representation of a project.

(03) rework discovery rate= Undiscovered Rework/Time to detect errors
Units: Tasks/Month
When errors occur they are not immediately discovered. Errors remain undetected until there is a review or integration activity that brings them to light. This is obtained based on 'undiscovered rework' and the 'time to detect errors'.

(04) time to detect errors= 3
Units: Month
Time to detect errors depends is a constant

(05) Undiscovered Rework= errors-rework discovery rate
Initial value: 0
Units: Tasks

(06) Work Accomplished= work flow
Initial value: 0
Units: Tasks

(07) work flow= MIN(Work Remaining,100)/speed of adaptation
Units: Tasks/Month
We are doing 100 tasks each month until the Work Remaining will be less than 100.

(08) Work Quality=0.9
Percentage of the tasks that are done correctly
Units: Dmnl

(09) Work Remaining= -work flow+rework discovery rate
Initial value: Initial Project Definition
Units: Tasks

(10) speed of adaptation=1
Units: Month

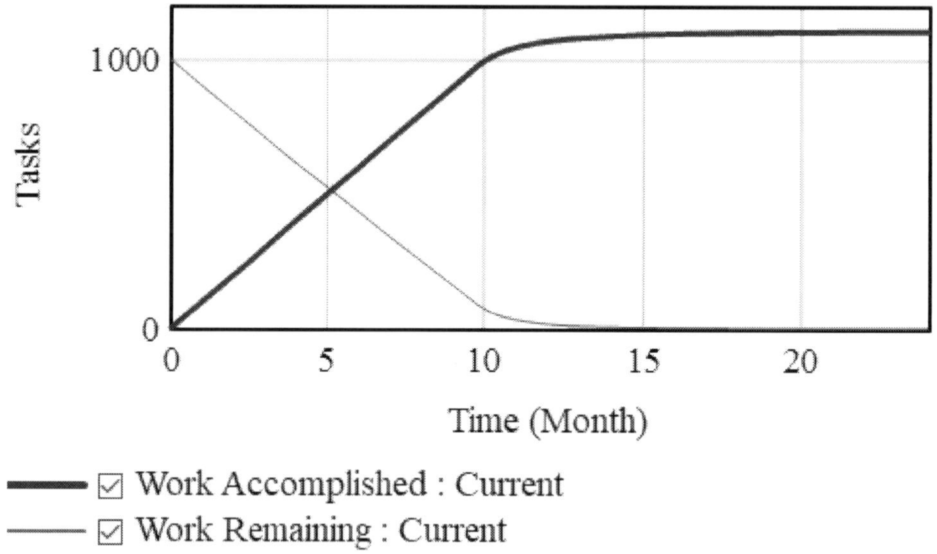

Click the icon Units Check

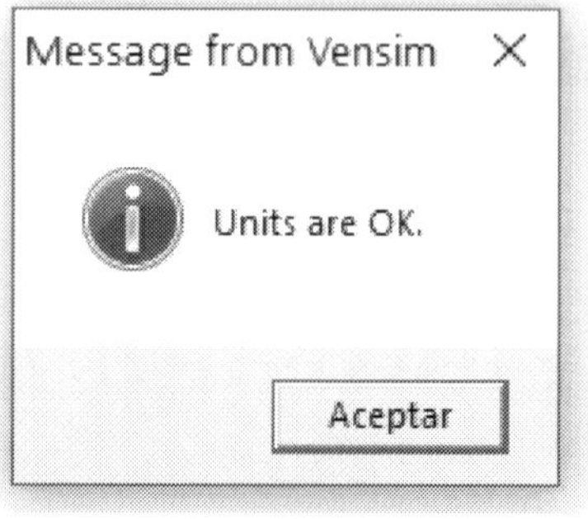

Model 4

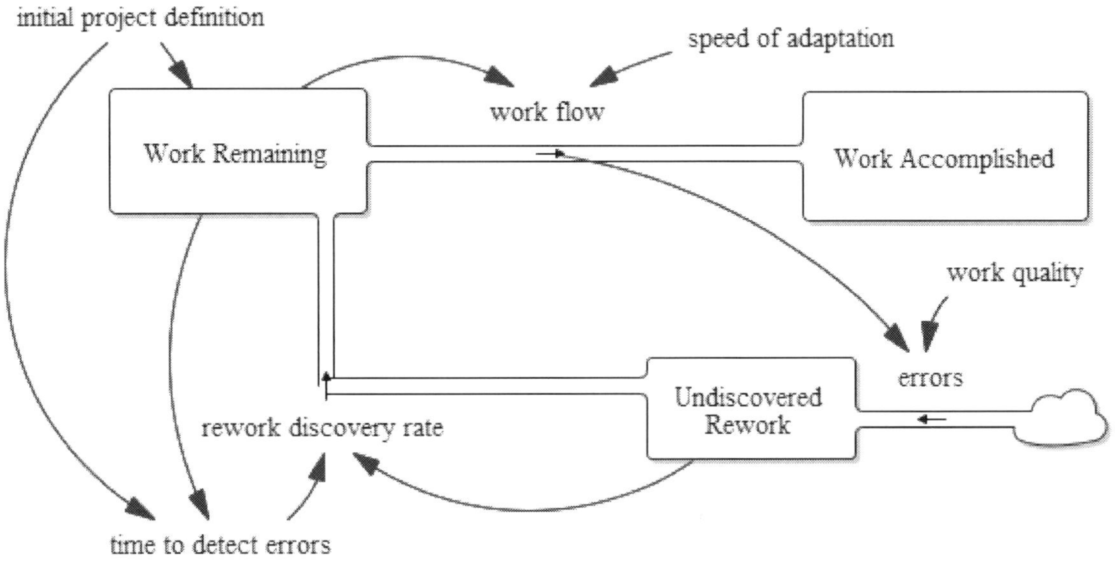

(01) errors=work flow*(1-work quality)
Units: Tasks/Month
In this model we will not assume that the tasks are done without errors. Errors are function of the tasks done (the work flow) and the percentage of tasks that do not accomplished the desired quality.

(02) initial project definition= 1000
Units: Tasks
We set 'Initial Project Definition' equal to 1000 tasks, and work flow 100. We are going to consider all the tasks are equal, because we will like to center the objective of the model in the study of the errors, not in the tasks. We are dealing with an aggregate representation of a project.

(03) rework discovery rate= Undiscovered Rework/time to detect errors
Units: Tasks/Month
When errors occur they are not immediately discovered. Errors remain undetected until there is a review or integration activity that brings them to light. This is obtained in base of the 'undiscovered rework' and the 'time to detect errors'.

(04) speed of adaptation= 1
Units: Month
This is the speed of adaptation between the workforce required to accomplish the required completion date with the current workforce

(05) time to detect errors = WITH LOOKUP (Work Remaining/initial project definition
(0,0.5),(0.2,1),(0.4,3),(0.6,6),(0.8,9),(1,12)
Units: Month
Time to detect errors depends of the state of the project. In this table we are using the idea that when we are in the beginning of a project, and there remains a lot of work to do, we are slower in detecting errors than when we are at the end of the

project. To define this variable, set Type=Auxiliary, Sub-Type = with Lookup, and As Graph=enter values in the table.

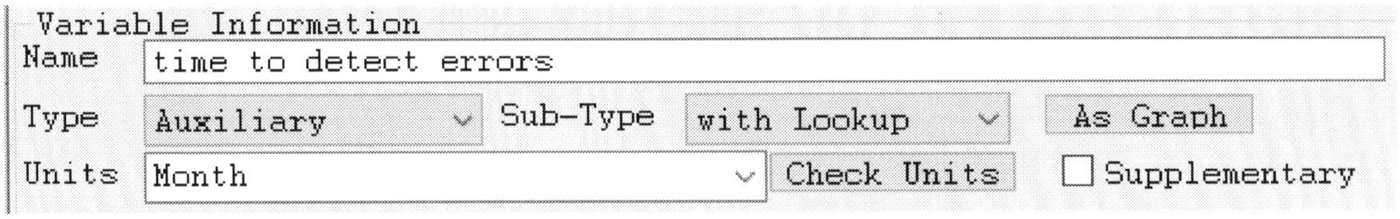

(06) Undiscovered Rework= INTEG (errors-rework discovery rate
Initial value: 0
Units: Tasks

(07) Work Accomplished= INTEG (work flow
Initial value: 0
Units: Tasks

(08) work flow= MIN(Work Remaining,100)/speed of adaptation
Units: Tasks/Month
We are doing 100 tasks each month until the Work Remaining will be less than 100.

(09) Work Remaining= INTEG (-work flow+rework discovery rate
Initial value: initial project definition
Units: Tasks

(10) work quality= 0.9
Units: Dmnl
Percentage of the tasks that are done correctly

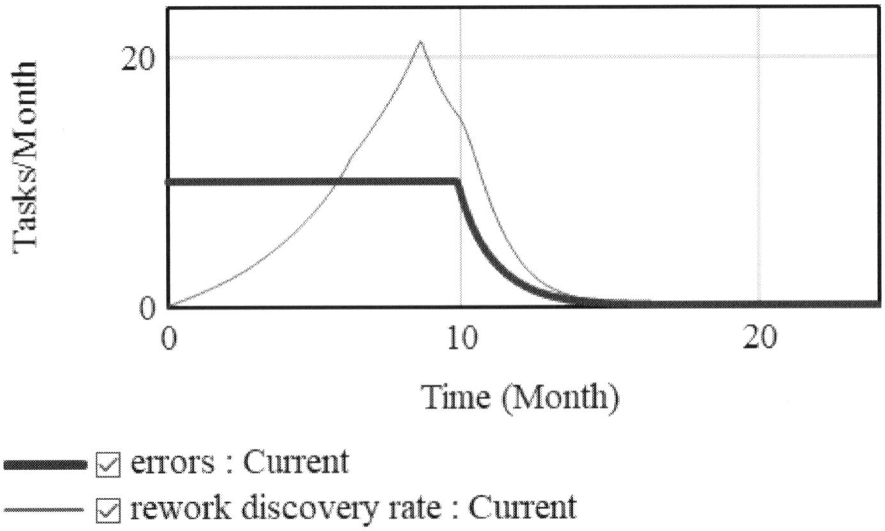

Model 5

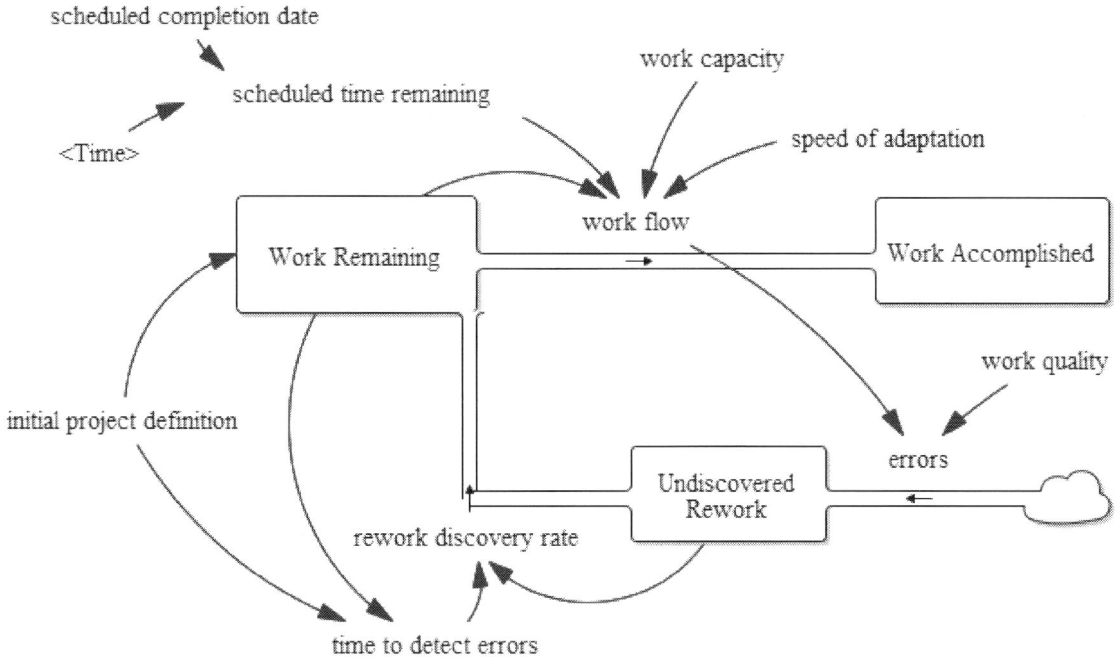

In every project there is a completion and delivery date, which will change the pace of work and the amount of resources used. The <Time> variable is a simple way to count the residual term at any time, starting from 10 months to 0. Select the <Time> variable using the icon Shadow Variable

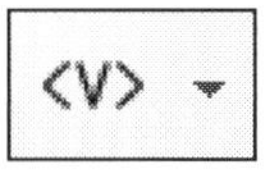

(01) errors=work flow*(1-work quality)
Units: Tasks/Month
In this model we will not assume that the tasks are done without errors. Errors are function of the tasks done (the work flow) and the percentage of tasks that do not accomplished the desired quality.

(02) initial project definition= 1000
Units: Tasks

(03) rework discovery rate=Undiscovered Rework/time to detect errors
Units: Tasks/Month
When errors occur they are not immediately discovered. Errors remain undetected until there is a review or integration activity that brings them to light. This is obtained in base of the 'undiscovered rework' and the 'time to detect errors'.

(04) scheduled completion date= 10
Units: Month
We have 1000 tasks and 100 tasks/month scheduled, so we will need 10 months to accomplish the project.

(05) scheduled time remaining=MAX(scheduled completion date-Time,0)
Units: Month
This is a easy way to count the remaining time to finish the work. Make sure you have added the <Time> variable using the icon Shadow Variable.

(06) speed of adaptation=1
Units: Month

(07) time to detect errors = WITH LOOKUP (Work Remaining/initial project definition,
(0,0.5),(0.2,1),(0.4,3),(0.6,6),(0.8,9),(1,12)
Units: Month
Time to detect errors depends of the state of the project.

(08) Undiscovered Rework= INTEG (errors-rework discovery rate
Initial value: 0
Units: Tasks

(09) Work Accomplished= INTEG (work flow
Initial value: 0
Units: Tasks

(10) work capacity= 250
Units: Tasks/Month
This is the maximum amount of tasks that our team can do in a month. We will consider it as a constant.

(11) work flow= MIN(work capacity, XIDZ(Work Remaining, scheduled time remaining, Work Remaining/speed of adaptation))
Units: Tasks/Month
The purpose of the project management is to keep projects on schedule. To do this it is necessary to know what the schedule is, and adjust resources to meet that schedule. The 'work flow' is the least (function MIN) between the 'work capacity' and the production capacity needed (obtained as: Work Remaining/scheduled time remaining) to accomplish the completion date. It is obtained based on 'Work Remaining' divided by the 'scheduled time remaining' except when this last value is equal to 0 (function XIDZ). In this case each month we will do the 'Work Remaining' divided by the 'speed of adaptation'.

(12) work quality= 0.9
Units: Dmnl
Percentage of the tasks that are done correctly

(13) Work Remaining= INTEG (-work flow+rework discovery rate
Initial value: initial project definition
Units: Tasks

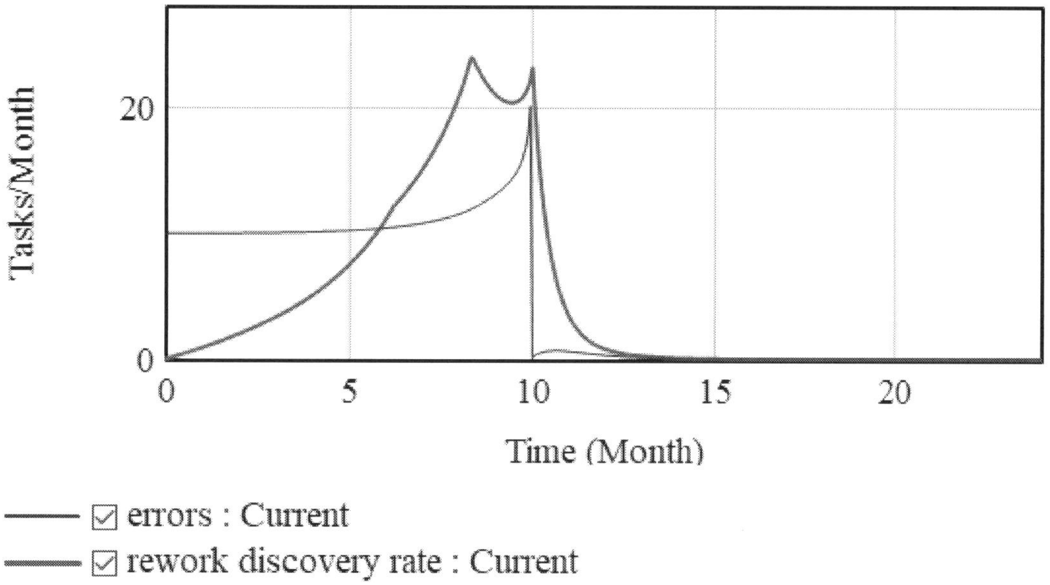
Tasks/Month
20
0
0
5
10
15
20
Time (Month)
errors : Current
rework discovery rate : Current

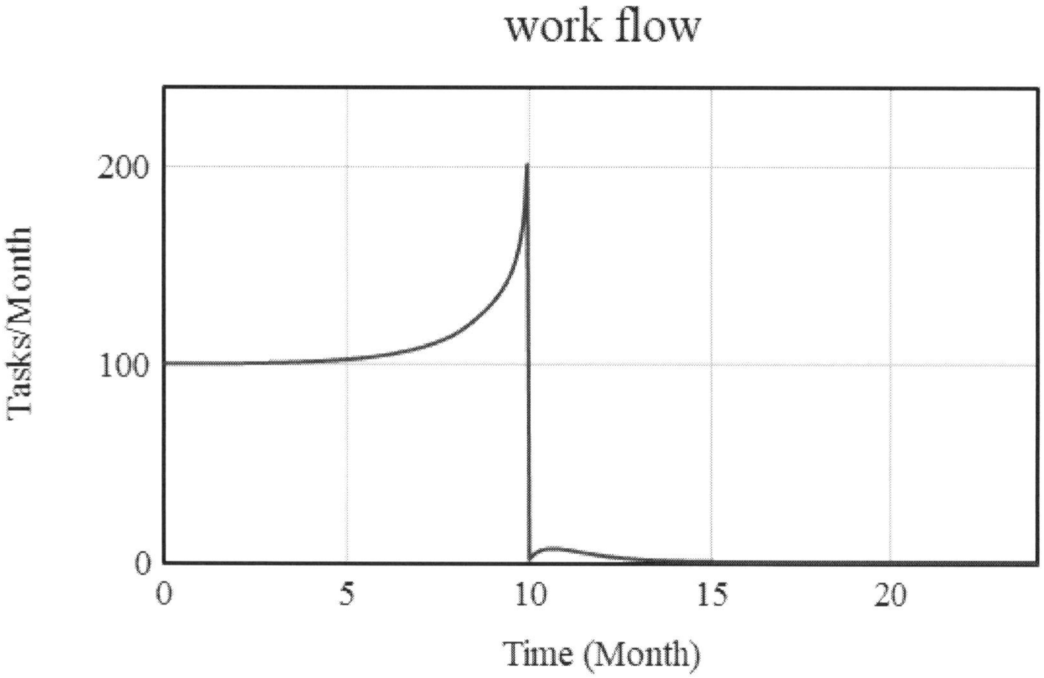
work flow
Tasks/Month
200
100
0
0
5
10
15
20
Time (Month)

3.15. Innovatory companies

In this work we can see the distinguishing characteristics of innovating companies and the steps followed in the construction of the model, the structure of the latest version, the behaviour obtained from this structure and the simulations and analyses of sensitivity that have been made. Lastly, the most suitable policies and the objectives which should be followed by innovating companies are explained.

The specialists and the bibliography that deal with this subject indicate that the distinguishing characteristics of innovating companies are: small size, environment with rapidly changing technology, high technological risk, need for investment in R+D, government aid policies and programs, difficulty of access to raw materials, difficulty in obtaining qualified staff, difficulty in the channels of distribution, markets without protective borders, attractive margins, financial difficulties, and need for government aid.

The next illustration shows the basic Causal Diagram

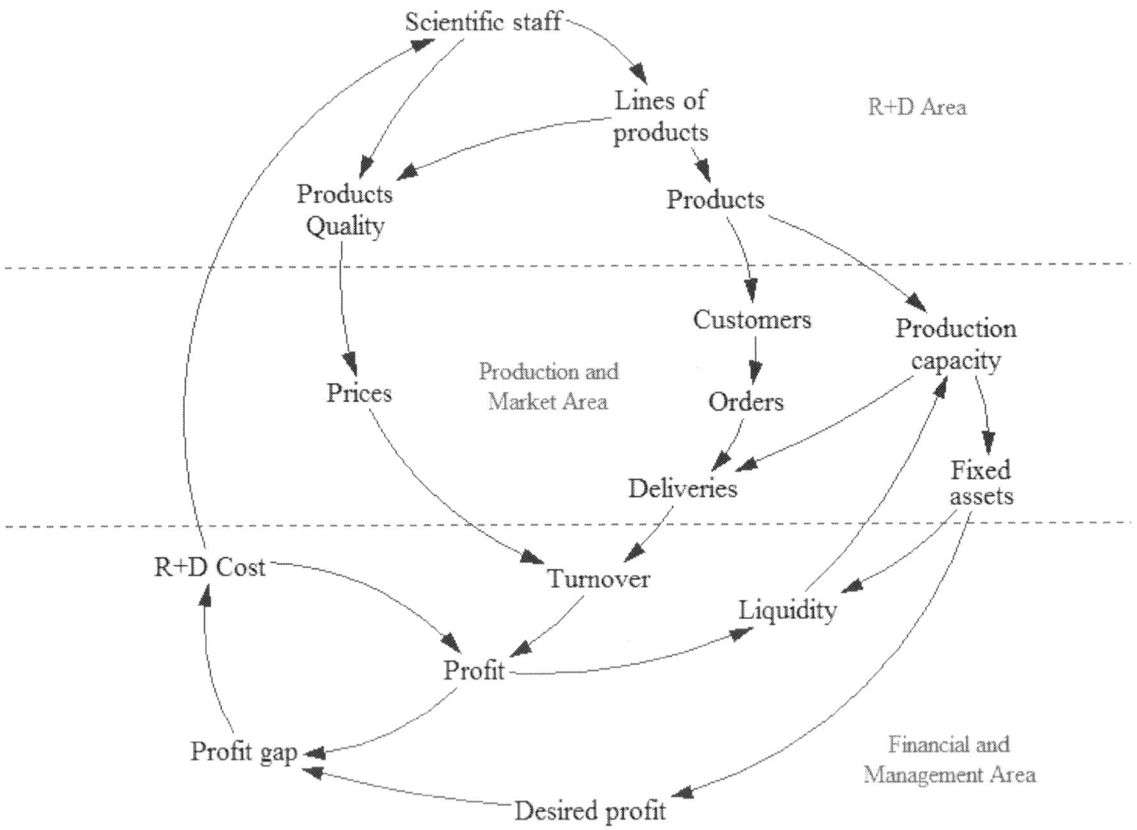

The model is divided into three main areas: the R+D Area, the Production and Market Area, and the Financial and Management Area. We can see how the R+D expenses make it possible to employ scientific staff who generate lines of production which lead to deliveries based on the current production capacity. Moreover, the quality of the products influences the price, and this, combined with the deliveries, defines the turnover. The turnover makes it possible to obtain a profit.

The cash situation limits the volume of the production capacity. In turn, the 'fixed assets', which is the accounting expression for the production capacity, influences the desired profit. When this is higher than the real profit, there tends to be cuts in the R+D expenditure. Although I will explain the main elements below, an overall view allows us to observe the existence of positive and negative loops. The former will by themselves lead the company to an exponential growth or a rapid collapse. The latter will act as stabilizers of the former.

We can clearly see the inputs and outputs of each area. Thus, the R+D Area receives the input of the R+D expenditure from the Finance and Management Area, giving as an output products of a certain quality. The Production and Market Area receives these outputs, together with the cash situation, and gives, as an output, deliveries at a certain price and production capacity. Lastly, on the basis of these inputs the Financial and Management Area offers a given R+D expenditure as an output. We will now see the main elements and relations of the model.

Research and development area

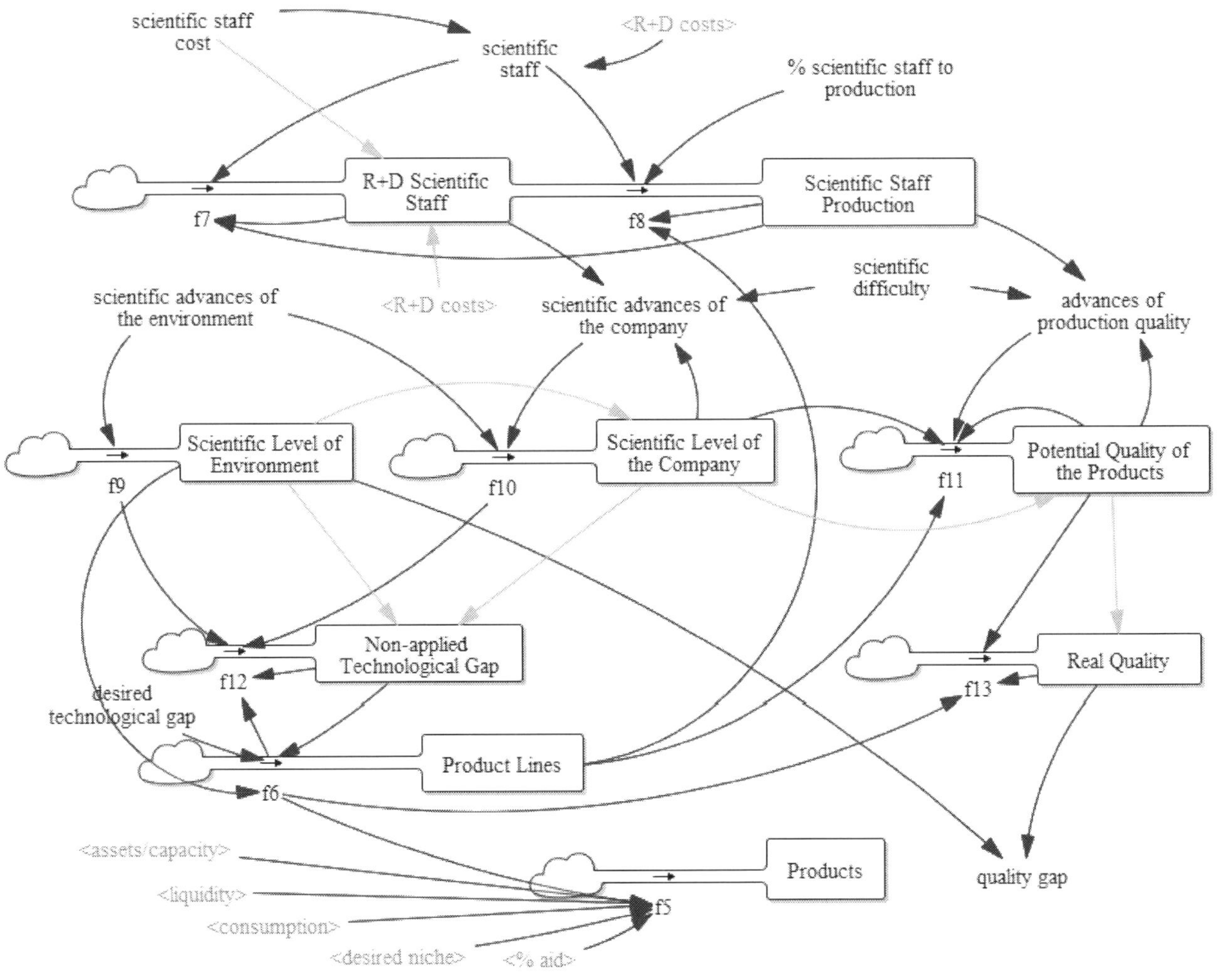

This is a key area in the model, showing the elements which affect the generation of products. It includes:

Scientific staff: These are persons who may be employed on the basis of the R+D expenditure. It is divided into two groups, the Scientific Staff devoted to R+D and the Scientific Staff devoted to production. Although initially all the scientific staff are devoted to R+D, after the launching of the first line of products they are divided into two groups.

Scientific progress: The scientific progress of the company develops on the basis of the number of employees devoted to research, and is hindered by the scientific difficulty of the chosen area of work and by the scientific level of the R+D work of the company.

Scientific difficulty: The degree of scientific difficulty of the area in which the company wishes to introduce its products, which may be high if it wishes to offer products with the latest technology, or low otherwise.

Scientific level: The scientific advances of the environment. Supposing there is a fluid relationship with the environment, the scientific level of the company increases according to both the scientific advances of the company and the advances in the environment. It is initially higher than that of its environment because it includes research previous to the creation of the company.

Potential quality of the products: This evolves according to the scientific level of the company and the advances in production quality. The increase in lines of products leads to diseconomies which lead to lower increases in quality.

Real quality: This is equal to the potential quality at the time the line of products is launched and is considered not to vary until the launching of a new line.

Quality gap: Is the percentage by which the technological quality of the company's products exceeds that of the environment. This gap has a clear influence on the price of the product.

Non-applied technological gap: This includes the research efforts devoted to creating a new line of products. It has not yet been applied as the desired technological advantage over the knowledge of the environment has not yet been reached. This increases according to the difference between the scientific level of the company and the scientific level of the environment. It diminishes as the product lines are applied.

Product lines: This indicates the number of lines which have been successful and serve as a basis for creation of products. In these companies the products are not generated in isolation, and one product type generates a diversity of products which meet the most common requirements of the customers. Product lines are generated when the Non-applied Technological Gap exceeds the scientific level of the environment by a desired value. This constant permits the simulation of different positions of entry in the market.

Products: After a line has been obtained, the products generated are no more than different applications of a single scientific base - the line of research - for the specific problems of the customers. This variable covers the products which are marketed, not those which may be marketed, as the volume of investment is limited. Thus, products are only launched while the cash available makes it possible to finance the new investments in fixed assets. Due to the dynamism of this sector, it is supposed that products not launched a certain time after the line of research has finished will not remain in the portfolio, and the company loses the opportunity to launch them. The flow of products shows a brief delay with regard to the appearance of the product lines.

Production and market area

This area contains the main aspects of production and sales.

Production capacity: This increases according to the launching of products. The production capacity determines the flow of deliveries, since production stocks are not considered to be significant.

Orders: These are calculated on the basis of the number of products, the number of customers and their consumption. A constant unit consumption per customer is assumed during the period studied.

Deliveries: This coincides with the volume of orders if the production capacity so permits. Orders which exceed the capacity are considered as lost. If the orders exceed the production capacity, this leads to what we will call Production Tension. The cash situation permitting, after a certain delay this tension will lead to an extension of the production capacity.

Price: It is considered that the company cannot significantly influence the reference price, but it can obtain a higher or lower price according to the quality gap of its products.

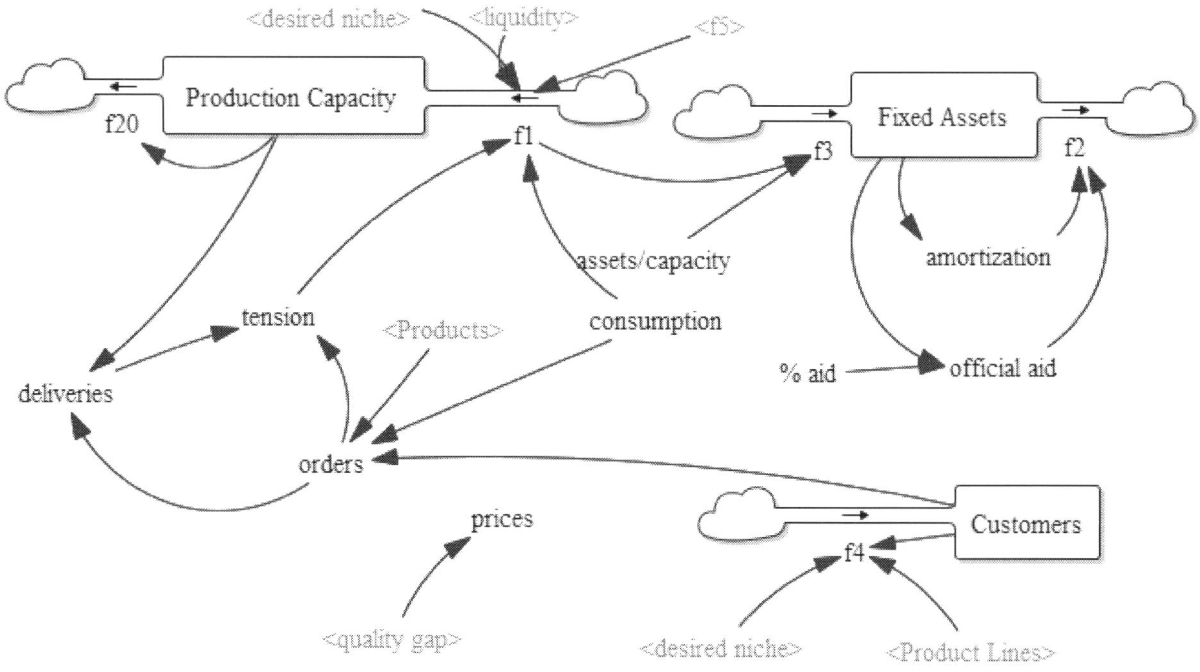

Customers: This deals with the number of customers of the company. The company starts to find customers in the desired niche after obtaining the first product line.

Fixed Assets: The value of the investments in fixed assets such as machinery and installations. It is difficult to find outside finance for these investments, and the company must therefore use its own capital. The fixed assets increase according to increases in the production capacity, and decrease according to amortization and the official aid received. They are of great importance in the calculation of the desired profit.

Finance and management area

This deals with the accounting and management aspects of the company. The basic aspects are:

Turnover: The deliveries multiplied by the price.

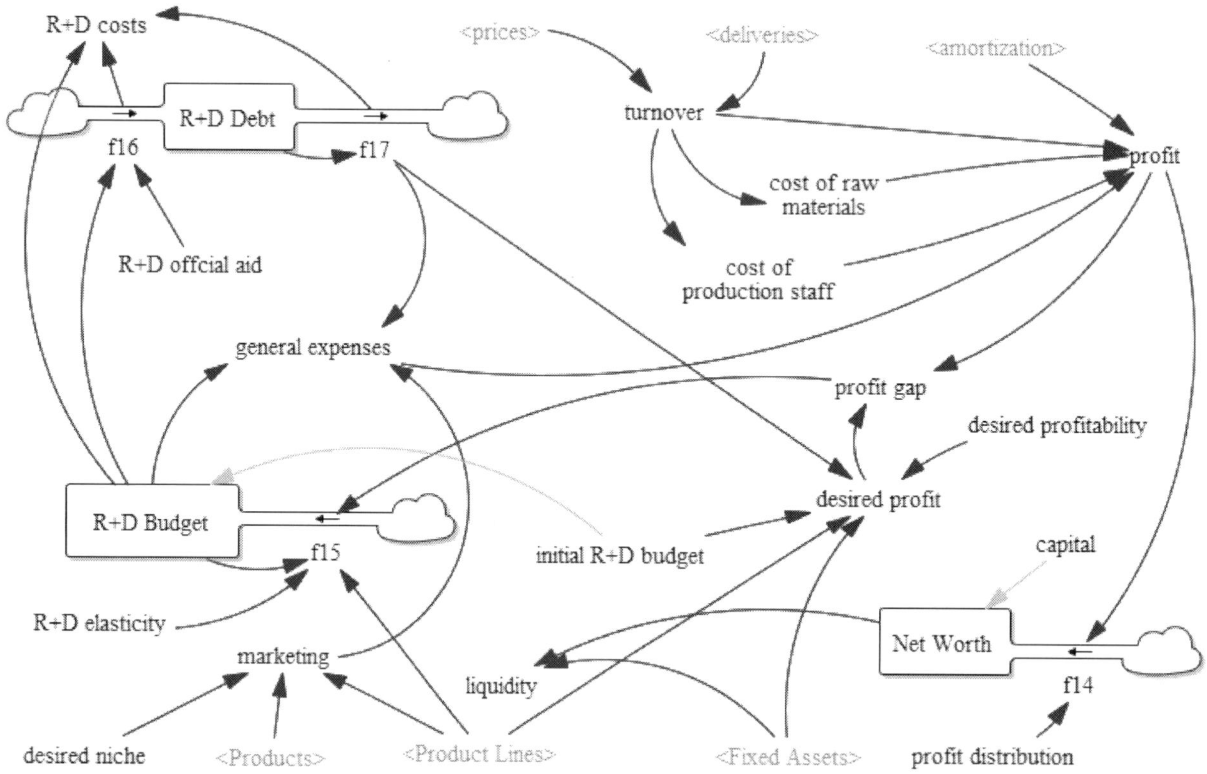

Profit: The concept of operating profit. It is calculated by subtracting from the turnover the cost of raw materials, the cost of production staff, general expenses - which include research and marketing - and depreciation. Neither cash income nor expenses are considered as they distort the indicators of the evolution of the company with external factors, without making a significant contribution. They are not considered as financial costs since this type of company is unlikely to find outside finance when they have to face liquidity problems.

Cost of raw materials and the cost of production staff: These are fixed percentages of turnover. It has been considered that two aspects come together at this point - firstly, a certain experience curve should permit a lower consumption of raw materials and labour; secondly, price of products reductions must be taken into account. The combination of these two aspects allows us to establish fixed percentages of turnover.

Net Worth: This is the net book value of the company at any given time, as the sum of the initial capital plus the annual profits, according to the policy of profit distribution. According to the profit distribution policy, the model calculates the profit to be carried to net worth when the profit is positive, and apportions losses when they occur.

Profit distribution: This deals with the company policy of distribution or reinvestment of profits.

Desired Profitability: This percentage is applied to the net worth of the fixed assets, in order to obtain the desired profit.

Profit Gap: The relative difference between the desired profit and the real profit.

Liquidity: The net worth not applied to fixed assets. According to the net worth, this establishes the maximum amount of fixed assets which the company can acquire, so that by financial orthodoxy the fixed assets are not financed with outside funds.

General expenses: Expenses of research and marketing. This also includes repayment of official aid received for research, since, unlike aids for investment, this is in the form of loans.

Loans for research and development: The amount that the company devotes to R+D. Added to the net official aid to R+D, this comes to the R+D expenditure. The management of the company may alter the R+D budgets with certain elasticity. The reasons for altering the R+D budget lie in the gap between the real and desired profits. Thus, when the real profit does not reach its objectives, the management tends to reduce the R+D budgets, and when it exceeds them, it does the opposite.

Marketing: The expenditure needed to publicize the products and inform potential customers of their characteristics. This depends on the number of products and the customer niche.

Desired niche: The group of customers to which the company wishes to aim its products. It may be very wide, with high marketing costs, distribution networks, etc., or very small, so the proximity of the company gives it a decisive competitive advantage over others.

Official aid to R+D: Five-year interest-free loans, up to 50% of the costs of the R+D project.

R+D costs: Includes the R+D budget, plus net official aid, i.e. the input of aid plus the return of the loans received.

Lastly, two indicators of the situation of the company are obtained:

Margin: This is calculated as a percentage of Profit/ turnover.

R+D costs/Turnover: As the study deals with innovating companies, the ratio of R+D to Turnover can give us an idea of the importance of this factor in the running of the company.

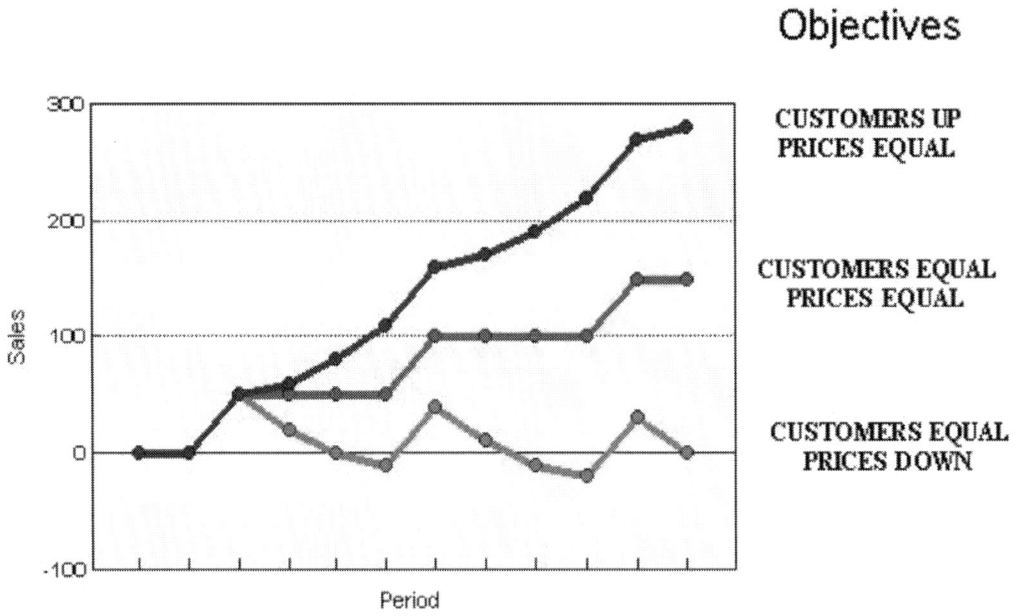

Model - Settings

INITIAL TIME= 0   FINAL TIME= 120   TIME STEP= 1   Units for time= Month

Equations

(01) % aid= 0.2

(02) % scientific staff to production= 0.5

(03) advances of production quality=Scientific Staff Production/(scientific difficulty *Potential Quality of the Products/100)

(04) amortization= Fixed Assets/(12*5)

(05) assets/capacity= 5000

(06) capital= 50000000

(07) consumption=1

(08) cost of production staff= turnover*0.25

(09) cost of raw materials= turnover*0.25

(10) Customers= f4
Initial value: 0

(11) deliveries= min(Production Capacity,orders)

(12) desired niche=1000

(13) desired profit= -initial R+D budget-f17+((Fixed Assets*desired profitability/12) +initial R+D budget+f17)*IF THEN ELSE(Product Lines>0.1, 1 , 0)

(14) desired profitability= 0.25

(15) desired technological gap= 0.5

(16) f1=((consumption*desired niche*f5)+smooth(tension,12))*IF THEN ELSE (liquidity>1,1 , 0)

(17) f10= scientific advances of the environment+scientific advances of the company

(18) f11= (advances of production quality+Scientific Level of the Company-Potential Quality of the Products)/(1+(Product Lines*Product Lines))

(19) f12=(f10-f9)-IF THEN ELSE(f6>0.1,Non-applied Technological Gap, 0)

(20) f13=IF THEN ELSE(f6>0.1,Potential Quality of the Products-Real Quality ,0)

(21) f14=IF THEN ELSE(profit>1, profit*profit distribution ,profit)

(22) f15=R+D Budget*R+D elasticity*IF THEN ELSE(profit gap>0.01, 1, -1) *IF THEN ELSE (Product Lines>0.1, 1 , 0)

(23) f16=R+D Budget*R+D official aid

(24) f17=R+D Debt/(12*5)

(25) f2= amortization+official aid

(26) f20= Production Capacity/(12*5)

(27) f3= assets/capacity*f1

(28) f4= DELAY3((desired niche-Customers)/12,3)*IF THEN ELSE(Product Lines>0.1, 1, 0)

(29) f5=smooth(f6*10,3)*IF THEN ELSE(liquidity>assets/capacity *consumption*desired niche*(1-% aid), 1 , 0)

(30) f6=IF THEN ELSE(Non-applied Technological Gap>Scientific Level of Environment*desired technological gap,1,0)

(31) f7=scientific staff-R+D Scientific Staff-Scientific Staff Production

(32) f8=IF THEN ELSE(Product Lines>0.1, scientific staff*% scientific staff to production, 0)-Scientific Staff Production

(33) f9=scientific advances of the environment

(34) Fixed Assets= +f3-f2
Initial value: 0

(35) general expenses= f17+marketing + R+D Budget

(36) initial R+D budget= 1500000

(37) liquidity=Net Worth-Fixed Assets

(38) marketing=desired niche*100*Products*IFTHENELSE(Product Lines>0.1,1, 0)

(39) Net Worth= f14
Initial value: capital

(40) Non-applied Technological Gap= f12
Initial value: Scientific Level of the Company-Scientific Level of Environment)

(41) official aid=Fixed Assets*% aid

(42) orders=Customers*Products*consumption

(43) Potential Quality of the Products= f11
Initial value: Scientific Level of the Company

(44) prices=1000*(1+quality gap)

(45) Product Lines= f6
Initial value: 0

(46) Production Capacity= f1-f20
Initial value: 0

(47) Products = f5
Initial value: 0

(48) profit = turnover-amortization-cost of production staff-cost of raw materials-general expenses

(49) profit distribution = 0.25

(50) profit gap = (profit-desired profit)/(max(desired profit,-desired profit)+1)

(51) quality gap = (Real Quality-Scientific Level of Environment)/Scientific Level of Environment

(52) R+D Budget = f15
Initial value: initial R+D budget

(53) R+D costs = R+D Budget+f16-f17

(54) R+D Debt = f16-f17
Initial value: 0

(55) R+D elasticity = 0.01

(56) R+D official aid = 0.5

(57) R+D Scientific Staff = f7-f8
Initial value:R+D costs/scientific staff cost

(58) Real Quality = f13
Initial value: Potential Quality of the Products

(59) scientific advances of the company=R+D Scientific Staff/(scientific difficulty *Scientific Level of the Company /100)

(60) scientific advances of the environment= 1

(61) scientific difficulty= 0.5

(62) Scientific Level of Environment= f9
Initial value: 100

(63) Scientific Level of the Company= f10
Initial value: Scientific Level of Environment*1.2

(64) scientific staff = R+D costs/scientific staff cost

(65) scientific staff cost = 500000

(66) Scientific Staff Production= f8
Initial value: 0

(67) tension = orders-deliveries

(68) turnover = deliveries*prices

Behavior

Leaving aside the period in which the founders of the company carry out the initial research and market research and manage to find the necessary capital, we can observe three clearly different stages.

The first begins when the company starts to operate, and lasts until the first product line is obtained. It is characterized by intensive R+D activity focused exclusively on solving scientific problems so that the company can obtain a certain line of products or services. The only challenge is to obtain the first line of products before the initial capital is exhausted.

The second stage is marked by the launching of the first line of products. This stage is characterized by the need to acquire production machinery, recruit new production staff, start production and transfer scientific staff from R+D to production. Marketing and sales costs must also be met. This is all dealt with in an environment of serious cash limitations and capital. The challenge thus lies in creating a production structure in an environment of serious financial difficulties.

The third stage is marked by the birth of new product lines. In this stage, there arise problems of distribution of the resources which the company is generating between the different agents who require them. The area of R+D will also request more funds to recover the initial levels of research, the area of production will request funds to increase its production capacity and close the gap between sales and production, and the representatives of the capital, after a long wait, will wish to obtain a high remuneration for the risky investment that they have made. Indeed, the final challenge consists in achieving an effective distribution of the resources.

We will now take a brief look at the evolution of the main elements of the model over a period of 120 months, or 10 years. Firstly, we can see the evolution of the scientific staff. Their behaviour is highly influenced by the obtaining of the first line of products. At this time two phenomena occur - some of them begin to be laid off due to the investments and expenses the company must make and others are redistributed, some of them being transferred from R+D to production.

Let us now see the evolution of the scientific level of the company, the scientific level of the environment, the potential quality and the real quality. The scientific level of the environment is approximately doubled in 10 years and as a result of the company's research, the scientific level of the company is always above that of the environment. The potential quality of the products is lower than the company's scientific level, since the products cannot incorporate all the company's knowledge. The real quality coincides with the potential quality at the time when new lines of products are launched.

We can see how the quality gap between the real quality and the scientific level of the environment is at a maximum after the launching of new lines of products, and then decreases. Lines of products appear successively when the scientific level of the company is higher than the scientific level of the environment, according to the concept of the Non-applied Technological Gap defined above. The products are generated after new lines are obtained. The number of these varies according to the availability of cash. The products which cannot be launched when the line is obtained are considered as lost.

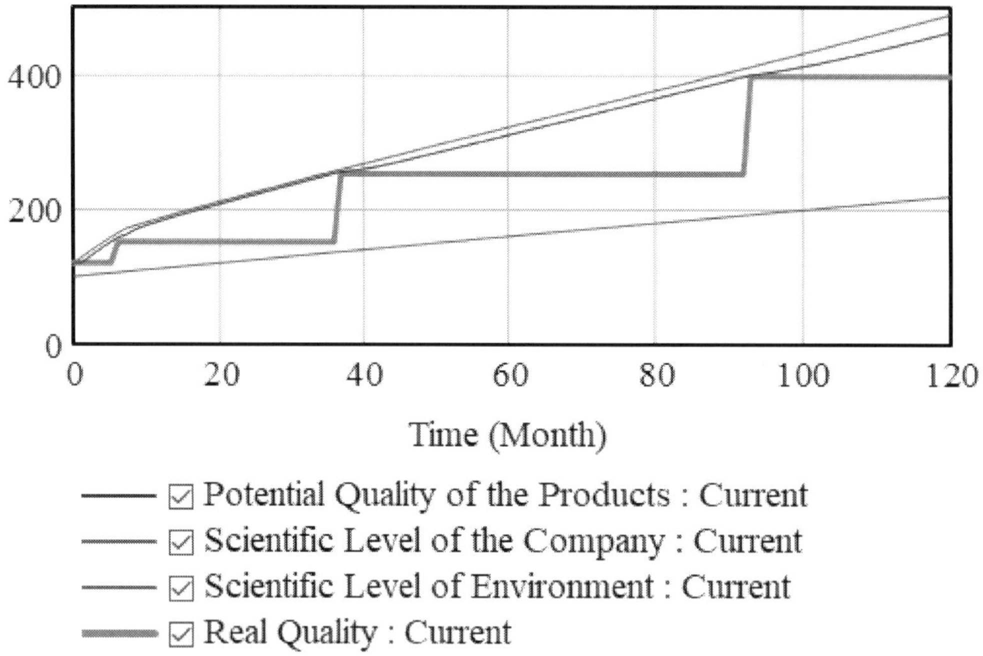

We can see here the evolution of production capacity, orders and deliveries. The evolution of the orders is similar to that of the products, since we assumed a constant consumption.

The increases in production capacity occur after the launching of new products and also when there are production tensions - they are always limited by the cash situation. The decreases occur as a result of the obsolescence of machinery.

As we have not assumed the existence of appreciable stocks, the real deliveries always coincide with either the orders or the production capacity, whichever is the lower. We can see below the components of turnover, which are deliveries and prices. The evolution of the deliveries is the same as that shown in the above table.

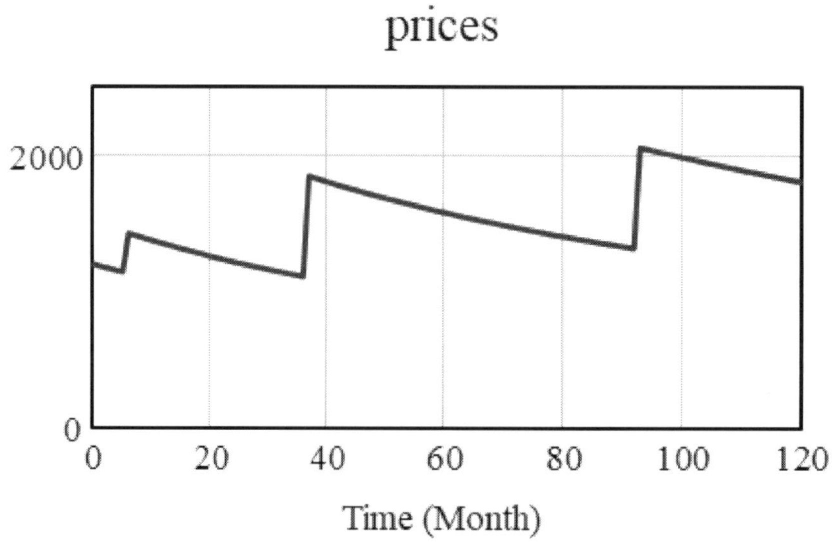

The prices show a saw tooth progression, since the maximum price is obtained on the launching of the lines of products, which is when the difference between real quality of the products and scientific level of the environment is at its maximum, and then immediately begins to decrease.

We can see a certain tendency to growth as a consequence of the company working in products with an increasing quality and price. As a result of the evolution of the deliveries and prices, we obtain the evolution of the turnover, which is highly influenced by the launching of new products and by gradual reductions in prices. One of the parameters which best characterizes these companies is the ratio of R+D expenses to turnover. As the R+D expenses have relatively little elasticity, their importance tends to decrease as the company increases its volume of turnover. To a certain extent we can see how the innovating companies, measured according to this ratio, tend to stop being innovative as they increase in size.

In the bibliography it is often stated that innovating companies are generally small in size, without giving an explanation for this. We show here the evolution of the real profit, which is highly dependent on turnover, and the evolution of the desired profit, which is highly dependent on the value of the fixed assets. We can see how scientific staff are laid off in the stages when the desired profit exceeds the real profit. We can see here the closely related evolution of net worth, fixed assets and liquidity. The initial net worth is the capital. This increases according to the profit obtained and the policy of capitalization. In the first phase the capital is the amount of cash, which decreases as a consequence of the expenditure on R+D. When new lines of products are obtained, a large part of this cash is transformed into fixed assets.The operating profit or profit margin of turnover shows a discontinuous evolution, with peaks on the launching of new lines of products and progressive decreases resulting from the price reductions we saw above.

Summary: The evolution of innovatory companies

In summary, we could say that the model of innovating companies we created has allowed us to observe behaviors which cannot be appreciated in the administrative charts, in the descriptions of processes or in the plant distribution. We can thus say that the evolution of these companies lies within a range of possibilities limited by the upper and lower curve of the next graph.

In this figure the central curve shows the theoretical evolution, measured in turnover of a company that manages to maintain a constant number of customers and a constant price. Its evolution in steps responds to the appearance of successive lines of products.

The lower line shows this same company in the case when it is forced to progressively reduce the price of its products, working with a closed niche of customers.

Finally, besides the step effect and the phenomenon of price reduction, the upper curve shows an increase in the number of customers, which makes it possible to compensate for price reductions.

3.16. Quality control

The company Torochip is a medium sized company dedicated to the fabrication of integrated circuits. Integrated circuits are small electronic components of semiconductor materials that have the same features as complete electric circuits (amplifiers, oscillators etc). Because of the delicate production process employed, only 30 – 50% of production is usable. For this reason all the units produced need to pass through quality control before being sold.

The company is concerned about the repercussion of the quality of the products on the image of the company. Torochip has observed that on occasions buyers have returned many defective products, while during other times, few products have been returned.

The company's solicitors are very aware of the situation and when they perceive an increase of the returns and reclamations of the product, they contract more personnel to work in the final stages of quality control to increase the effectiveness of this process. The amount of new hiring is conditioned by the existing number of controllers and the frequency of the reclamations received (06).

(in brackets, the number of the equation of the model)

The major difficulty of the final control process requires some months of preparation and training, even though some controllers learn more quickly than others. The controllers on the training course do not check integrated circuits that are going to be sold because Torochip doesn't want to run the risk that the inexpert controllers let defective components pass the control.

The new employees who receive training are instructed by the expert controllers. An experienced controller assigned the task of teaching new employees should dedicate half of his time to this task and the rest of the time should go towards the usual controls of finished products.

Torochip doesn't have a specific policy to reduce the number of controllers, but the natural dynamic of this department corrects the excesses in personnel that can be produced (02).

Currently, the strong demand obliges the controllers to continue in their work in the same rhythm as the production process. For this reason the time dedicated to the final checks depends on the volume of production (10).

The clients appear to perceive fluctuations in the quality of Torochip products. In this way during some periods, clients return many defective products and during other periods they receive very few returns.

The previous implies that a good model should demonstrate a trend towards the fluctuation in the quality of Torochip's products. When the number of controls per employee increases due to an increase in sales, the final quality decreases. Sometime later, new controllers are contracted. Initially, this increase in new controllers reduces the number of effective controllers until the new controllers are capable of helping in the control. The new training tasks decrease further the observed quality. Nevertheless, when the period of training is over, the number of controls per employee decreases and the observed quality increases.

(remember in brackets the number of the equation of the model)

We have the following qualitative information:

- There are currently 80 controllers with 20 controllers in training (07) (09)
- Each controller stays in the company for an average of 16 months. (02) (19)
- Four months of training is necessary for a controller in training to become a qualified controller. (12) (20)
- The value of acceptable quality is 1 (03)
- Currently, 7000 chips are produced each month. This amount corresponds exactly with the number of orders.
- A test will be carried out simulating that from period 10, an increase of 700 chips will be ordered each month. (21)
- Each of these units will be considered as a reclamation.
- The returns will be received with a delay of 3 months. (18)
- The observable quality has a delay of 3 months over the value of the current quality.
- The production is adjusted to the average of the orders in the last 6 months. (15)
- The orders that are received are due to the observable quality.
- We will analyse a period of 60 months.

Based on this information:

1) Study the situation described and give a concise summary of the problem and of the behaviour of the model. Identify the management policies that could cause this behaviour.

2) Develop a brief causal diagram will be developed based on the analysis.

3) Create the corresponding Stock and Flow Diagram SDF and compare it with the diagram that we are about to see. In the diagram the variables that are going to incorporate the internal tables underlining the name of the variable are drawn and the delays have been signalled in the arrows.

4) Write the equations in the format that will be presented shortly. Take special note in this practical case of the use of Internal Tables, Delays, Functions, and an element of the test to examine different scenarios. The test is activated if the time of the function Step is inferior to the horizon of the simulation.

5) Execute the model on the computer. Compare the behaviour obtained with the behaviour expected based on your analysis.

6) Experiment with structural changes or with the parameters employed that can alleviate the problem. Define which modifications in the management policies employed can offer a more stable quality of the products. We want to minimise the fluctuations of quality in order to maximise the volume of orders.

Note:

Internal Tables can be used to reproduce the relationship that exists between two variables with the help of a table, define the dependent variable 'Auxiliary with lookup' and enter the values of the points of the table using the key 'As Graph', and signal the independent variable (the cause) between the options that appear on the screen.

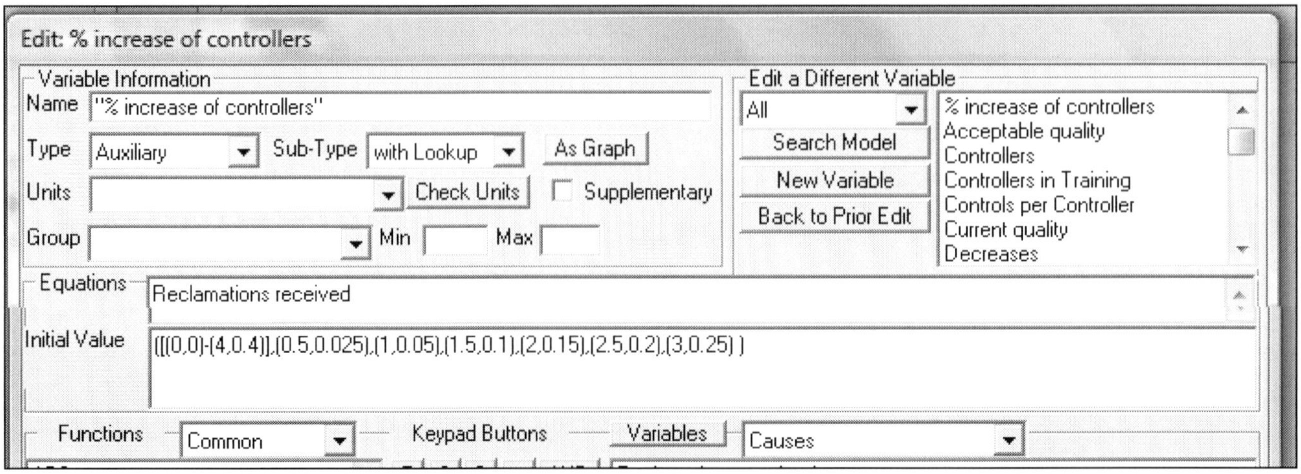

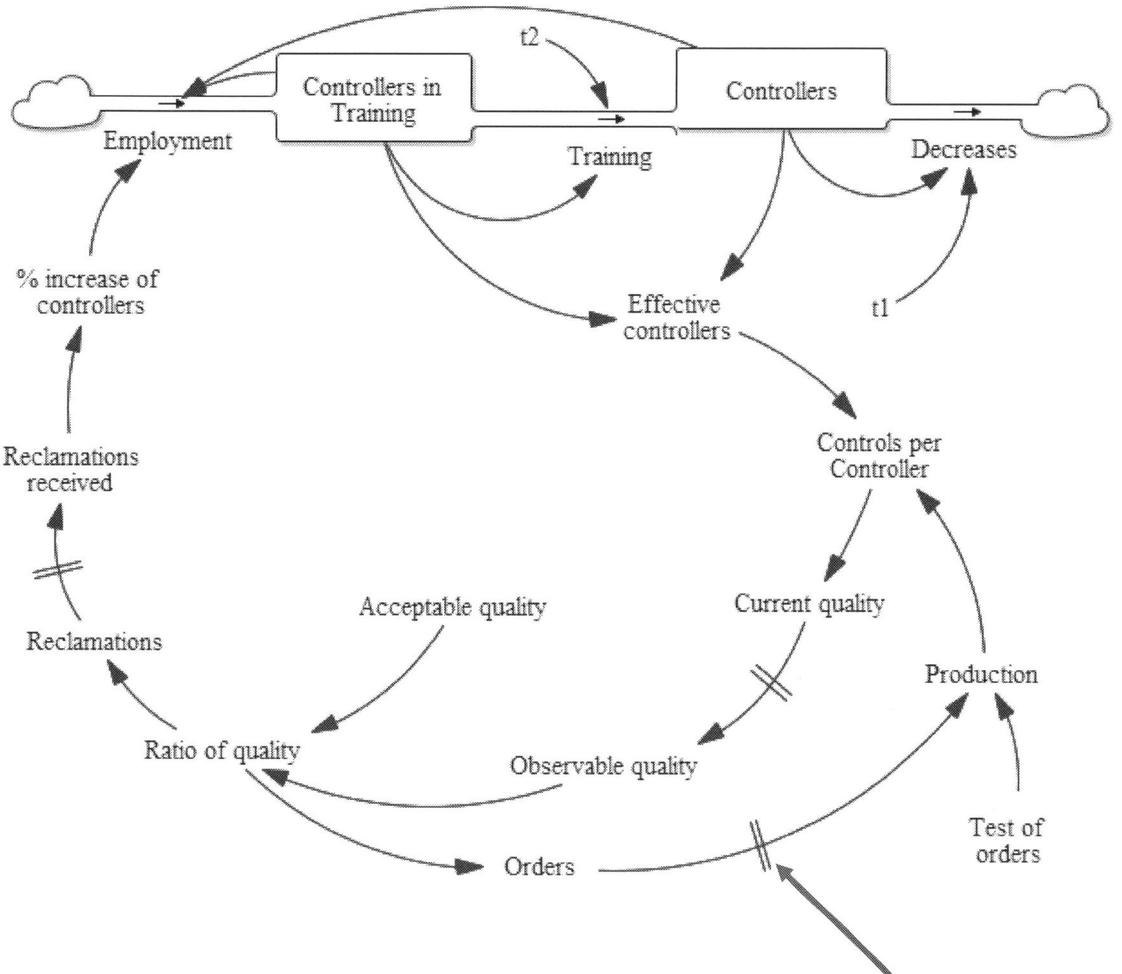

To show that there is a delay function, click in the arrow head with the mouse right button and select Delay mark.

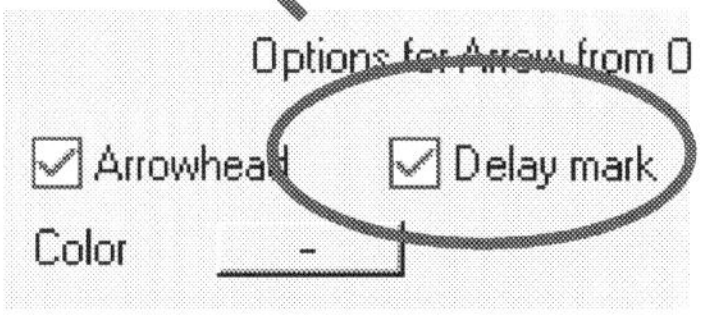

Model – Settings:
INITIAL TIME= 0 FINAL TIME= 60 TIME STEP=1 Units for time=Month

Equations 

(01) % increase of controllers = reclamations received
Lookup: (0.05,0.025),(1,0.05),(1.5,0.1),(2,0.15),(2,0.15),(2.5,0.2),(3,0.25)

(02) Decreases = Controllers/t1

(03) Acceptable quality = 1

(04) Current quality = Controls per Controller
Lookup: (50,1.25),(75,1.1),(100,1),(125,0.7),(150,0.6),(175,0.5),(200,0.4)

(05) Observable quality = SMOOTH I (Current Quality, 3,1)

(06) Employment = % increase of controllers * (Controllers + Controllers in training)

(07) Controllers = + Training – Decreases
Initial value: 80

(08) Effective controllers = Controllers – 0.5* Controllers in Training.

(09) Controllers in Training = Employment-Training
Initial value: 20

(10) Controls per Controller = Production/Effective Controllers

(11) Training = Controller in Training/t2

(12) Orders = Ratio of quality
Lookup: (0,0),(0.5,4000),(1,7000),(1.5,11000)

(13) Production = SMOOTH (Orders, 6) + test of Orders

(14) Ratio of quality = Observable quality/Acceptable quality

(15) Reclamations = Ratio of quality
Lookup: (0.5,4),(0.75,2),(1,1),(1.25,0.5)

(16) Reclamations received = SMOOTH I (Reclamations, 3,1)

(17) T1 = 16

(18) T2 = 4

(19) Test of orders = STEP (700,10)

Remember to enter the data in all the tables using the AsGraph button. This is the easiest way.

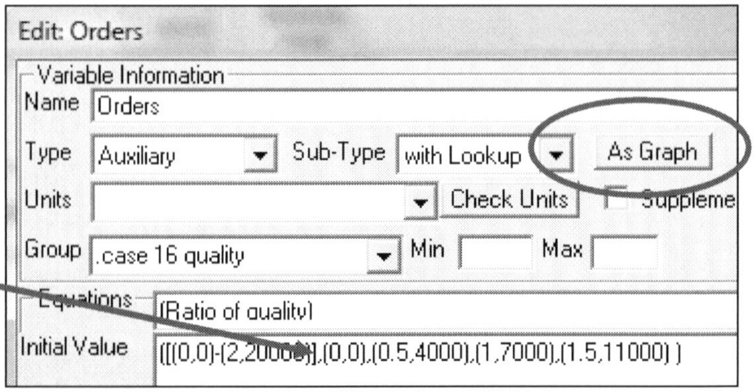

Detail of the initial values of the variables

These values reproduce an initial situation of balance in the system. When possible, it is important to get a model that reproduces a situation of balance of the system. In this way we can better analyse the cause of the variations and oscillations that appear in the system.

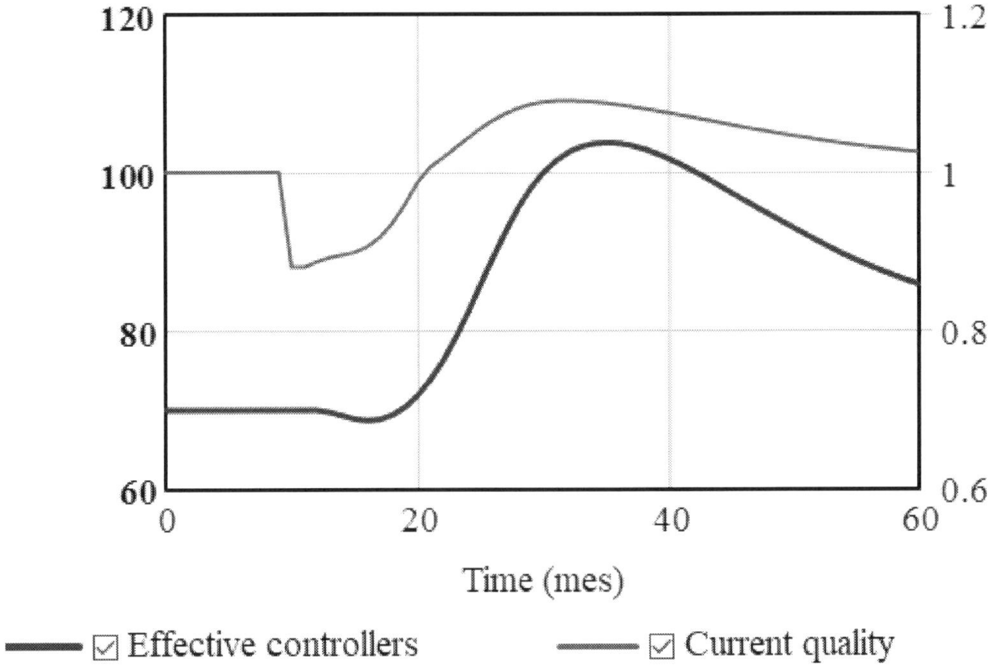

3.17. The impact of a Business Plan

The purpose of a business plan is to facilitate the achievement of the most important objectives and goals of a company, organization or corporation. This takes on a particular importance in a turbulent, uncertain and competitive world. The use of a business plan can minimize risks at all levels of business management, especially for new businesses, considering that 90% of these do not last beyond the first three years of existence. Therefore, it is important to be able to determine the endogenous factors that can affect the growth of a business.

The objective of all business plans is to obtain a profit, which may be social, environmental or purely economic. As a minimum, this objective requires the company to achieve a reasonable return on investment (ROI), either at managerial or institutional level, in order to generate resources for further use.

In the economic world, especially in small and medium enterprises (SME), a detailed ROI calculation can minimize the risk of failure, especially for new business during the first three years of existence.

A business plan is an interactive exploration in which we create possible future scenarios, both external and internal, in an attempt to design action plans in advance that may potentially lead us to our goals.

A business plan will not turn bad ideas into good business but it will allow us to discover these errors in time and avoid greater mishaps. Good ideas which are converted into a good business plan generate better business and can be used as the foundation for a more ambitious strategic reconsideration when circumstances require.

The definition of the basic qualitative variables such as environment, economic situation and market, as well as that of the basic quantitative variables such as capital, revenue or sales necessary to formulate a business plan, can permit small businesses to make an initial evaluation of their economic viability in terms of their Return on Investment (ROI) over the first few years and, therefore, assure their implementation.

Throughout this framework, we repeatedly come across the inevitable conflict of interests and opinions between managers and owners, which lead us to the creation of the model presented below.

Model – Settings:

INITIAL TIME=0 FINAL TIME=100 TIME STEP= 1 Units for time = Month

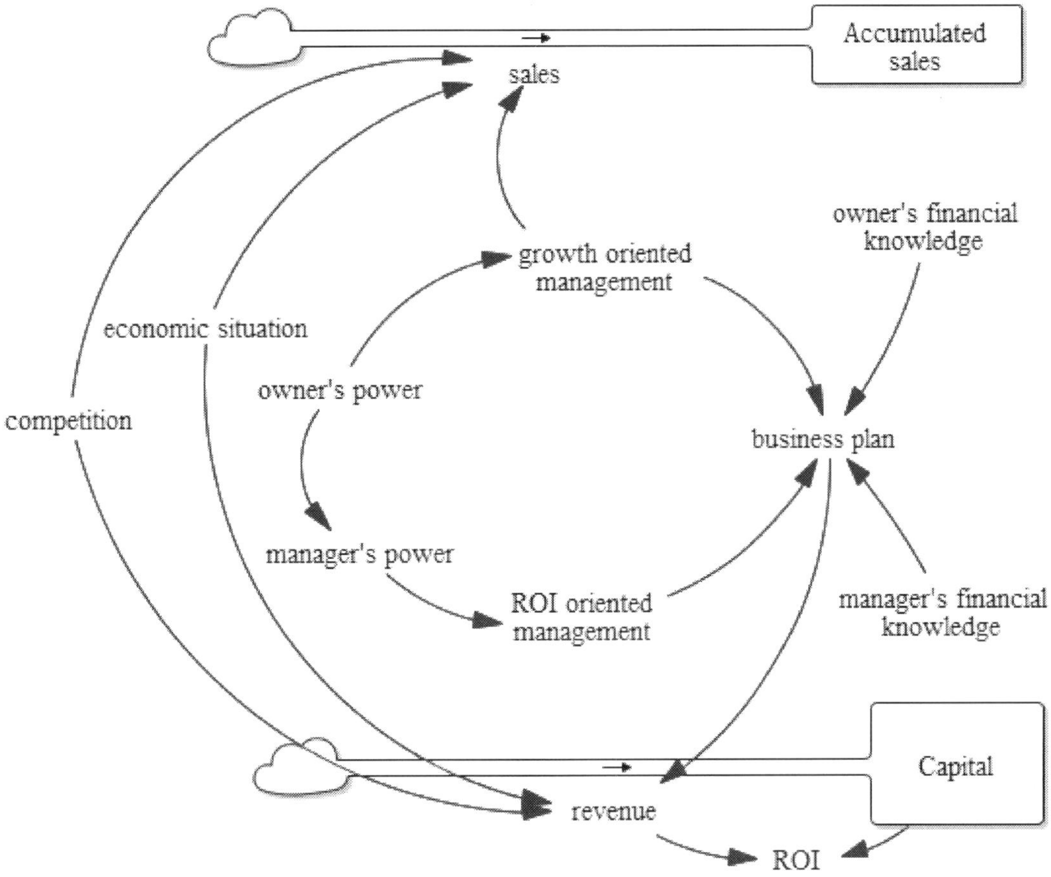

(01) business plan = (ROI oriented management*manager's financial knowledge) + (growth oriented management*owner's financial knowledge)
This variable will show us the quality of the company's business plan. If the company has a strongly ROI oriented management strategy (the only requirement is is the willingness to implement this strategy) and a manager with good financial knowledge, we will obtain an excellent business plan. However, if management is more oriented towards growth at any cost, the business plan either will not materialize or will be of very low quality. The sum of the variables ROI oriented management and growth oriented management is 1, allowing this equation to represent the relative weight of the two management strategies.

(02) Capital = revenues
Initial value = 10
We will define the variable Capital as the simple accumulation of net revenue and assign it an initial value of 10.

(03) competition = RANDOM NORMAL(-1, 0 , -0.5 , 0.5 , 777)
This is a negative or detrimental factor in the company's evolution from the perspective of both sales and revenue. Its value is to be set between -1 (very negative impact) and 0 (no significant impact), with an average of -0.5 and a standard deviation of 0.5. This variable takes into account the permanent pressure put on the company by competitors.

(04) manager's financial knowledge= 1
We assume that the manager has the necessary financial knowledge to carry out the responsibilities his position requires.

(05) owner's financial knowledge=0.5
We will set an average level of financial knowledge, as the owner is not usually an expert in finance, but rather a person with an idea about a product or service and the desire to use his company to mass distribute it.

(06) economic situation = RANDOM NORMAL(-1, 1, 0 , 0.5 , 777)
We will define the values between -1 and 1 to indicate unfavorable situations (-1) or favorable situations (1). The average is 0, as it is halfway between the minimum and maximum, and the standard deviation is 0.5. This variable takes into account the impact of environmental factors (customers, suppliers, interest rates, exchange rates, etc.), which are favorable or unfavorable to the company's activity in an alternating and random way.

(07) growth oriented management = owner's power
We define the extent to which a company is oriented towards growth as dependent on the extent of the owner's power. This is because the owner is usually a person who is in love with his company's product, and partly to feed his ego, wishes to achieve the greatest possible turnover and growth at the cost of profits.

(08) ROI oriented management = manager's power
We define the extent to which a company is oriented towards ROI as dependent on the extent of the manager's power. Unlike the owner, the manager's main goal is to keep his job and salary, which requires the company to be profitable.

(09) revenue = business plan+economic situation+competition
This variable takes into account both the positive and negative effects of the existence of a business plan, the economic situation and the pressure from the competition.

(10) manager's power = 1-owner's power
The manager's power is given to him by the owner through the delegation of tasks. This power can be total (value=1) or purely symbolic (value=0).

(11) owner's power = 0.8
As the maximum value is 1, we will set a very high level of power for the owner.

(12) ROI = revenue / Capital
The definition of Return on Investment is the revenue obtained divided by the capital invested, and is equivalent to the number of years necessary to recover the capital invested.

(13) sales = (growth oriented management+economic situation+competition)*100+100
This variable takes into account the positive and negative effects of the management strategies oriented towards sales growth, the economic situation and

the competition. It is multiplied by 100 to obtain a figure significantly greater than that of the revenue, and 100 is added to avoid sales resulting in a negative number during any period. However, there are other ways to achieve the same result, such as using the MAX function.

(14) Accumulated sales =sales
Initial value: 1000
We will define the initial value of sales as 1000 to represent the sales that were agreed upon before the official start of company activity.

Results

On the same screen, we can compare how the company's capital evolves using the values indicated in the model (owner's power is defined as 0.8) and how it evolves when the owner's power takes on a value of 0.2.

The following image shows us that in the first scenario, the capital remains stable with some fluctuations, while in the second scenario it shows a significant and steady growth. This allows us to conclude that it is better for the company if the owner has less power.

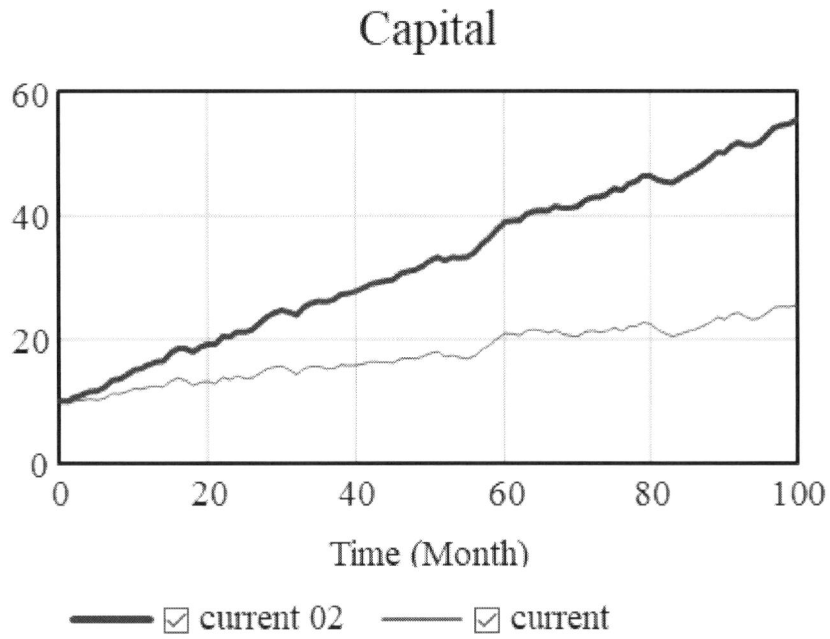

Social Area

3.18. Filling a glass

In every house there is an automatic mechanism that fills the water deposit in the bathroom every time we empty it. This mechanism doesn't need our intervention to fulfil its function. There is a buoy that acts as a measurer of the level of water - a mechanism that automatically regulates the amount of the water that enters so that the deposit is filled. Once full, the flow of water is retained. This process functions correctly if one condition is met - that there is gravity.

Now we will analyse a similar subject as we are going to make a simulation model in which a person fills a glass of water. We suppose that one condition is met - that the person acts with logic.

To create this model we will make a **Causal Diagram**. We know that:

- The more water that enters, the higher the level of water in the glass.
- The higher the lever of water in the glass, the lower the empty volume will be.
- The more empty volume, the more water will enter.

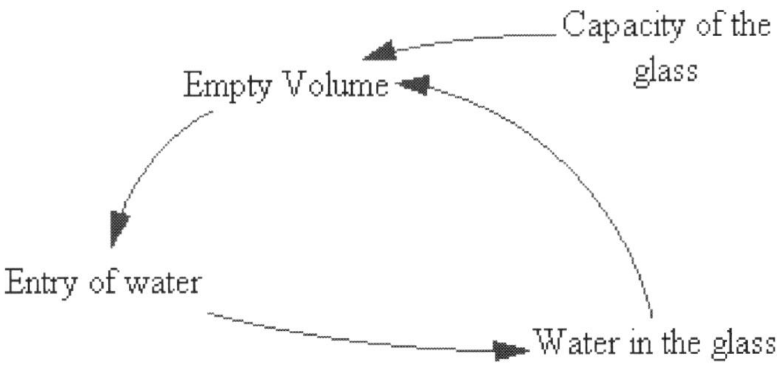

To make a flow diagram, we will:

- Make a mental photo of the system. What appears in the photo is a Stock.
- Define as Flows the elements that vary the Stock.
- The rest of the elements of the system are Auxiliary variables.
-

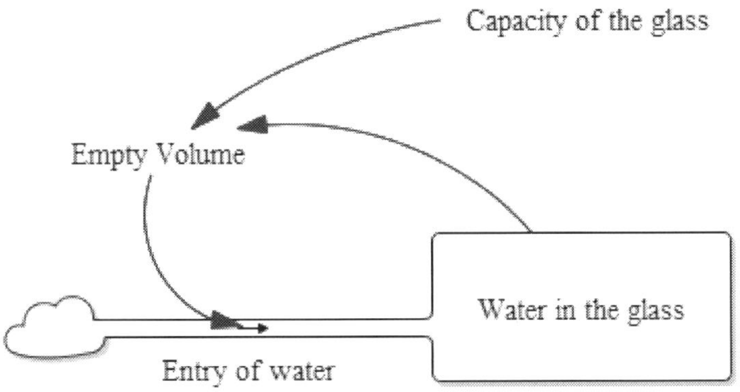

Model -Settings

INITIAL TIME=0 FINAL TIME=60 TIME STEP= 1 Units for time = Seconds

Equations 

(1) Water in the glass = entry of water
Units: cm3
(2) Capacity of the glass = 250
Units = cm3
(3) Entry of water = Empty volume
Lookup (0,0),(50,10),(100,30),(150,50),(200,50),(250,50)
We will use a table to define the behaviour of letting the water enter depending on the empty volume of the glass. In this way when the glass is empty we will let 50cm3/second of water into the glass, point (250,50) and when the empty volume is 0 the entry will be 0, point (0,0)
Units = cm3/seconds
(4) Empty Volume = Capacity of the glass – Water in the glass
Units: cm3

To define the variable with the table, select Type: Auxiliary and 'with Lookup' and click AsGraph to enter the values.

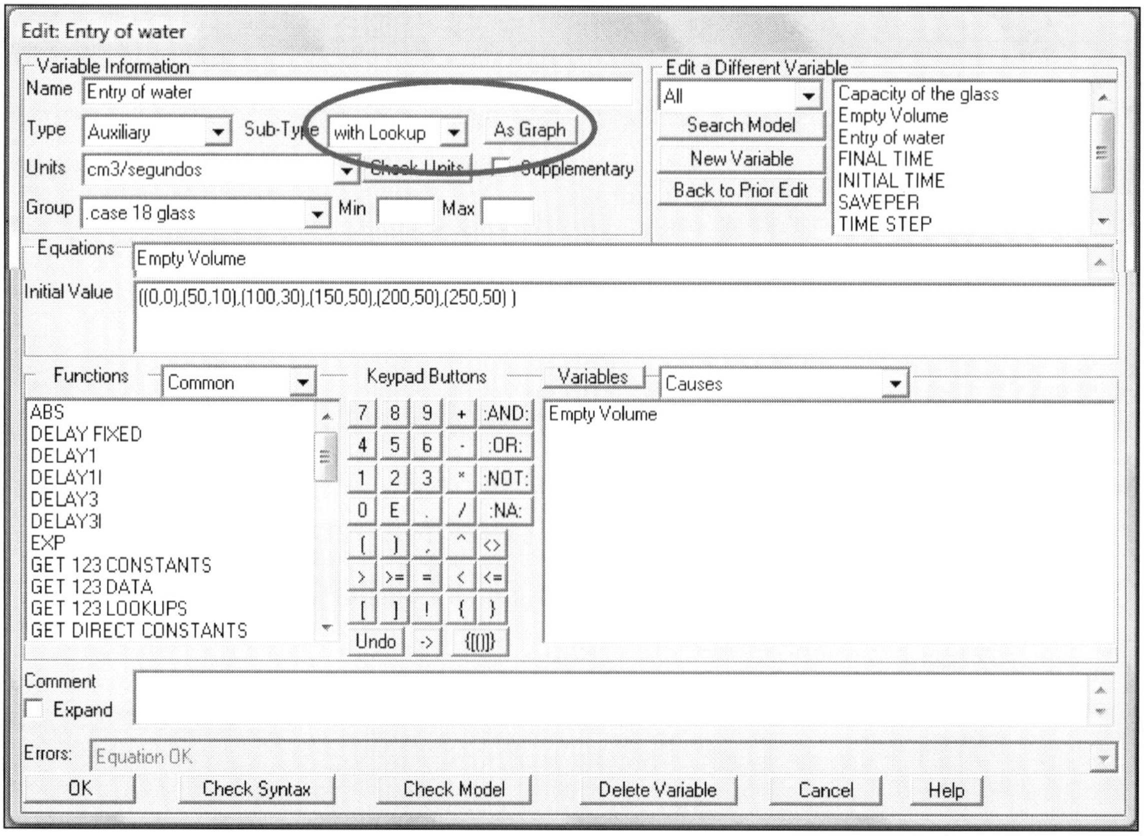

Simulate

The first execution of the model done with the icon receives the name Current.

Simulate different behaviours modifying the table that can be found in Entry of Water. Remember that the points signal how much water enters per second (output) depending on the quantity of water there is in the glass (input).

To display all the simulations together, respond No when the following screen appears and signal a name for the simulation (fast, slow, etc.)

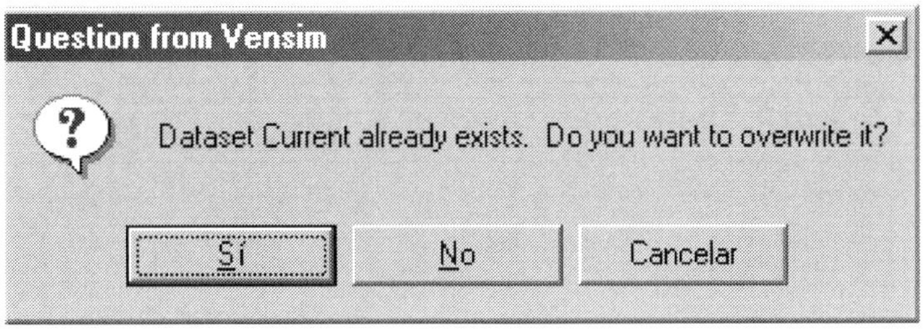

The result has to be similar to the following:

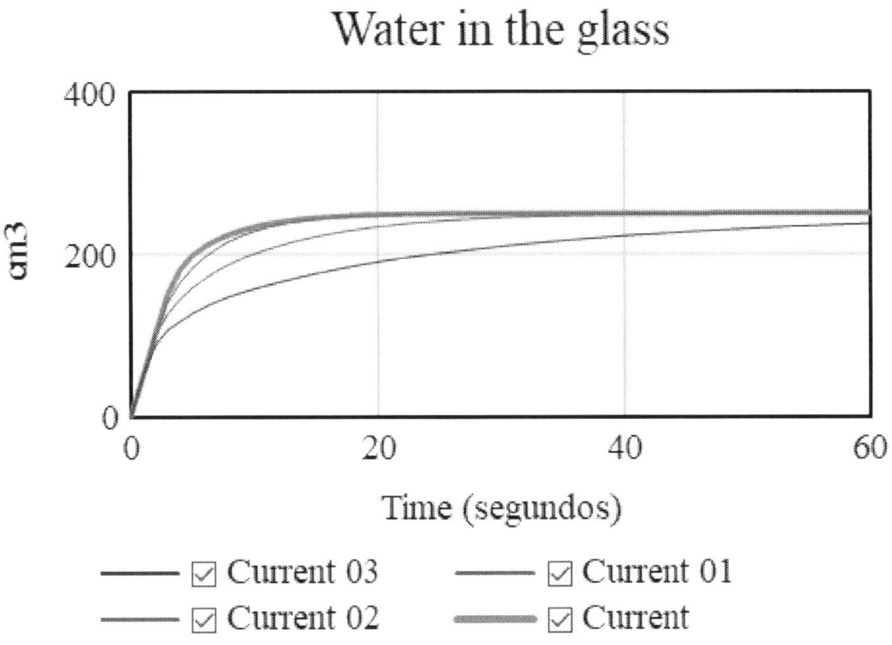

In a way that we can observe graphically, this shows the result of different behaviours using a Real state and a Desired State.

3.19. A catastrophe study

Woodland is a prosperous region located in a temperate area with abundant vegetation. Its inhabitants lead a happy and comfortable life. Its total population is 1,000,000, a figure which has remained stable in recent years.

Currently 40% of its population is young people under 20 years old, 50% are adults between 20 and 70 years old and the others are elderly people.

Furthermore, we know its population parameters to be the following - the life expectancy is 80 years, the birth rate of the adult population is 6% per year and the mortality rate is 2.5% in young people and 2% in adults. All of these parameters have remained stable and are not expected to change in the future.

On an ill-fated night in the summer of 2005 a terrible fire breaks out and over the course of a long week, the fire advances uncontrollably, devastating everything in its path. Emergency services manage to help all the young and elderly people to safety, but when the fire is put out, they discover that the total number of victims is 100,000 people, all of whom are adults.

We would like to make an estimation of how the total number of people in Woodland will evolve following the catastrophe, knowing that the population parameters (life expectancy, birth rate and mortality rate) will remain constant. More specifically, we must determine the number of years it will take the region to return to its population of 1,000,000 people.

To carry out this study, we will use a simple simulation model in which the start date will be set as the year 2000 and the time frame will be 50 years.

Taking into account the parameters we have been given, we will construct an initial version of the model based on a stable population from 2000 to 2050. Then, in a second version, we will introduce the death of 100,000 adults in the year 2005 and evaluate the impact that it will have up to the year 2050.

In the next page we have a diagram that allows us to reproduce the situation of a stable population.

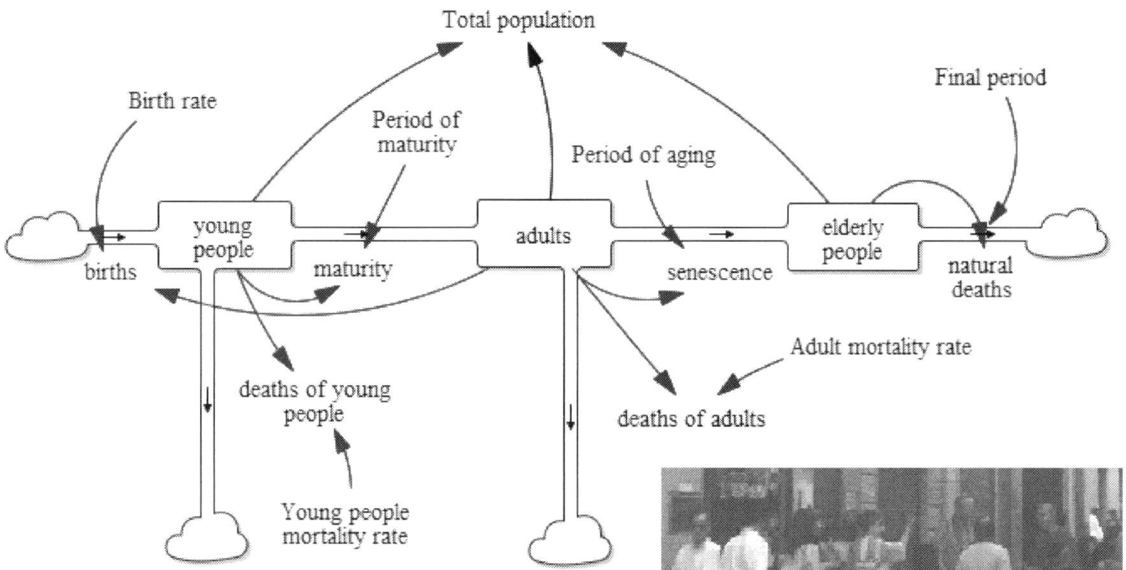

 Note: To add an image to the diagram go to your photo editor, open an image, select the region of the image you wish to add and copy it. Go to the Vensim screen, click on the 'Comment tool' icon in the toolbar and click on the model screen. Select the Metafile option (click on the circle that appears to the left of the word Metafile) and then click OK. The area of the image you have copied will be imported.

Another way to add an image is by clicking Import in the window that opens when you click on the 'Com' icon and then click on the model screen. With this option, you must select a file containing an image in bmp format. The full image contained in the file will be imported.

When click on File – New Model and a window appears, enter:
INITIAL TIME=2000 FINAL TIME=2050 TIME STEP= 1 Units for time =Year

Stocks 

(01) Adults = +maturity-senescence-deaths of adults
Initial value: 500000
Units: persons
(02) Young =+births-maturity-deaths of young people
Initial value: 400000
Units: persons
(03) Elderly = +senescence-natural deaths
Initial value: 100000
Units: persons

Flows 

(04) maturity= young people/period of maturity
Units: persons/year
(05) deaths of adults= adults*adult mortality rate
Units: persons/year
(06) deaths of young people= young*young people mortality rate
Units: persons/year
(07) natural deaths=elderly/final period
Units: persons/year
(08) births= adults*birth rate
Units: persons/year
(09) senescence=adults/period of aging
Units: persons/year

Auxiliary variables 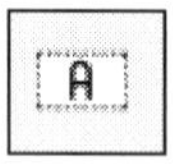

(10) period of maturity= 20
Units: year
Number of years it takes young people to become adults
(11) period of aging= 50
Units: year
Number of years it takes adults to become elderly people
(12) final period=10
Units: year
Number of years in the life of elderly people
(13) adult mortality rate= 0.02
Units: 1/year
(14) young people mortality rate= 0.025
Units: 1/year
Percentage of young people that die before becoming adults
(15) birth rate=0.06
Units: 1/year
(16) total population= adults+young+elderly
Units: persons

If you then select the icon Units Check, you should get an information box saying:

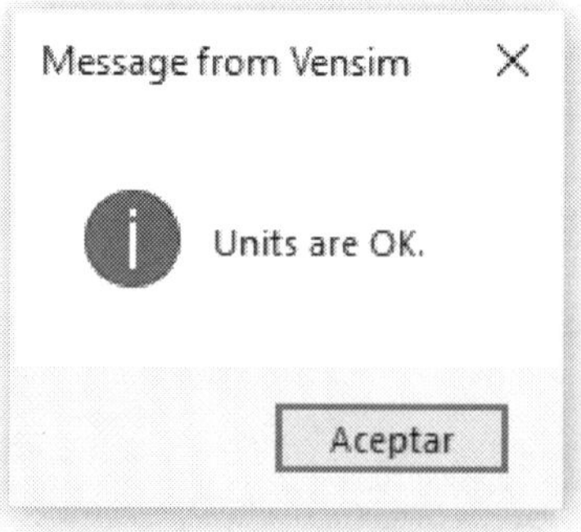

This model can be used to reproduce a stable situation. Below, the model will be modified to simulate the catastrophe. The initial effect will clearly be a reduction of the population from 1,000,000 to 900,000 people, but the future evolution will not yet have been determined. Naturally, our intuition will give us an indication of how this evolution will progress. Before continuing, we should note the approximate number of years we predict it will take the population to return to its initial figure of 1,000,000 people.

To simulate the effects of the 2005 catastrophe resulting in the deaths of 100,000 adults, we must create a new adult output flow. Using the PULSE function, we indicate that the period will be 2005 and the duration will be 1 period. Furthermore, it will be necessary to modify the adult equation to include the new flow.

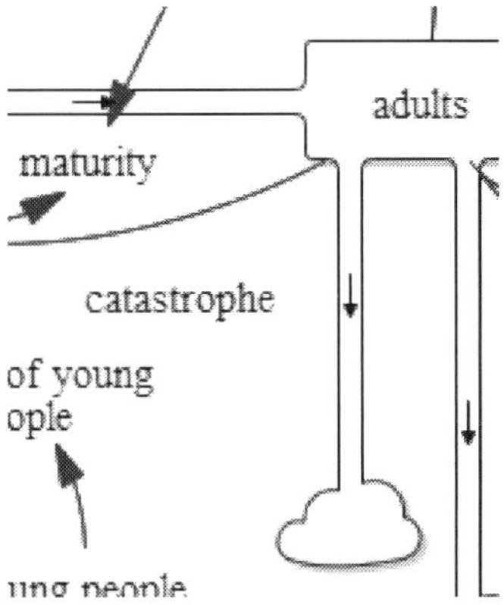

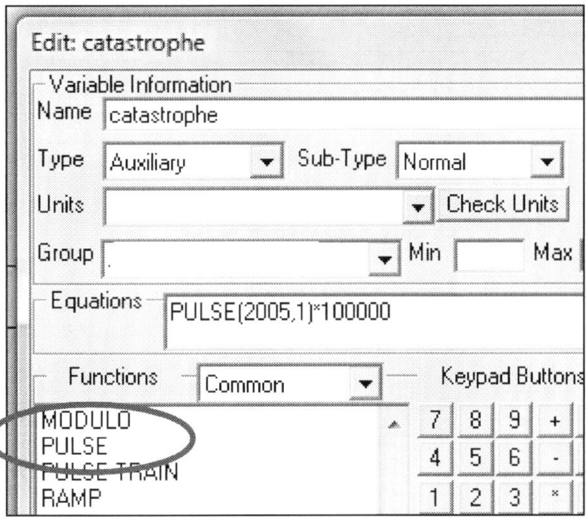

We can see the details of the PULSE function in the Functions, Tables and Delays chapter, but basically it takes on the value of 0 until the period t (2005 in this case), and for n periods (1 in our model) it takes on the value of 1. We want to simulate the loss of 100,000 people, and for this reason we multiply the result of the PULSE function (which is 1) by the figure 100,000.

Before simulating the model we should ask ourselves what type of situation we expect to see in 20 years time, in other words:

a) The population will remain stable at 1,000,000 minus 100,000 = 900,000
b) The population will recover its initial value within a few years
c) The population will recover its initial value over the course of many years

It is important to make a prediction about the evolution we expect to observe in the model before running the simulation because if we obtain results which confirm our prediction, it gives us a foundation on which to build future studies.

However, the results may contradict our prediction. In this case, if there are no errors in the model (which must be verified), some thought should be given to the assumptions on which we have based our predictions and they should be reconsidered.

In this study, the evolution of the population is not consistent with the situation we expected, given that the population neither remains stable at its new level nor returns to its original level, but decreases even further before stabilizing. The model clearly shows that the evolution of the total number of people is not consistent with our expectations.

The population neither remains stable at 900,000 people nor returns to the original figure over the course of the extensive time frame we have defined but stabilizes at a figure clearly lower than 900,000 people. We can monitor what happens in the different collectives using the graphs which show us the evolution of one variable and the others which influence it.

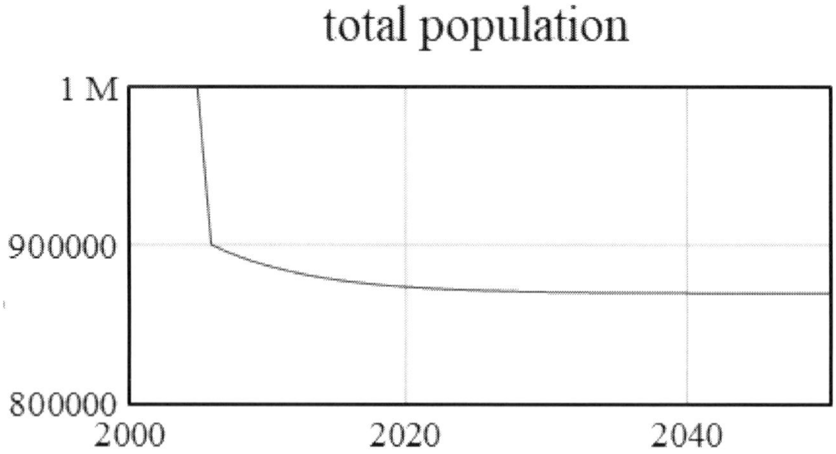

Select the variable Total Population and then the Causes Strip icon to see the detail.

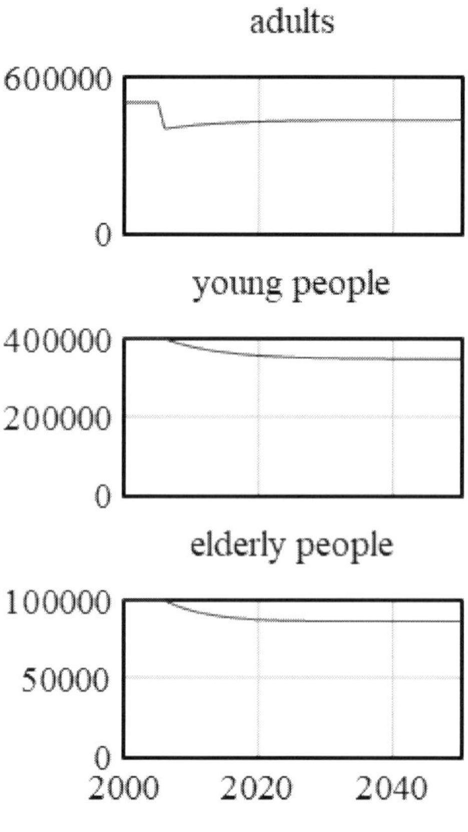

3.20. The young ambitious worker

An ambitious young worker of 20 years, after completing his basic studies, comments that he has heard his father say many times that he has 'killed himself working since he was a young man' and this has given him ulcers and heart problems. The young worker is determined not to end up like his father.

From our conversation with him, we can make the following points (somewhat disordered):

- His desired income is 5,000 € a year.
- Working hours: he supposes that he can find work, as he is healthy and has the enthusiasm to work.
- Salary per hour: with only basic studies he calculates 5€/hour.
- Real Income: this depends on the salary and the hours worked.
- He needs some hours of rest – as everyone does. For him, rest is everything: sleeping, eating, reading etc.
- Health: he doesn't have any problems.
- Fatigue: Even though he is young, he is not a machine. It will depend on the hours he works.
- Spending: At the moment he lives with his parents (spending = 0).
- Bearing in mind his desired income (5000€) and the salary per hour (5€/hour), we calculate that he will need to work 1,000 hours a year (4 hours a day).
- Having seen the experience of his father, he knows that if he ends up working 12 hours a day his health will suffer.
- He hopes to marry and have children at 35, then he will need a larger income, maybe 20,000€/year.

Before making a decision on his life and to get his ideas into order, **he asks us to make a simple model that will allow him to plan his future better**

Comments:

- The model should include all his working life (from 20 to 65 years)
- There are two aspects that worry him: health and income.
- He needs us to propose an alternative, not to guess his future.
- The elements that we believe are necessary can be incorporated into the model to propose alternatives.

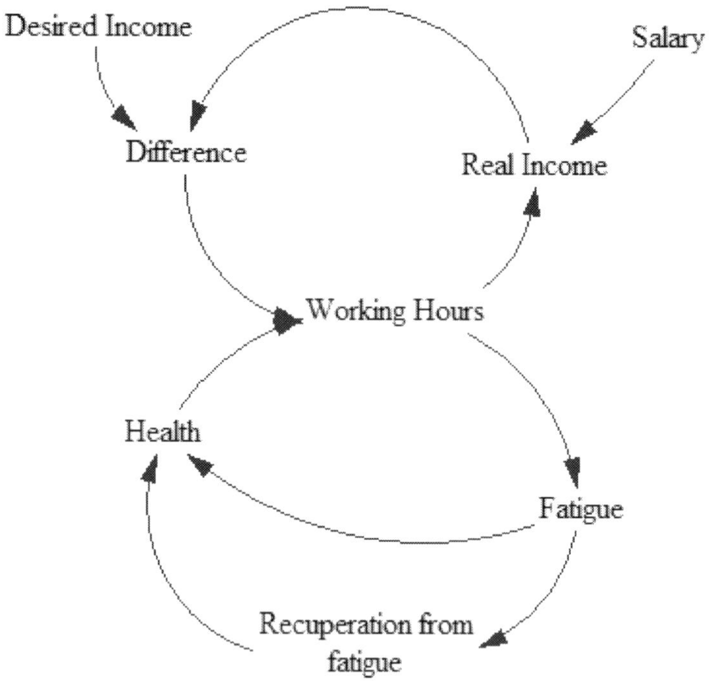

We will structure all the information received in the following causal diagram that has two zones - the economic zone and the health zone. The translation of the causal diagram to the Stock and Flow Diagram SDF is necessary to be able to work on the computer and it is not an automatic process but a process that sometimes requires the creation of new elements called Flows.

One possible solution is that which is indicated. The working hours have been defined in the flow diagram as a stock. This is a simplified model. Later in this exercise we will make a more complete version of this model.

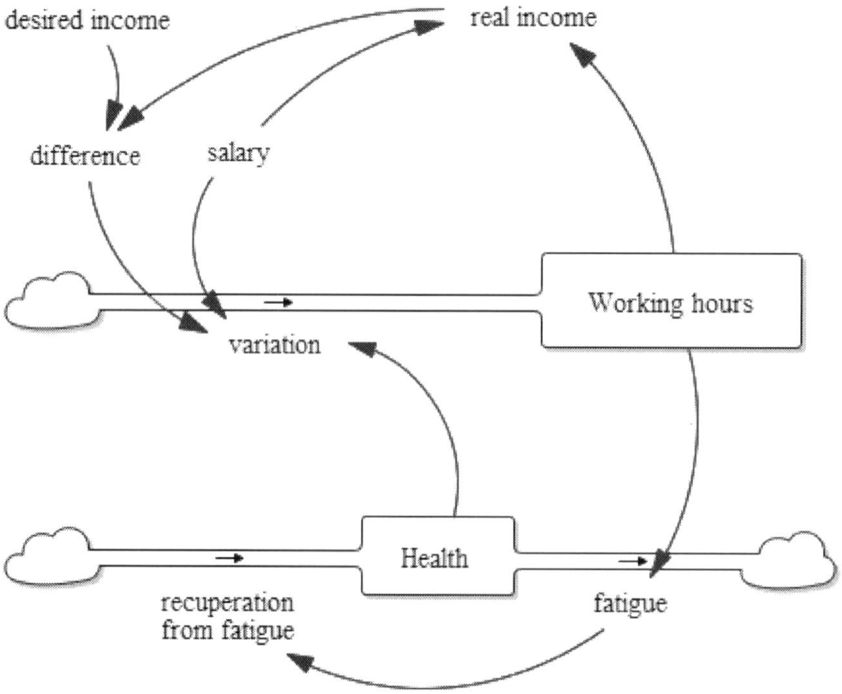

Model - Settings

INITIAL TIME= 20 FINAL TIME= 65 TIME STEP=1 Units for time = Years

Equations

(01) Difference = desired income – real income
It is the difference between the desired income and the real income that he presently receives.

(02) Fatigue = WITH LOOKUP (Working hours)

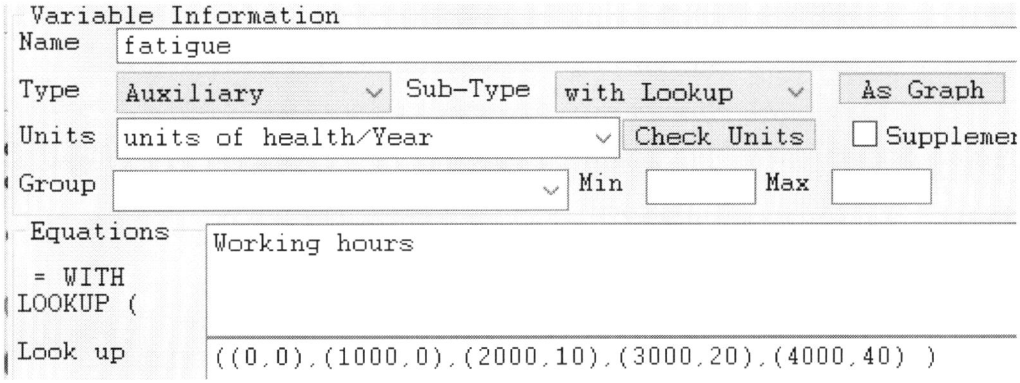

This depends on how many hours he works annually according to the table. In the formula, we will indicate that the independent variable is the Working hours, the dependent variable is the fatigue. The relationship between the two is defined in the table: (0,0),(1000,0),(2000,10),(3000,20),(4000,40)
As input we have the number of hours worked each year and as output the fatigue, measured in units of health. Given that the usual number of working hours is 1,600-1,800 hours, when the hours are 1000, the fatigue will be 0, when the working hours are 2,000, the fatigue will be 10 and it will keep increasing accordingly. We will define the table as Type Lookup, and it is important to bear in mind that it reflects the relationship between the two elements but it is NOT an element of the system even though the aspect of the Flow Diagram can appear confusing.

(03) Working Hours = Variation
Initial Value: 0
They are the working hours annually that he completes. Initially zero.

(04) Desired Income = 5000 + STEP (15000,35)
Initially 5,000€/year and from the age of 35 20,000€/year, due to family responsibilities.

(05) Real Income = Salary* Working Hours
The real income is the product of a salary per hour of work.

(06) Recuperation from fatigue = WITH LOOKUP (fatigue)

We will consider that the recuperation depends on the fatigue in the sense that if the fatigue doesn't exist (as a threat to health) throughout the day, there is nothing to recuperate. This function is in the table (0,0),(10,10),(20,15),(30,20),(40,20)

We will start from the point (0,0) when there is no fatigue the is no recuperation and a moderate fatigue corresponds to a complete recuperation (10,10). When the fatigue is elevated (20,30,40) the recuperation doesn't totally cover the loss of health due to fatigue.

(07) Salary = 5 + STEP (0,35)

The salary is 5€/hour. The evolution of the variables of the initial model shows us that at 65 years old, he will be a person with bad health and low income. But we can include a 2€/hour increase in salary when he turns 35 => STEP (2,35). If he follows the recommendation of studying what he likes from the age of 20 to 35, (it appears to be a very moderate increase), the result changes completely both for his health and his income.

(08) Health = Recuperation from fatigue – fatigue

Initial value = 100

We can take the scale from + 100 to – 100, in the way that in the beginning, a young person aged 20, his health is 100. It will decrease depending on fatigue and will increase with recuperation.

(09) Variation = (difference/(salary*5))*(health/100)

Depending on the Difference (euros) and on the salary (euros/hour) we calculate how many more hours he needs to work. As this variation is not instantaneous, we will divide it by 5. From this we can deduce the idea that it will take 5 years to find the job he wants. We can vary this hypothesis. Also we consider that he will increase his working hours depending on what his health (+100 to -100) will allow. If his health is 100, he can increase his working hours as much as he wants.

Results

The results of the model are not very satisfactory as he ends up poor (the income doesn't reach 5000 €) and very ill.

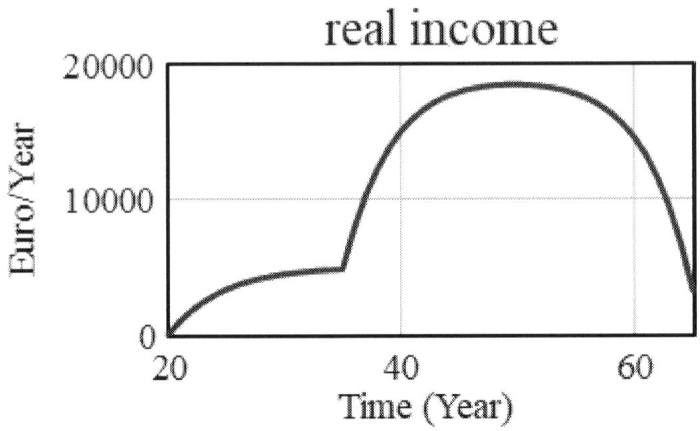

Proposal

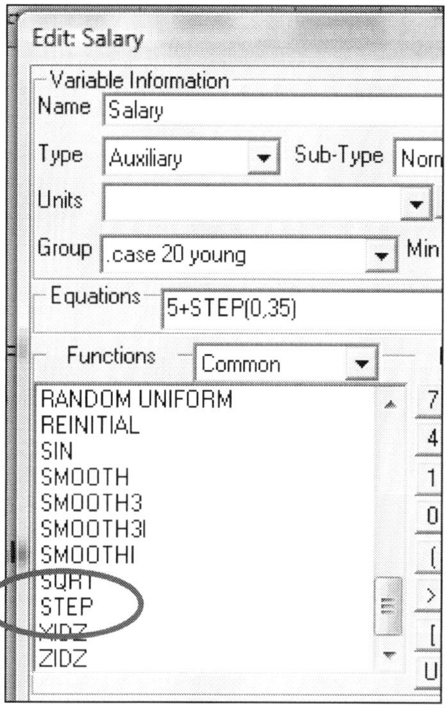

He decides to study further until the age of 35 (1000 hours/year) in order to be able to obtain a higher salary (2 €/hour more).

We will introduce this change in the salary equation = 5+STEP(2,35)

The results of the simulation show that he will retire with less health that he has now but without having 'killed himself working', and with a desired level of income.

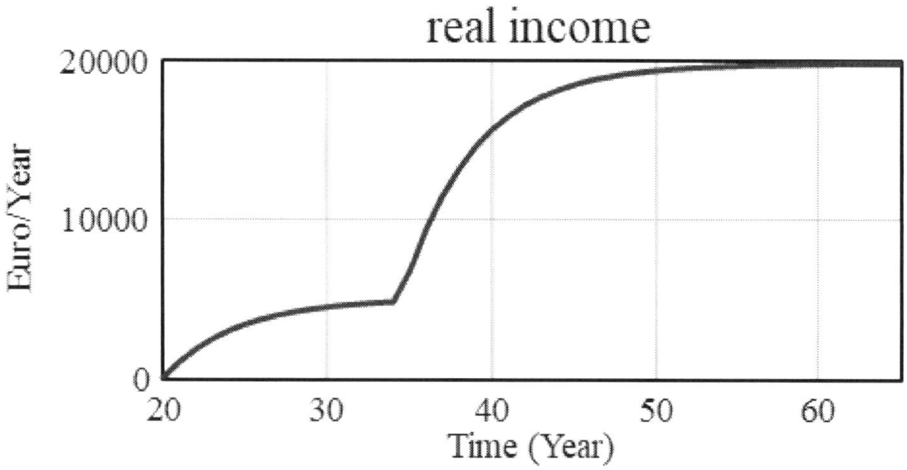

Our final recommendation based on the results of the model is that he should study while he is young to achieve a higher salary from the age of 35.

Final Comments

In this model we have seen how to use qualitative elements, like health and fatigue, together with quantitative elements like working hours and level of income.

3.21. Development of an epidemic

Let us imagine a population that is initially healthy. In this population a number people infected with a contagious disease appear. An individual could transmit or catch the illness from other individuals. The transmission of the illness is due to the physical proximity. During the infectious process the individuals can pass through some or all of the following states:

- Susceptible (S), the state in which the individual can catch the illness from another infected person.
- Infected (I), the state in which the individual finds himself infected and can infect others.
- Recuperated (R), or cured, the state during in which the individual can not infect or be infected because he will have acquired an immunity (temporary or permanent) nor can he infect others as he has recuperated or has passed through the contagious stage of the illness.

In the various infectious diseases, we can find two principle groups:

- Those with immunity (temporary or permanent) who have been infected and have since recuperated. The majority of these illnesses are of viral origin (sarampion, varicela, poliomyelitis).
- Those that, once recuperated, may turn again to susceptibility. These illnesses are mainly caused by bacterial agents (venereal disease, pest and some forms of meningitis) or protozoos (malaria).

Bearing in mind the different states in the infectious process, the epidemiological models can be divided into three big groups:

SIR: The model susceptible-infected-recuperated, related to illnesses that produce permanent immunity and a typical cycle that includes the three states. This does not mean that all the individuals of the population must pass through these stages, some will not be infected and remain healthy, in other words, they will remain in state (S). Others will be immunised artificially by vaccination or another method and will pass directly to state (R) without having been infected.

SIRS: The model susceptible-infected-recuperated-susceptible, the same as the previous model but applicable in cases where the immunity is not permanent and the individual is again susceptible after a certain amount of time, such as the 'flu.

SIS: The model susceptible-infected-susceptible is used in cases where the illness does not produce immunity, the individual can pass from being infected to being susceptible again skipping the recuperation stage completely.

A model can take into consideration the vital dynamics of the population (births, deaths, migratory movements) depending on the time period studied, the characteristics of the illness and the population being studied.

The model

We are going to use one of the most well known epidemiological models in biology as a reference for our model. It is the Kermack and McKendrick model that is expressed as:

$$\begin{cases} S' = -\beta SI \\ I' = \beta SI - \gamma I \\ R' = \gamma I \end{cases}$$

S is population who is susceptible to catching illness, I is the population infected and R the population that has passed through the illness and has recuperated. There are two constants, the rate of infection and the rate of recuperation.

To draw a flow diagram we need to follow the following steps

We need to create 3 stocks

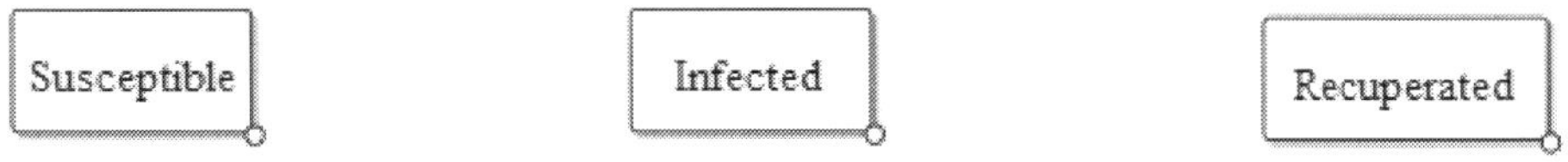

Add the flow of the vaccinated individuals

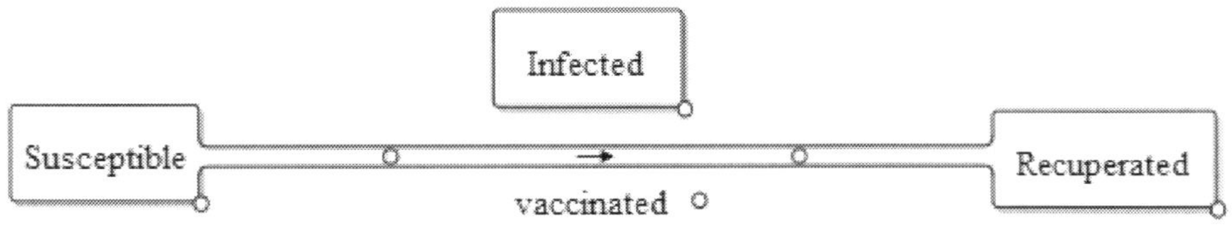

Displace the flow by dragging vaccinated downward.

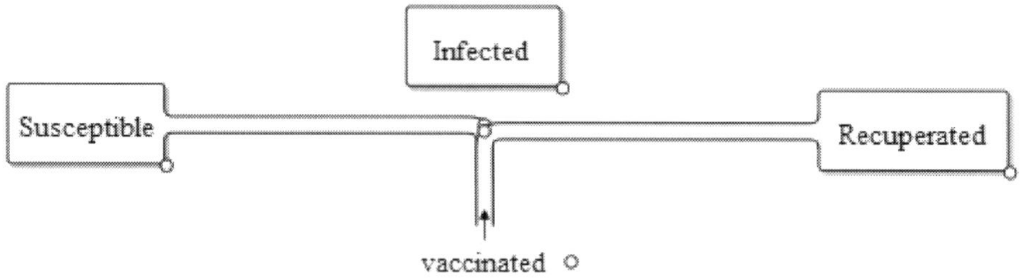

Click in the circle situated in the canal of the flow and move right and downwards to draw the entry of the recuperated individuals

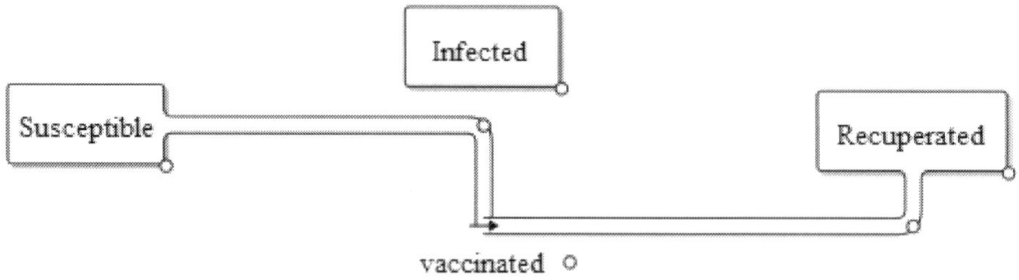

It is possible to draw a Flow with corners, rather than a straight line, as seen below for "exposed". This is done by holding down the Shift key and with the cursor clicking first inside the starting level, (Recuperated) then in each corner of the channel and finally in the level of arrival (Susceptible).

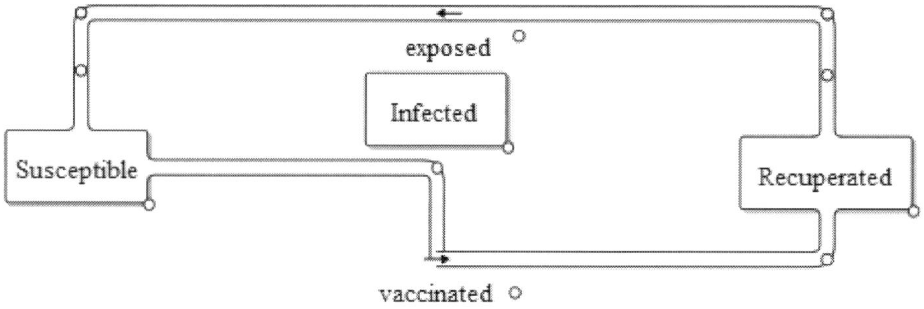

Complete the model by adding the flows of make ill, cure, and death.

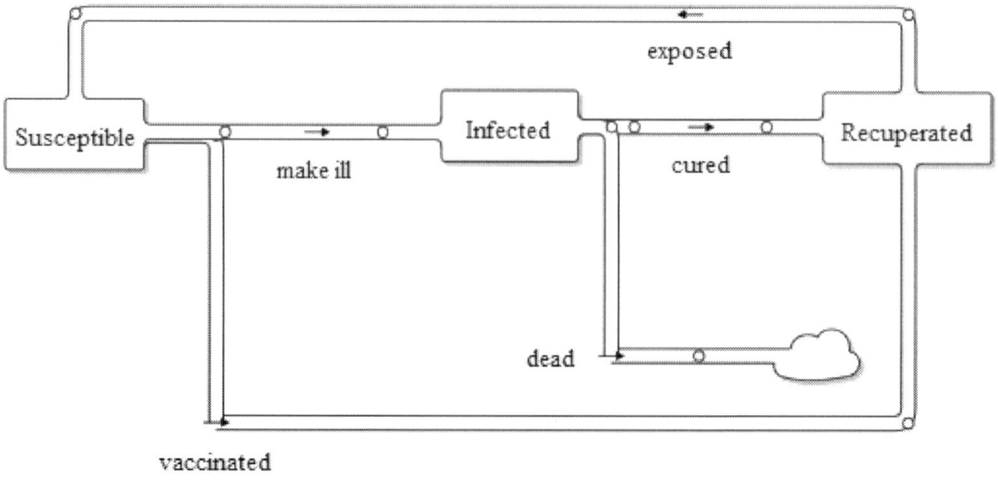

Clicking the right button of the mouse on the flow of death symbol we can change the situation of the text and its colour.

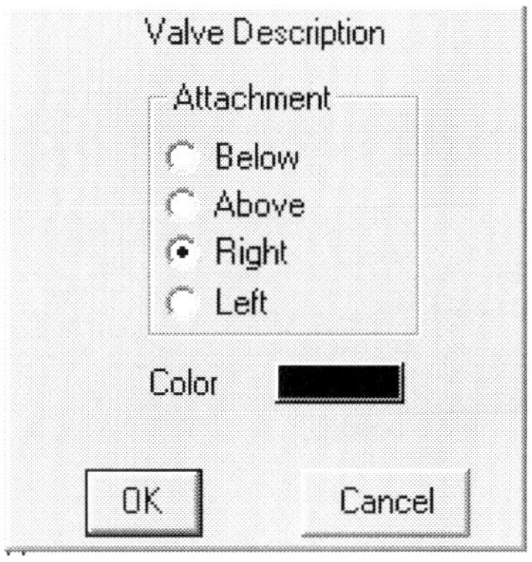

Clicking the right hand button of the mouse on the circles in the arrows or the flows, it is possible to modify their colours to emphasize certain relationships in the model.

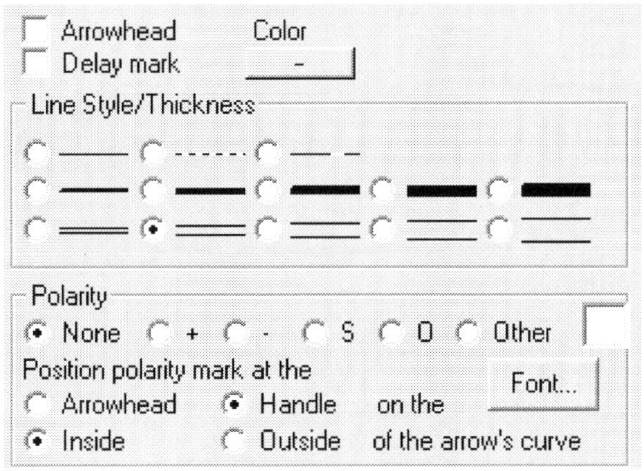

To add the remaining elements of the system, introduce the equations and simulate.

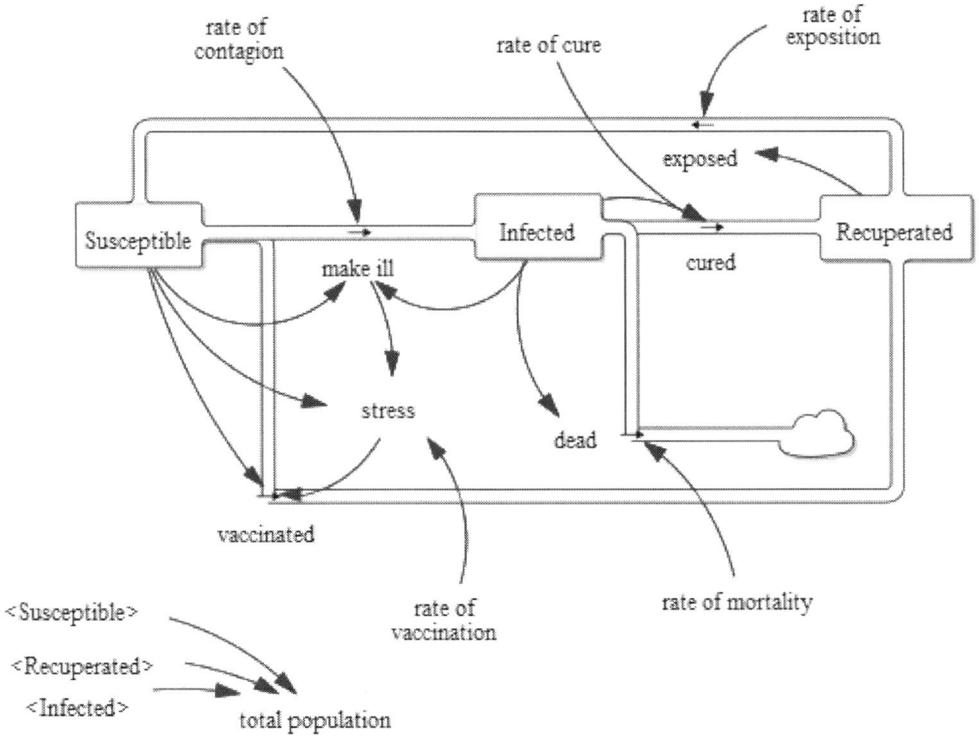

To add shadow variables click the key: a menu opens and we can choose the variable which we want to copy. The usefulness of these copies is that we can avoid producing excessive crossing of arrows in the model. In our case, we will create the variable Total Population as the sum of the three possible states.

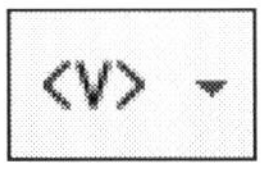

The software pre-writes the equations of the stocks based on how we have drawn the flows.

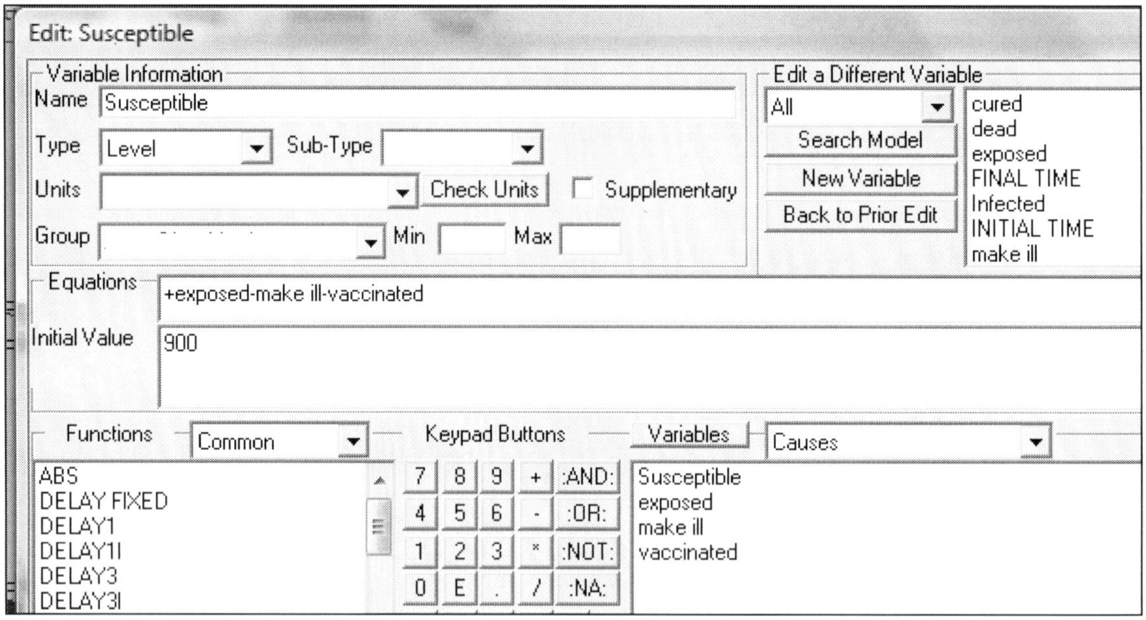

The equations are simple as the stocks vary depending on the entries and exits that we have assigned to them with their corresponding sign and the flows are, in general, a product of the value of a stock by the value of a rate.

Nevertheless, there are formulas that are slightly more complicated - those of the infection:

'make ill'

We can apply the equation formula from the model by Kermack and McKendrick. According to this model, the number of people who infect can be calculated as the product of the number of people susceptible by the number of people infected by the rate of contagion.

'stress'

We implement the concept of stress to account for the fact that number of people who are capable of infecting in relation to the number of people susceptible to infection will influence the number of 'vaccinated' people by having become ill.

Model - Settings

INITIAL TIME = 0 FINAL TIME = 20 TIME STEP = 1

In this model we won't worry about units because, while the units in the Kermack and McKendrick equation are valid, matching units in this case is not important.

Equations 

Stocks 

(01) Susceptible = + exposed – make ill – vaccinated
Initial Value = 900

(02) Infected = + make ill – cured - dead
Initial Value = 100

(03) Recuperated = cured + vaccinated –exposed
Initial Value = 0

Flows 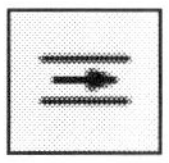

(04) cured = Infected * rate of cure

(05) make ill = Susceptible* rate of contagion* infected

(06) exposed = Recuperated* rate of exposition

(07) dead = Infected* mortality rate

(08) vaccinated = Susceptible* stress

Auxiliary variables 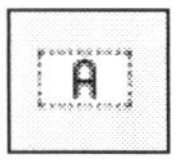

(09) rate of contagion = 0.001

(10) rate of cure = 0.4

(11) rate of exposition = 0.05

(12) rate of mortality = 0.1

(13) rate of vaccination = 0.5

(14) stress = (make ill / Susceptible) * rate of vaccination

(15) total population = Infected + recuperated + susceptible

Simulations

Clicking the automatic simulation icon we can observe that in a simultaneous form the evolution of each variable in the model, and also simulate the effect of changes on the variables in the rates that are constant by displacing the cursor to the cursor to the left and right.

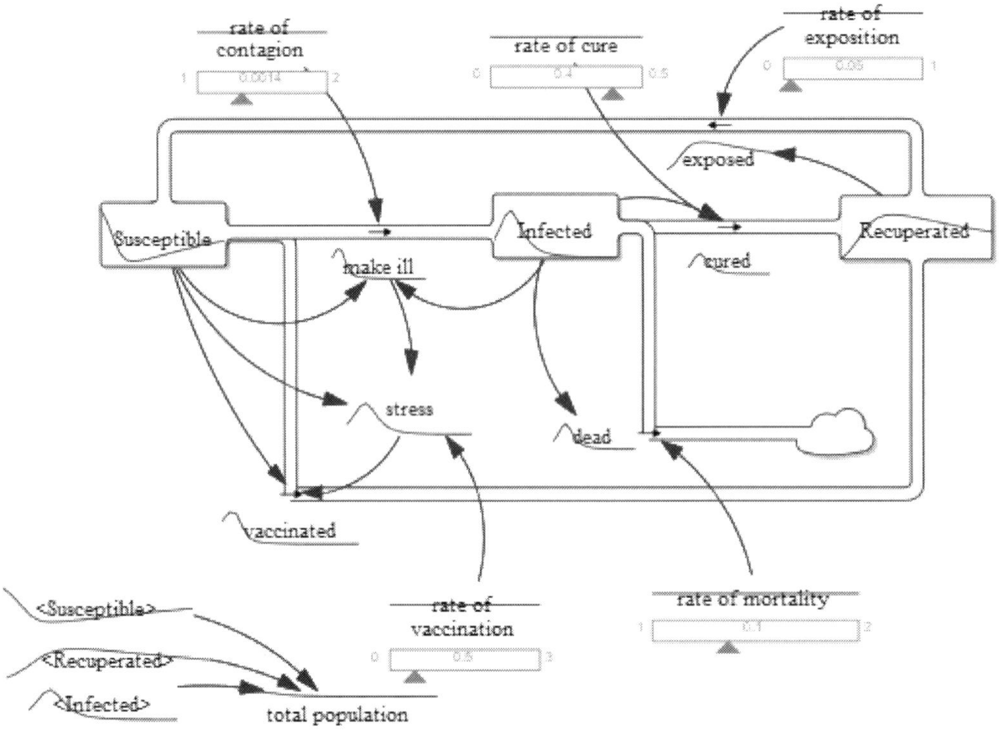

Pressing the icon Control Panel – Custom Graphs we can see the behaviour of specific variables in more detail.

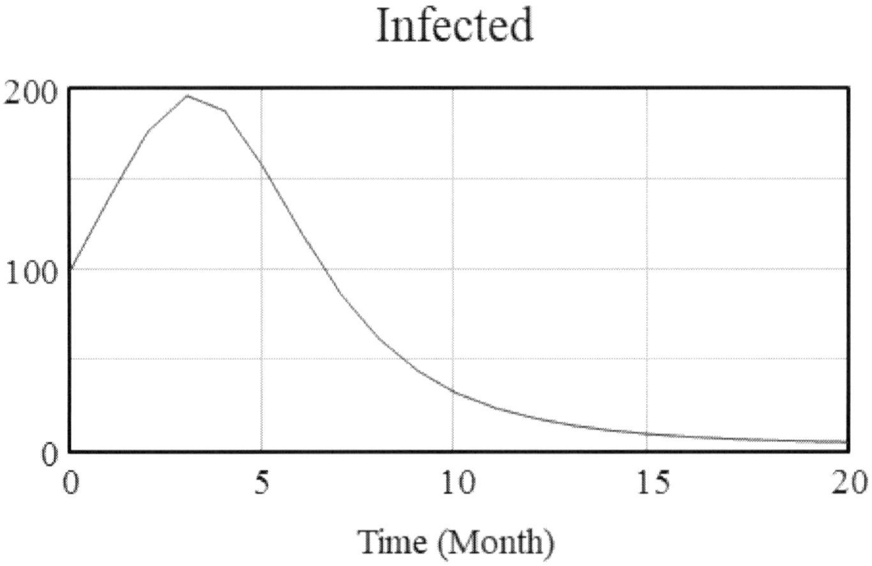

3.22. The dynamics of two clocks

A soldier stopped in front of a clock shop every day at 8 o'clock in the morning and adjusted his watch, comparing it with the chronometer situated in the window display. One day the soldier went into the shop and congratulated the owner on the exactness of the chronometer.

'Are you in line with time from Westminster?' asked the soldier.

'No' answered the owner 'I adjust it according to the canon fire at 5 o'clock in the castle. Why do you stop here every day and check the time of your watch?'

The soldier answered, 'I am the one who fires the canon'.

Knowing that the chronometer from the clock shop is delayed about one minute every 24 hours, and the clock of the soldier is delayed 1 minute every 8 hours, what is the total error in the time of the canon in the castle after 15 days?

We can make a summary of the situation in the following causal diagram:

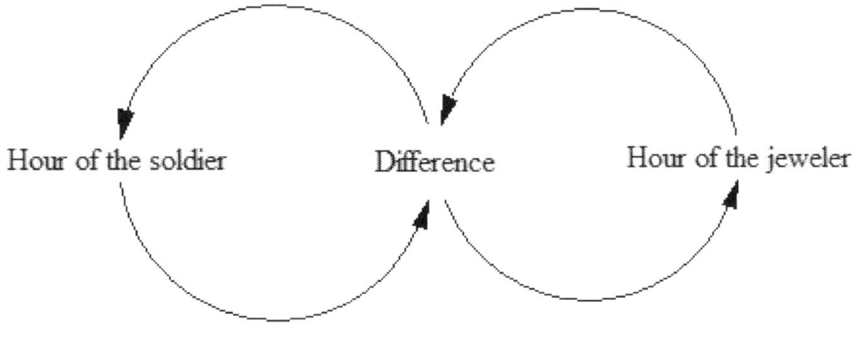

To introduce this scheme into the computer we have to translate it to the format of the Stock and Flow Diagram SFD. This model presents the particularity that the time is a Stock. We have to differentiate the real time (24 hour/day, 1440 minutes) from the time measured by clocks, this being something almost physical that we can manipulate.

We can imagine 1 minute measured = 1 tennis ball, being the watches of both boxes from which the tennis balls enter and leave.

Taking the indicated flow diagram as a reference, we can define the equations.

Parameters:

- Each day contains 24 hours, in other words, 1440 minutes.
- The clock from the shop counts (enter) 1440-3 minutes and the soldier counts (enter) 1440-1 minutes.
- Each watch marks 1440 minutes (leaves).
- Everyone adjusts their clock (enter or leave) to the difference of the minutes that one has in respect to the other.
- We have to see the state of the system at the end of 15 days.

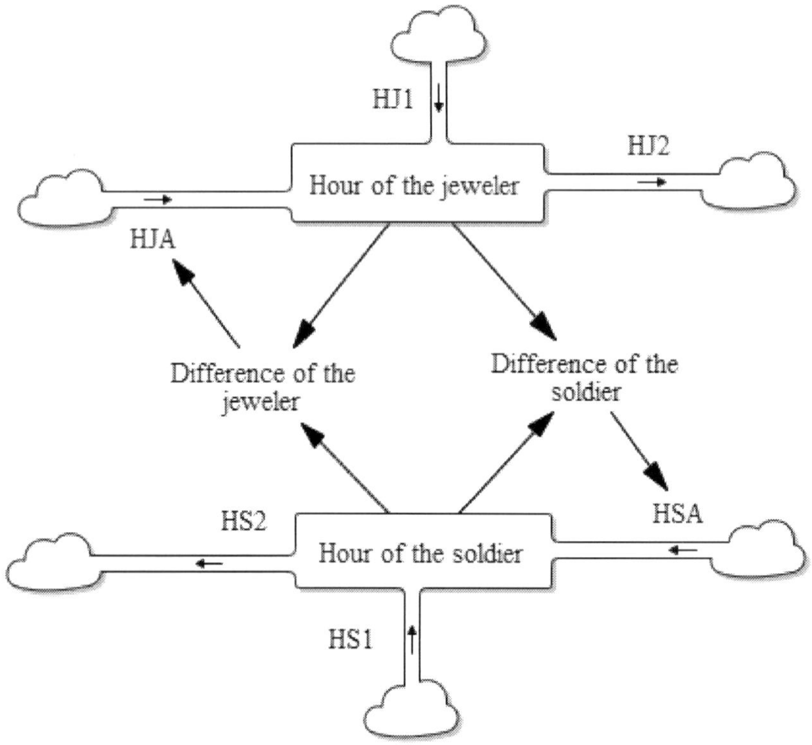

Equations

Model – Settings: INITIAL TIME = 0 FINAL TIME = 15 Units for time: Day

(01) Difference of the jeweler=Hour of the soldier-Hour of the jeweler

(02) Difference of the soldier=Hour of the jeweler-Hour of the soldier

(03) HJ1= 1440-1

(04) HJ2= 1440

(05) HJA= Difference of the jeweler

(06) Hour of the jeweler= INTEG (HJ1+HJA-HJ2
Initial time: 0

(07) Hour of the soldier= INTEG (+HSA+HS1-HS2
Initial time: 0

(08) HS1=1440-3

(09) HS2=1440

(10) HSA=Difference of the soldier

In this case the result in graphical form is not very useful as a date is demanded, not its historical evolution.

We will use the icons that can be found in the left side of the screen.

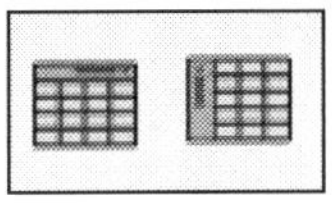

The results of the simulation in vertical table form indicate that the difference at the end of 15 days is 31 minutes.

Table Time Down

Time (Day)	"Hour of the soldier" Runs: Current	Hour of the soldier
0		0
1		-3
2		-4
3		-7
4		-8
5		-11
6		-12
7		-15
8		-16
9		-19
10		-20
11		-23
12		-24
13		-27
14		-28
15		-31

Mechanical Area

3.23. Dynamics of a tank

We want to simulate the behaviour of a tank containing liquid, which contains only one entry and one exit, both of which are always open. The tank has a capacity of 100 litres, and initially contains 50 litres of liquid. We want to know the dynamic of the tank's content before making changes in the entry and exit flows. In other words, we want to make sure that the tank will not get overfilled and that it won't get empty.

The entry is regulated in a way that, to avoid overfilling, the flow increases when the tank is getting emptier, and the flow decreases when the tank gets filled up. Initially we equilibrate the entry in the tank in a way that enters a fraction of 1/10 of the empty volume in the tank.

On the other hand, we regulate the exit in a way that, to avoid emptying, we will increase the exit flow when the tank is almost filled, and it will decrease when the tank becomes empty. Initially, we regulate the exit in a way that exits a fraction of 1/10 of the volume in the tank.

Questions:

1. What behavior would you expect, if initially, as we described earlier, the tank gets filled up at a rate of 1/10 of it empty volume, and gets empty at the same rate?
2. What would happen if initially, the tank contains 80 litres?
3. What would happen if the entry is regulated at 1/20 of the empty volume of the tank, and the exit at 1/10 of the volume?

We want to create a model to understand the dynamics of the content of the tank to be able to answer those questions.

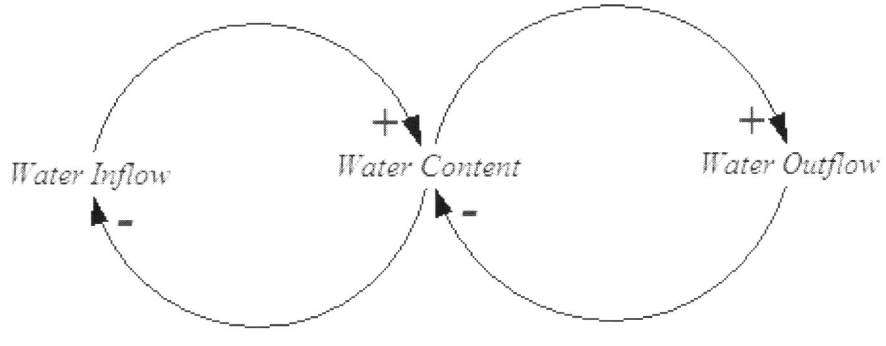

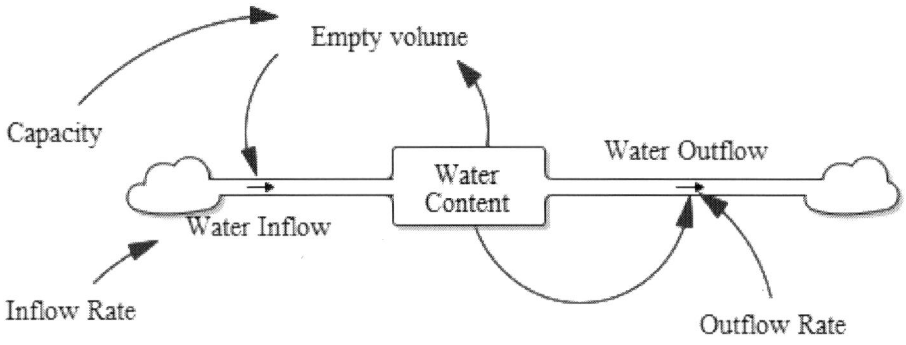

Model - Settings:
INITIAL TIME=0 FINAL TIME=100 TIME STEP=1 Units for time=Hour

(01) Capacity = 100
Units: Litres (write "Litres" because it is not on the menu)

(02) Empty volume = Capacity-Water Content
Units: Litres

(03) Inflow Rate= 1/10
Units: 1/Hour
The tank is filled every hour a 10% of its empty volume.

(04) Outflow Rate = 1/10
Units: 1/Hour
Every hour the tank is empty out a 10% of its contents.

(05) Water Content= Water Inflow-Water Outflow
Initial value: 50
Units: Litres

(06) Water Inflow= Empty volume*Inflow Rate
Units: Litres/Hour

(07) Water Outflow=Water Content*Outflow Rate
Units: Litres/Hour

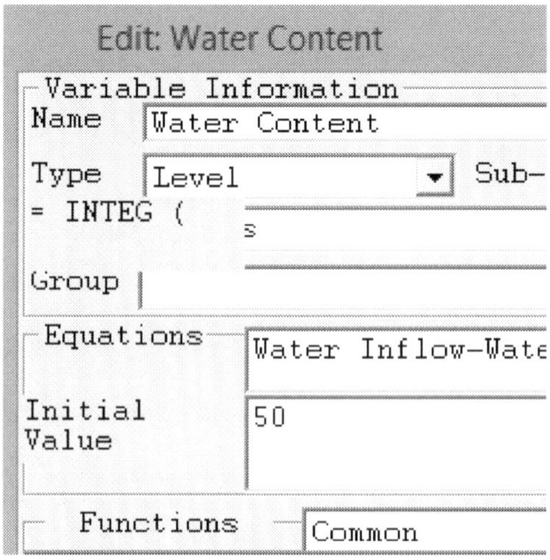

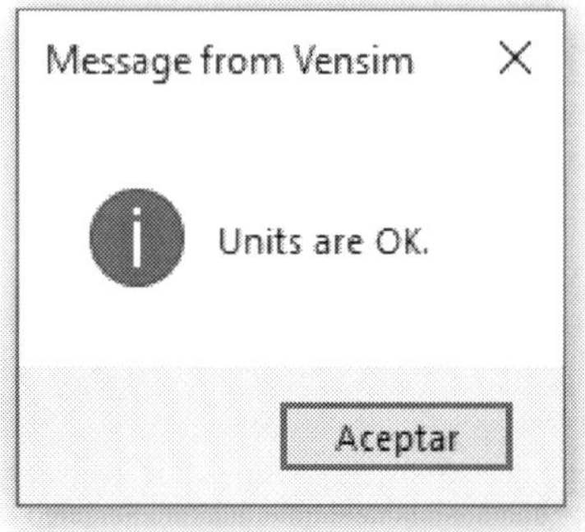

Click the icon Units Check. You should see the message.

Now, run the model and make the simulations. 

1. What behavior would you expect, if initially, as we described earlier, the tanks gets filled up at a rate of 1/10 of it empty volume, and gets empty at the same rate?

The water content would be constant at 50 litres and water inflow and outflow would also be constant at 5 liters.

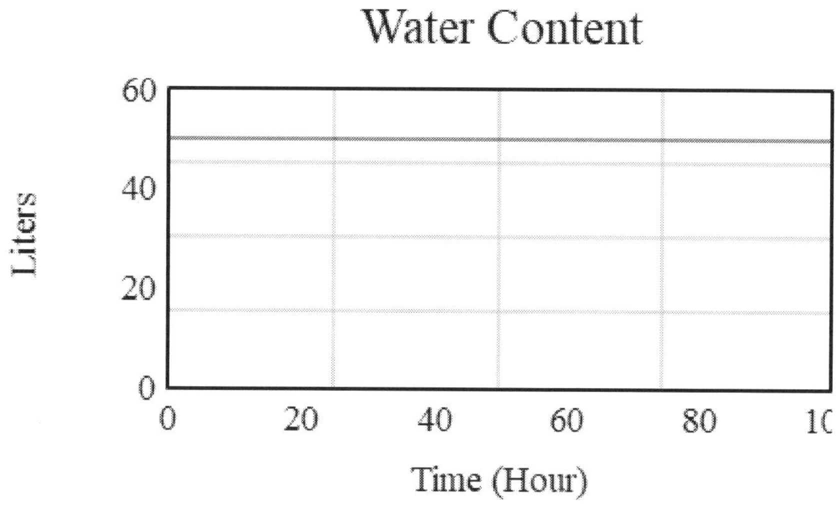

2. What would happen if initially, the tank contains 80 liters?

Change the initial value of the Stock and Simulate. Click Yes to override.

During the first period the inflow rate is (100-80)/10=2 liters, and the outflow rate is 80/10=8 liters, this make that the contents drops until it reach the value of 50 liters, at this moment the inflow and outflow rates are 50/10=5 and the water content is stable.

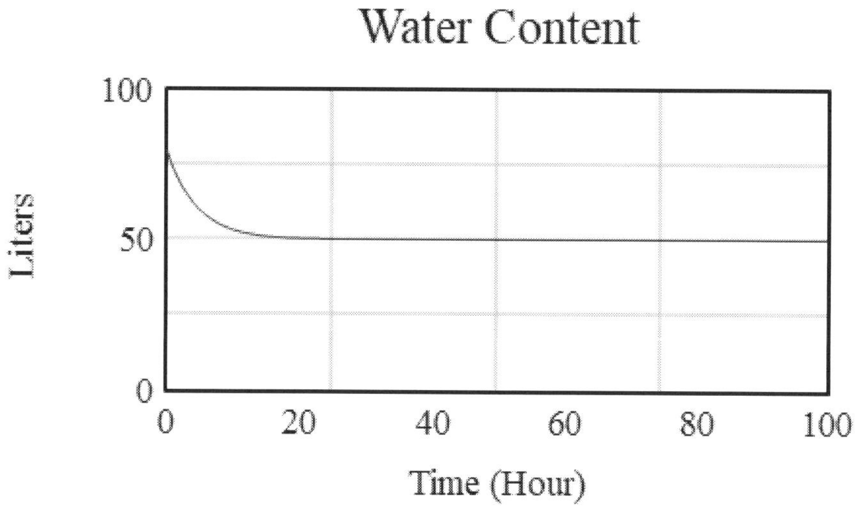

3. Change the initial value of the Stock to its initial value of 50. What would happen if the entry is regulated at 0.05 of the empty volume of the tank, and the exit at 0.10 of their content?

The water content at the first will fall down rapidly until 20 hours, after which it remain fairly constant at 33 litres. The water inflow will increase and the outflow will decrease at the beginning, and then remain constant at about 3 litres.

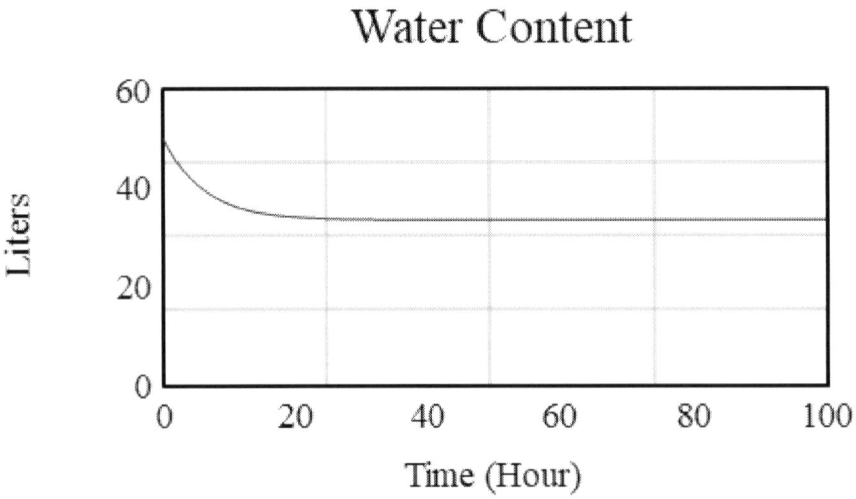

When you run a simulation you can save it as a dataset. These datasets can be managed from the Control Panel. The default name for a simulation dataset is current.vdfx.

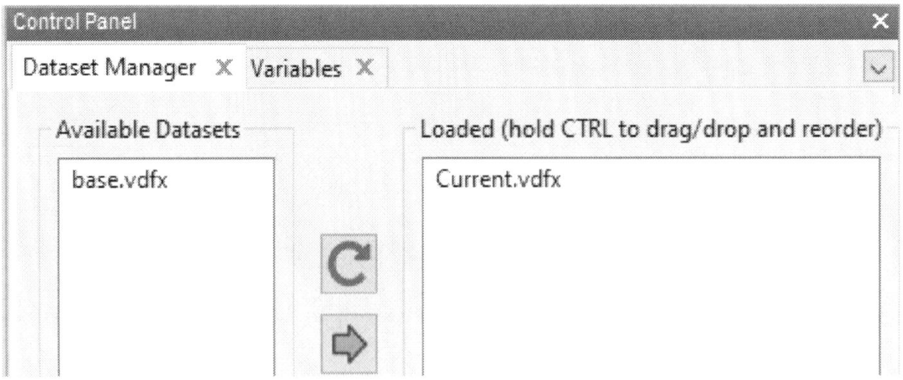

3.24. A study of the oscilatory movements

The study of oscillatory movements has always been motive for conflict, especially for students. What is this conflict? From the mechanical courses to the study of oscillators, we see systems whose complexity is given in cases with constant acceleration. From here, the said cinematic variable also changes its value depending on the time which can be disconcerting for students.

This uncertainty increases when the students solve the differential equations to find the cinematic variables (position, speed and acceleration) and/or the dynamic variables (quantity of movement and strength), and they lose the capacity to analyse the behaviour of the variables that have an influence - or not - in the movements.

To try to solve this situation, an alternative study of the different Oscillators is proposed using System Dynamics as a tool.

The Forced Oscillatory Movement

We will study the oscillatory movement in a dimension that describes a system formed by a pendulum, a body of mass m, suspended from the extreme of an ideal spring with an elastic constant k-, when this is disturbed – by elongating or compressing it – in respect to its position of balance x0, a distance x.

The forced oscillatory movement is produced by an elastic force, –k.x, a friction force, -b.v, and a harmonic force:

$$F(t) = F \,.\, \cos (wt)$$

Where w is the frequency of variation of the force, the time, and F is its maximum intensity acting over a body.

The equation of movement is:

$$-k \,.\, x - b \,.\, [dx / dt] + F(t) = m \,.\, [d^2x / dt^2]$$

The model of the forced oscillatory movement is represented below:

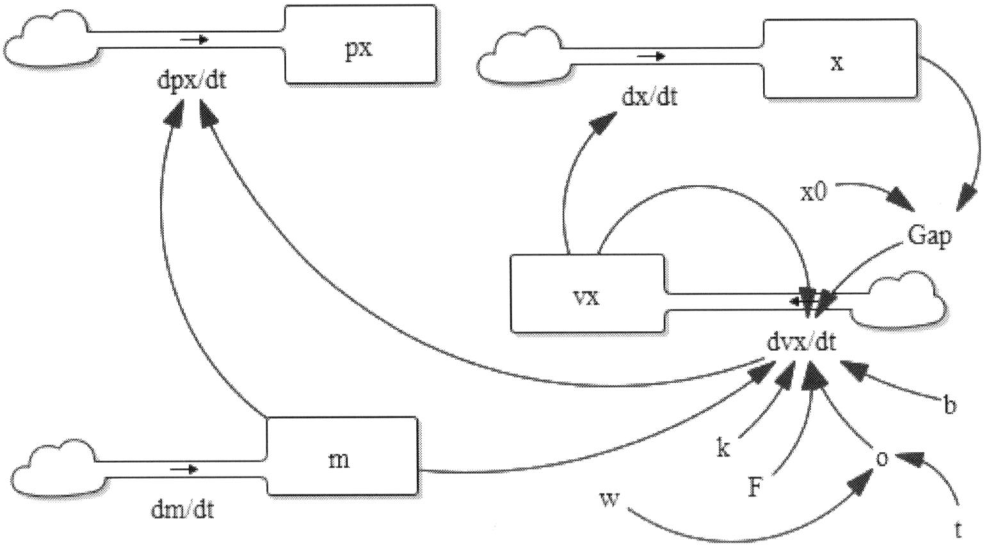

This model has 4 stocks:

a) the quantity of movement px, that is modified by the flux dpx/dt (the strength in x)
b) the position x that is modified by the flux dx/dt (the velocity in x);
c) the velocity vx, affected by the flux dvx/dt (the acceleration in x);
d) the mass m, affected by the flux dm/dt (in this model the mass doesn't vary in time).

The acceleration is influenced by the following auxiliary variables:

a) the constant elastic of the spring k;
b) the mass m
c) the difference or Gap between the position x (stock) and the position of balance x0
d) the constant of deadening b;
e) the strength F;
f) the phase o, which is the same as the product of the frequency w and the time t.

The model can be used to analyse the cinematic variables (position, velocity and acceleration) or dynamics (quantity of movement and strength).

Although the variables can be analysed in an individual form, they can be grouped together in two visual representations - one that involves the cinematic variable, and the other for quantity of movement and strength.

As a result of the simulation, the following graphics are obtained:

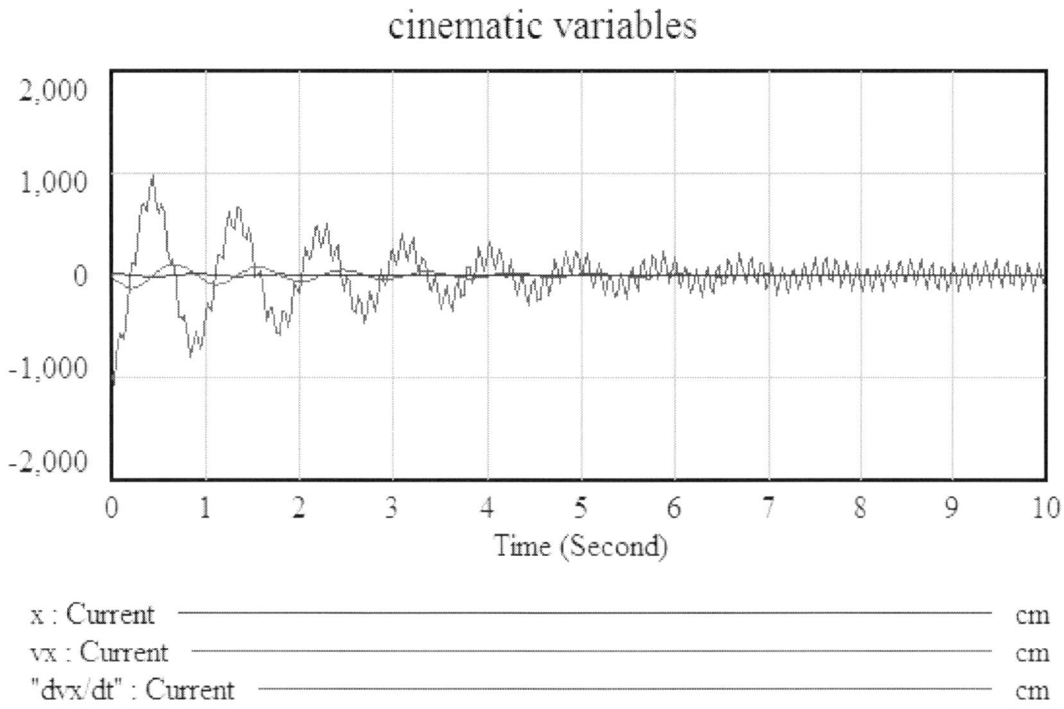

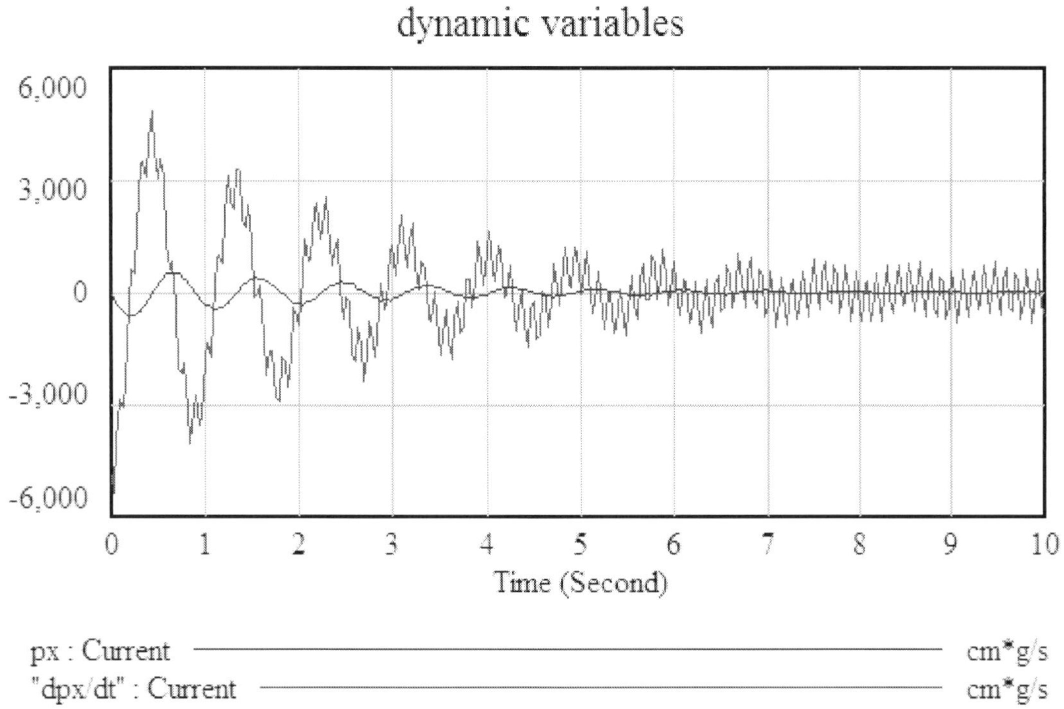

If we give F a value of zero in this model, a deadened oscillatory movement will be produced. The following graphics are obtained:

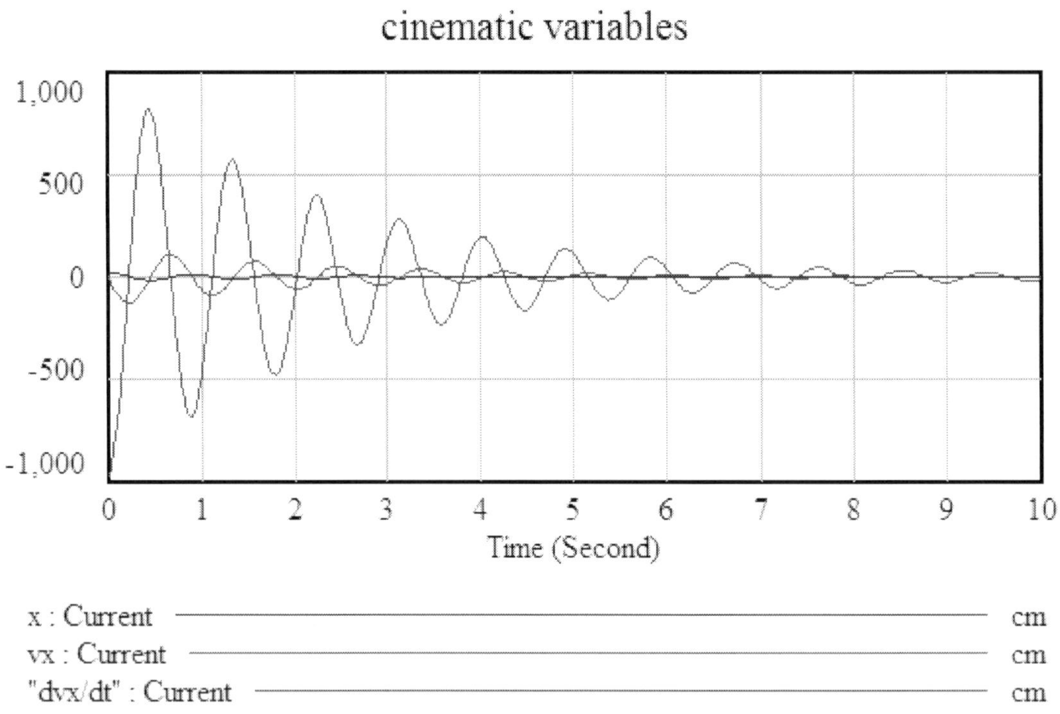

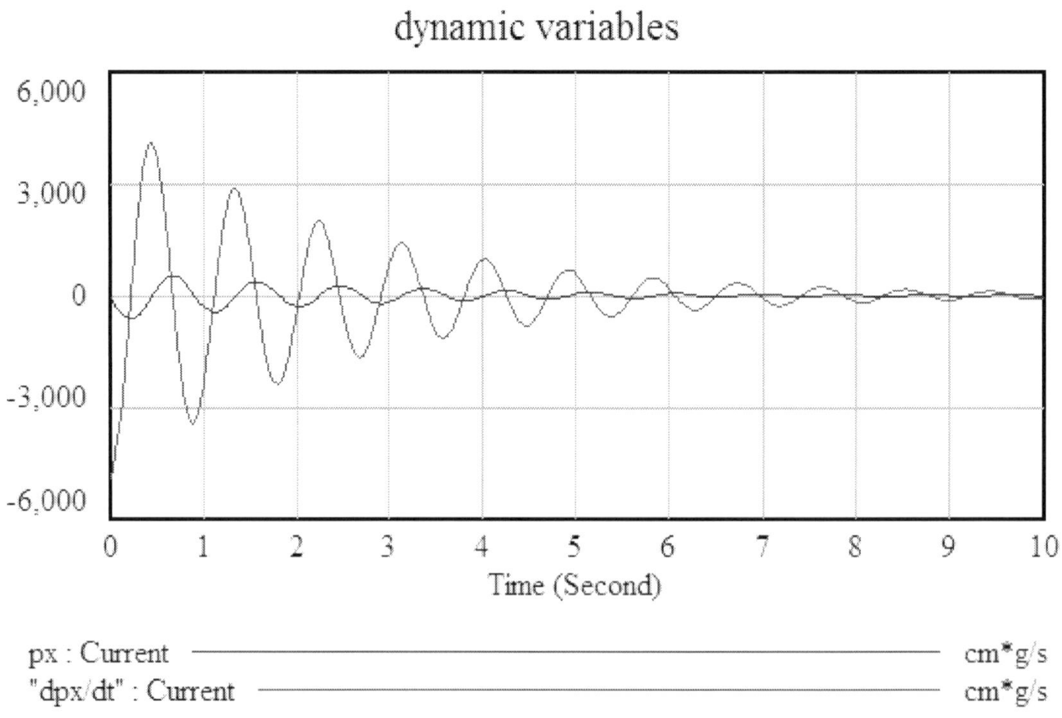

It also has an additional advantage - it is possible to analyse if deadened oscillatory movement exists. How is this done? Simply by changing the value of the constant b and doing the simulation. The following graphic shows us the results.

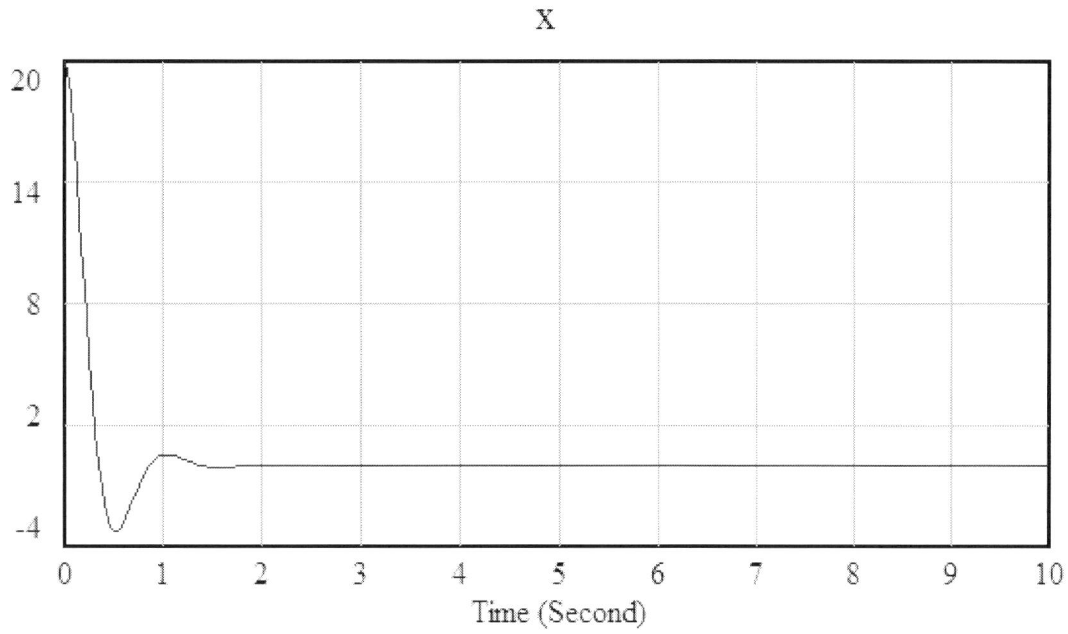

In conclusion, with this model it is very easy to analyse if oscillations exist for other values of b.

If the model of the deadened oscillator is given the value of b = 0, a deadened oscillatory movement will be produced. This shows clearly the great educative usefulness of System Dynamics.

The graphic representations obtained for the chosen variables will be shown next.

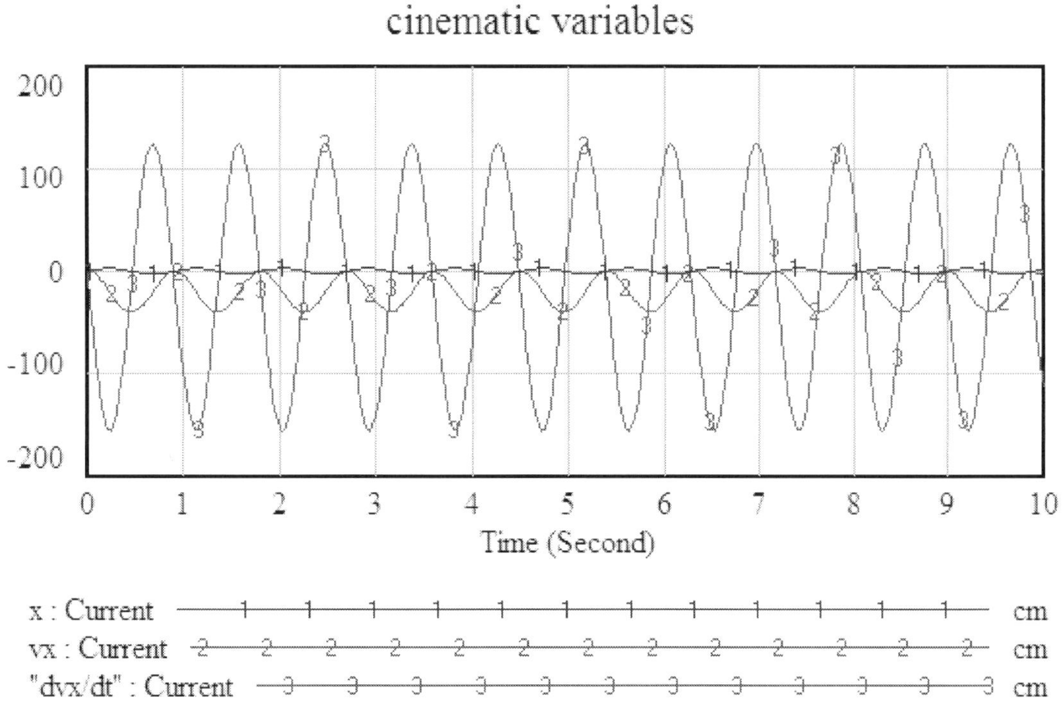

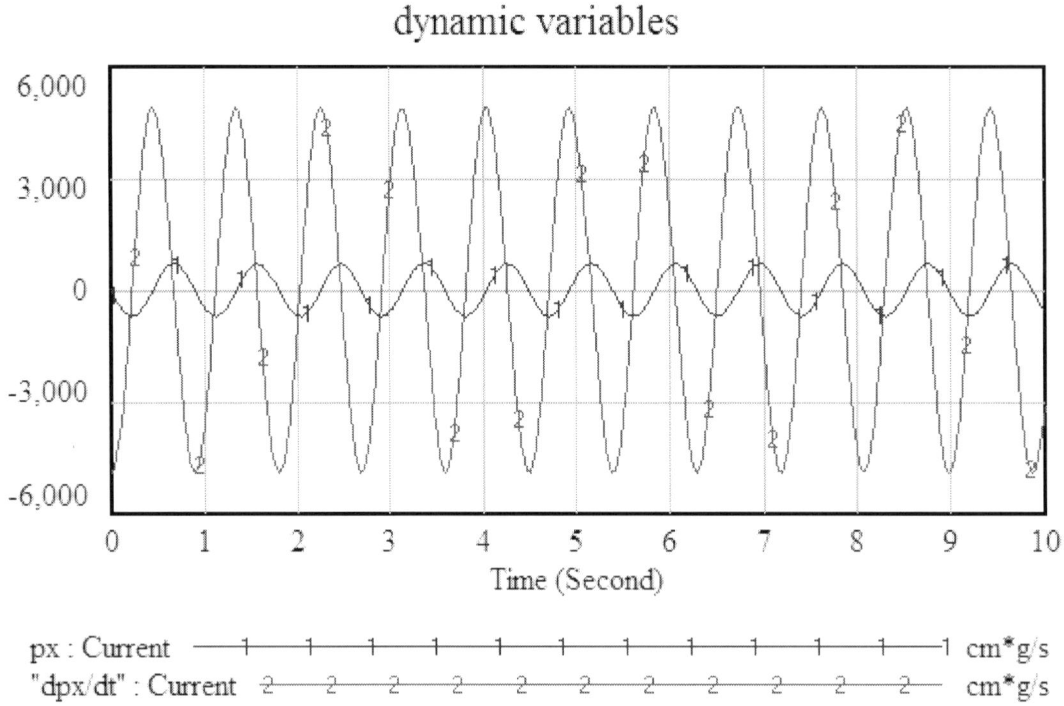

By changing the values of the variables that influence the movement, it is possible to analyse the behaviour of the variables. For example, for the simple deadened oscillator, it is possible to view the change in the amplitude, the longitude of the wave, and the frequency (or the period, that is its reciprocal) when the values of k or m change in the model.

The following graphics show the values of the position x depending on the time, according to the values of the original model:

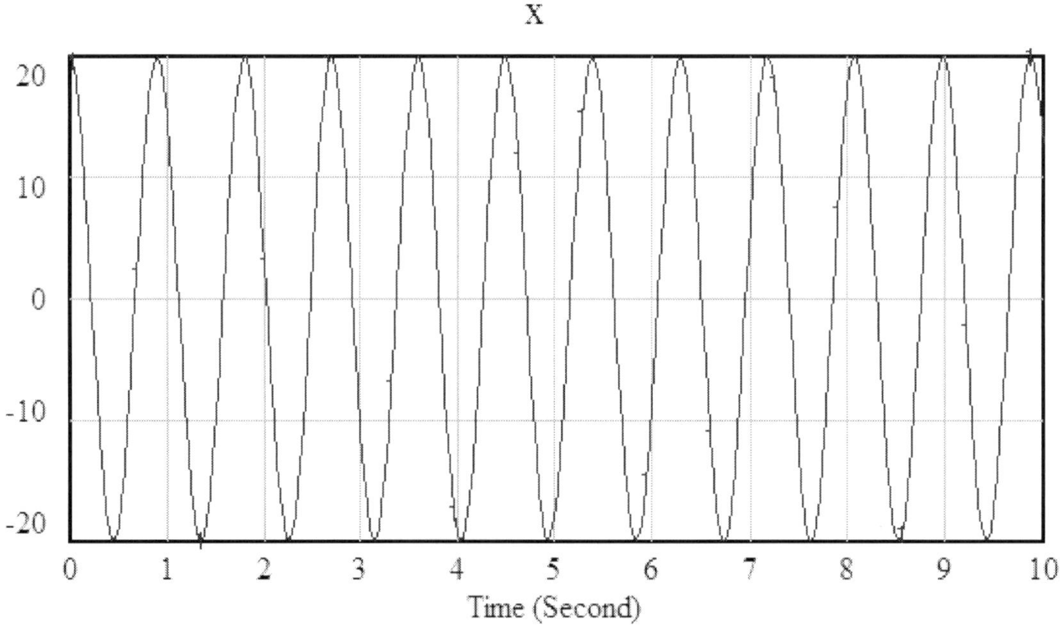

If we modify the numerical value of the mass m, the following graphic is produced:

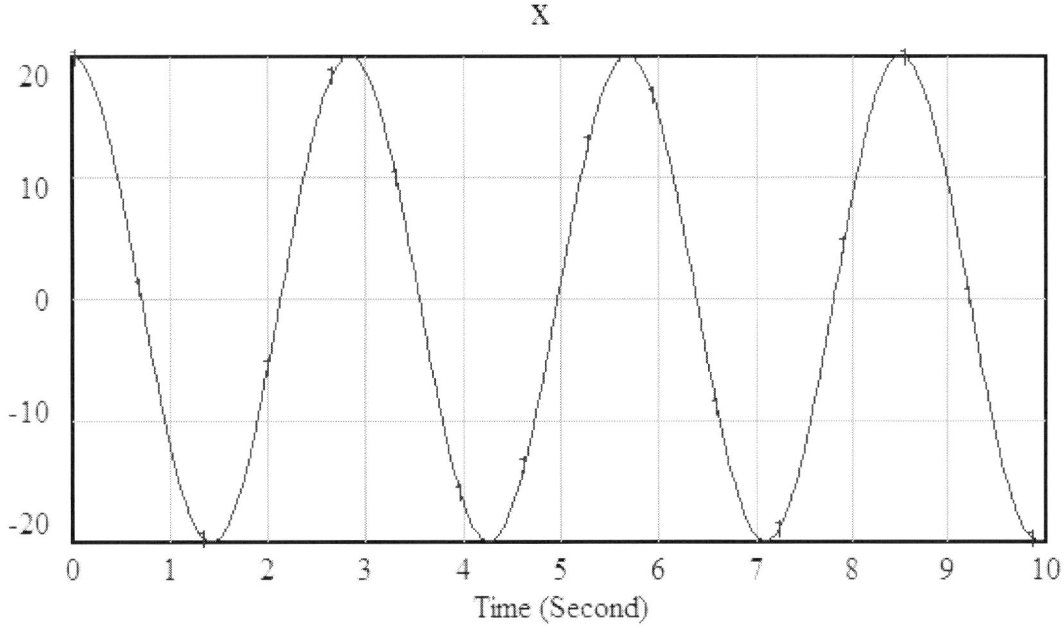

But if the numerical value of the variable k is changed, the following results are obtained:

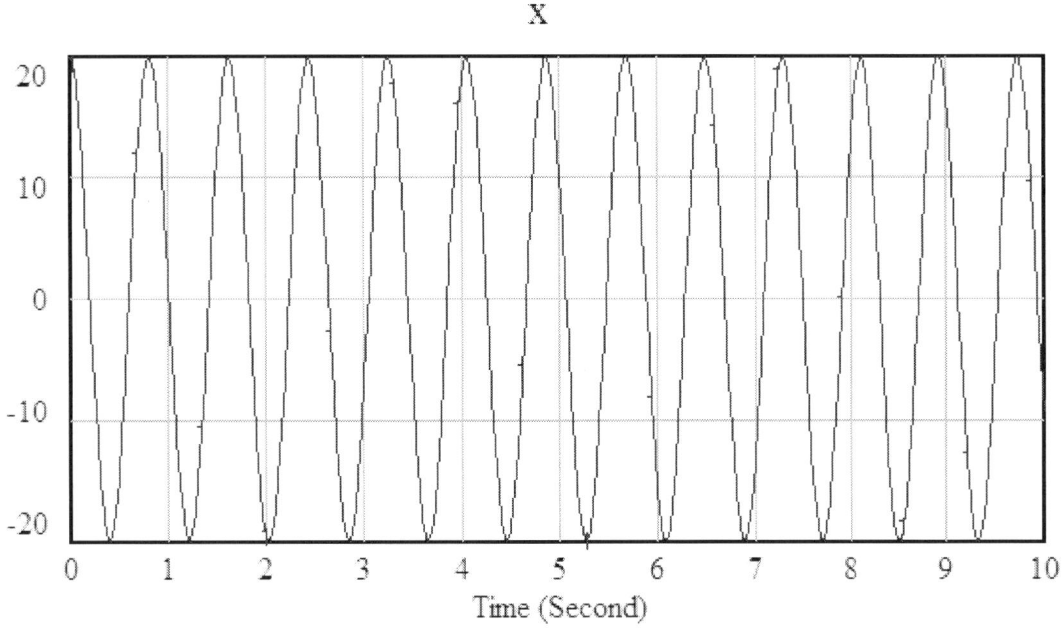

Model - Settings

INITIAL TIME = 0   FINAL TIME = 10   Units for time= Second   TIME STEP= 0.03125

The equations of the model of the forced oscillator are the following:

(01) b=4
Units: g/s

(02) dm/dt=0
Units: g/s

(03) dpx/dt=m*dvx/dt
Units: g *cm/s*s

(04) dvx/dt=Gap*k/m-(b/m)*vx+(F/m)*SIN(o)
Units: cm/s*s

(05) dx/dt= vx
Units: cm/s

(06) F=700
Units: dyn

(07) Gap= x0-x
Units: cm

(08) k=245
Units: dyn/cm

(09) m=dm/dt
Initial value: 5
Units: g

(10) o=w*t
Units: Dmnl

(11) px=dpx/dt
Initial value: 0
Units: g*cm/s

(12) t=RAMP(1, 0 , 10)
Units: s

(13) vx=dvx/dt
Initial value: 0
Units: cm/s

(14) w=750
Units: 1/s

(15) x= dx/dt
Initial value: 20
Units: cm

(16) x0= 0
Units: cm

The previous graphics have been produced in the model modifications in the following parameters:

- for the graphic where deadened oscillatory movement doesn't exist, b = 35 g/s
- for the graphic of x depending on the time where the value of the mass m = 50g is modified
- for the graphic of x depending on the time where the value of the elastic constant of the spring is modified, k = 500 dyn/cm.

We can obtain the following **conclusions** from the model created:

- when the mass increases, the period of deadened oscillatory movement increases;
- when the constant k increases, the period of deadened movement decreases.

As a consequence, a direct relationship between the period of oscillation and the mass of the pendulum, and an inverse relation between said period and the elastic constant of the spring, exists. As a consequence it is possible to write:

$$T = m / k$$

The mathematical expression of the period depending on the mass and the elastic constant of the spring can be found using processing of the numerical data of the simulation, or using differential equations that are the result of the application of the second law of Newton.

This study attempts to highlight the great educative advantages of using System Dynamics in analysis of Forced Oscillatory Movements both deadened and harmonic.

Employing the same model and modifying the values of the variables according to the case being studied, graphics are produced that greatly facilitate the comprehension and the discussion of the system analysed.

3.25. Design of a chemical reactor

One distinctive characteristic of industrial processes is its complex structure, generally constructed by many phases, each consisting of numerous sub-components. Its mathematical description generates equations ranging from simple to very complicated.

This is added to the lack of an exhaustive knowledge of the characteristics of the sub-components and the elevated grade of interrelation between the variables, generator of non-lineal results that complicate the exact resolution of the equations.

In this way, the design of new plants and the optimisation of those that already exist, two inherent tasks in the activity of engineers of process and project usually require intuitive criteria. Neither for economic reasons nor security is it possible to carry out these excessive laboratory simulations or experiments with existing plants.

As a result, the systematic analysis and the creation of dynamic simulation models appear to be a very attractive solution to the problem.

Physiochemical characteristics of the process considered.

The present model simulates the behaviour of a discontinued reactor, perfectly mixed, of a constant volume, with a system of temperature regulation using an interior coil where a reaction in the gaseous phase of the first order is completed, of type:

$$A \rightarrow B + C$$

In the figure, a simplified scheme of the reactor is shown, with the agitator that ensures it is mixed perfectly and the cooling coil, around which a cooling current circulates to Tm with enough flow in order not to experience a significant change in temperature.

The equations that represent the balances of material and energy are:

Balance of material for component A:

$$dn_A / dt = - k \, n_A$$

Where
nA = number of mols of A in a time of t
k = constant of the velocity of reaction in the first order, where 1/hr and depending on the temperature in the form k = A1* exp (-B1/T), where A1 and B1 belong to a certain reaction and T is the absolute temperature.

Balance of energy:

$$Cv\ (dT/dt) = \Delta H_R\ (-\ k\ n_A) - U\ A\ (T - Tm)$$

Where
Cv = calorific capacity of the mix contained in the reactor, to the constant volume Btu / mol lb ºF
AHr = heat of the reaction Btu/mol-lb A
U = global coefficient of the transmission of heat Btu/h ft2 ºR
T = temperature of the reactor, ºR
Tm = average cooling temperature, ºR
A = area of transference of heat supplied by the coil, ft2

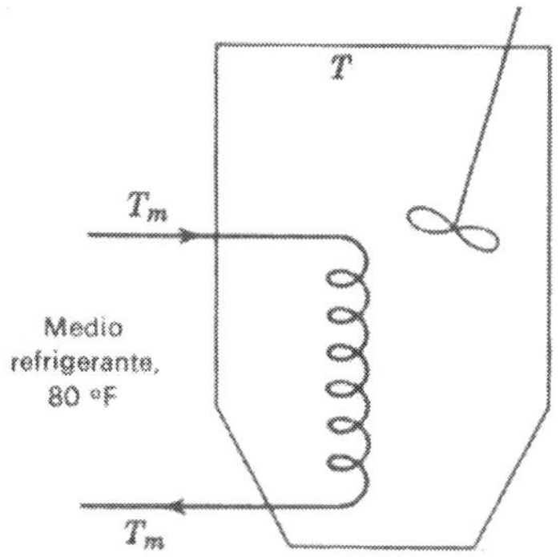

The reaction of conversion from A in products B and C is exothermic. During its development, it generates a quantity of heat that can be calculated starting from the heat of its reaction, HR. In this way, not applying a cooling system, the temperature of the reactor will increase constantly. There are also other factors that should be considered.

1) The elevation of the temperature in the reactor is related to the global calorific capacity of the mix contained and this continues to vary with the advance of the reaction since A is substituted by B and C. For a certain composition in the reactor it will be:

$$Cv = n_A\ C_A + n_B\ C_B + n_C\ C_C$$

Where the n_i are the mols of each component to a time t, and the C_i are the respective calorific capacities, considered constants for which ever value of the Temperature. Given the certain stequiometry of reaction, the expression can be simplified expressing it only depending on NA.

2) The constant K of the reaction depends on the temperature and in another way, the generation of heat of the number of mols of A present in the reactor to a time t, as it can be seen in the energy balance equation. This generates a double link between the equations representative of the process and to introduce difficulties for the solution.

For the construction of a dynamic simulation model, two stocks have been considered. In one, the disappearance of A is produced by reaction, in the other, the accumulation of heat by reaction and its dissipation through refrigeration that is measured using the temperature. The symbols of the equations have been conserved, the usual equations in the field of chemical engineering.

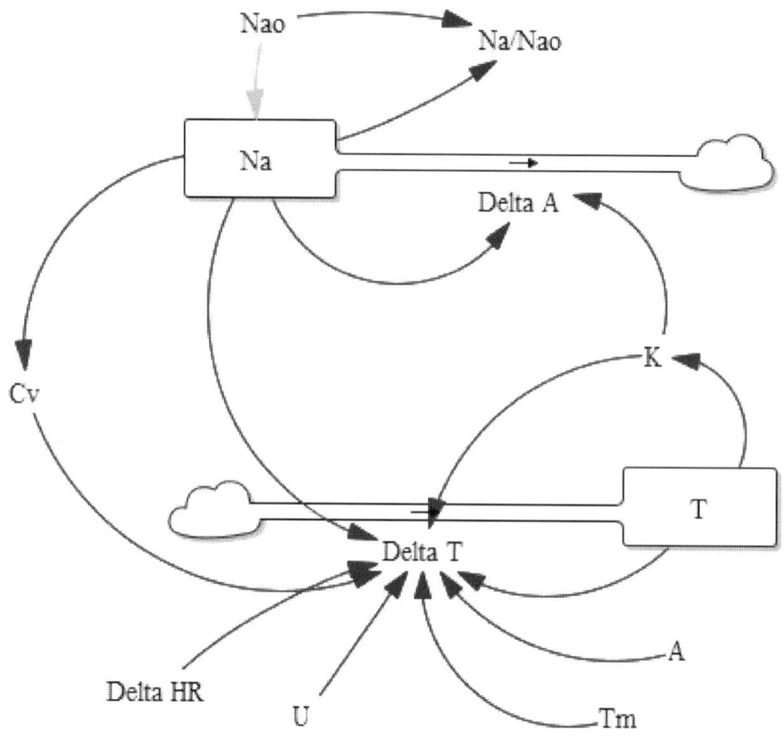

The data used in the model are:

nAo = 0.1 (number of mols present at the beginning of the reaction)
T = 600º R (initial temperature of the mass contained in the reactor)
k = $0.5 * 10^{10} * \exp(-1.394 * 10^4/T)$
ΔH_R = -2500 Btu/mol-lb A
U = 3 Btu/h ft^2 ºR
Tm = 540 ºR
A = 0.2 ft^2
C_A = 30 Btu / mol lb ºF
C_B = 25 Btu / mol lb ºF
C_c = 25 Btu / mol lb ºF
Cv = 5 – 20*n_A (Considering the stoichiometry of the reaction)

Model - Settings

INITIAL TIME=0 FINAL TIME= 90 Units for time= Second TIME STEP= 0.0625

Equations

(01) A = 0

(02) Cv = 5-20*Na

(03) Delta A = K*Na/60

(04) Delta HR = -2500

(05) Delta T = (Delta HR*(-K*Na)-U*A*(T-Tm))/(Cv*60)

(06) K = 1.712*1e+010*EXP(-1.394*10000/T)
Units: 1/hr

(07) Na = -Delta A,
Initial Value = Nao
Units: mollb

(08) Na/Nao = Na/Nao

(09) Nao = 0.1

(10) T = Delta T
Initial Value: 600
Units: Grados R

(11) Tm = 540

(12) U = 3

The model reproduces the behaviour obtained using the application of the most complex analogical models like those of Himmelblau and Bischoff. This allows us to see clearly the behaviour of this reactor under different conditions of operation, given that it is possible to change the temperature of the refrigerator, the area of transference of heat, the initial temperature of the process and the global coefficient of the transference of heat.

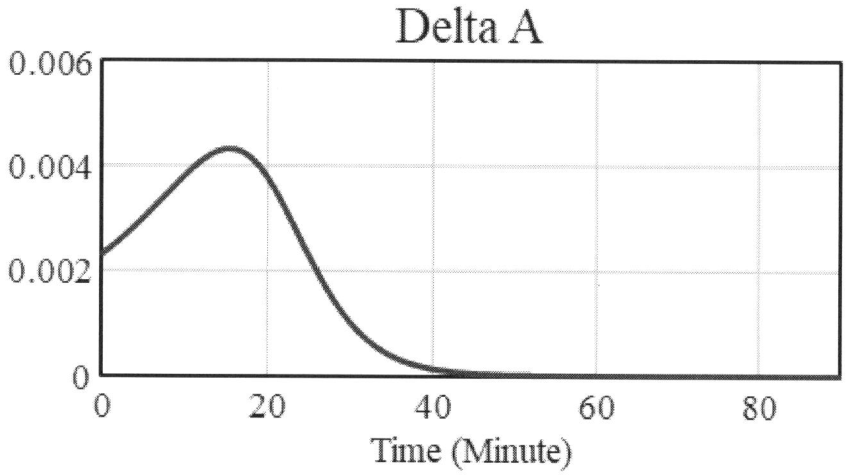

3.26. The Butterfly Effect

When a simulation model is created, the system's elements often behave in a surprisingly and even totally unexpected way. The changes we make to the initial conditions can also result in opposite or very different effects from those anticipated. Moreover, small changes in initial values may result in considerable differences in the behaviour of the system's elements.

Perhaps unknowingly we create a simulation model with a structure and a relationship among variables such that, under specific conditions, it behaves in a way known as chaos. A definition of chaos is 'an aperiodic behaviour in a deterministic system that is very sensitive to initial conditions'.

The simulation model does not have to be very complex with a lot of variables, parameters and feedback. Numerous studies have shown that with three differential equations and with one of them being non-linear, we have the conditions needed for the system to have chaotic behaviour under certain conditions.

In the last few decades of the twentieth century, the Chaos Theory attracted considerable interest since it showed the interconnected reality that surrounds and fills us with feedback loops, where each constituent element acts to modify the behaviour of the environment that surrounds it but does not do so independently. Instead, it obeys an integrated behaviour of the whole. This theory is particularly useful to tackle the study of social phenomena, which are always complex and difficult to solve in terms of linear cause-effect relationships.

Fortunately, there are examples of physical phenomena or purely mathematical systems, such as the forced pendulum - a physical phenomenon or a third order differential equation - a mathematical model that makes it easier to understand chaotic behaviour before modelling much more difficult situations, such as social phenomena. We have an even more famous example, because of its cinematographic repercussions, from the work of meteorologist Edward Lorenz who over forty years ago constructed a three differential equation system to model meteorological behaviour in a simple way. He obtained a surprising as well as eye-catching response commonly known as the 'Butterfly Effect'.

In the 1960s, Lorenz began a series of investigations focused on solving the problem of meteorological prediction. Working in a two-dimensional rectangular atmosphere whose lower zone is at a higher temperature than the upper zone and using continuity equations, amount of movement and thermal balance, he developed a simplified system formed by three differential equations that can be written as follows-

$$dx/dt = a\,(y - x)$$
$$dy/dt = r.x - x.z - y$$
$$dz/dt = x.y - c.z$$

It is important to observe that there are three differential equations that have two non-linearities, the products 'x.y' and ' x.z '. That is why this system meets the conditions for chaotic behaviour to appear in its state variables ('state variables' are called Stocks in System Dynamics).

We can represent these equations with a dynamic simulation model. Nevertheless, it is important to keep in mind that the previous equations are the result of a regular process when analysing physical and chemical phenomena. It consists of adimensionalising variables.

The result of this process is the appearance of groups of parameters (for example, characteristic density, viscosity, longitude) known as adimensional numbers, which establish relationships between the driving forces of change in the system being studied, in other words, its dynamics.

The created model is made up of three Stocks called Convective Flow, Horizontal Temperature Difference and Vertical Temperature Difference, which depend on their respective Flows, which are: Convective Flow Variation, Horizontal Temperature Variation and Vertical Temperature Variation.

In addition, there are three adimensional parameters - the Prandtl Number - which establishes a relationship between the viscosity and thermal conductivity of the fluid - the Rayleigh Number - which quantifies heat transmission in a fluid layer through internal heat radiation and - Height - which represents the thickness of the layer being studied.

The model basically establishes the relationship between the convective flow and the variations in temperature of the air mass, which is complex in itself since a difference in temperature produces a convective flow and this flow, in turn, modifies the difference in temperature, all this contingent on the properties of the medium studied, such as viscosity, density or thermal conductivity, which are grouped together in adimensional numbers which appear as model parameters.

A simulation model with the following values thus described-

Prandtl Number = 10
Rayleigh Number = 28
Height = 8/3

And the following initial Stocks values:

Convective Flow = 0
Horizontal Temperature Difference= 1
Vertical Temperature Difference= 0

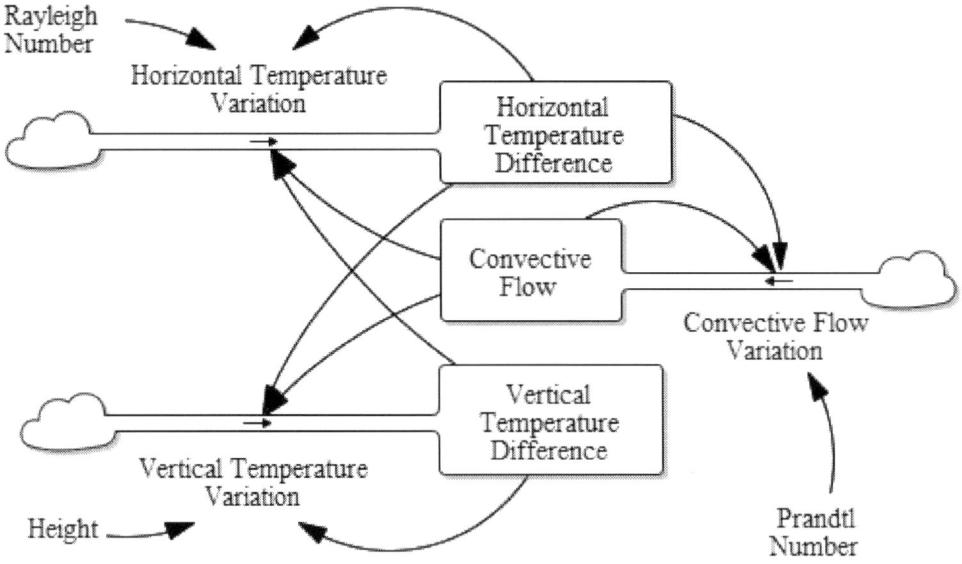

Model - Settings

INITIAL TIME= 0 FINAL TIME= 50 TIME STEP= 0.0078125 Units for time=Month

Equations

(01) Height = 8/3

(02) Horizontal Temperature Difference= Horizontal Temperature Variation
Initial value = 1

(03) Vertical Temperature Difference= Vertical Temperature Variation
Initial value = 0

(04) Convective Flow = Convective Flow Variation
Initial value = 0

(05) Prandtl Number = 10

(06) Rayleigh Number = 28

(07) Horizontal Temperature Variation= Convective Flow*(Rayleigh Number-Vertical Temperature Difference)- Horizontal Temperature Difference

(08) Vertical Temperature Variation= Convective Flow*Horizontal Temperature Difference-Height *Vertical Temperature Difference

(09) Convective Flow Variation= Prandtl Number*(Horizontal Temperature Difference-Convective Flow)

Integration method

The models add a small error to each simulation period when calculating Stock condition. When we want to minimize this error, we can use the Runge Kutta integration method. Click Model - Settings to select this integration method once you have constructed the model. The tool bar will appear as follows. Click on the word Euler until RK4 appears.

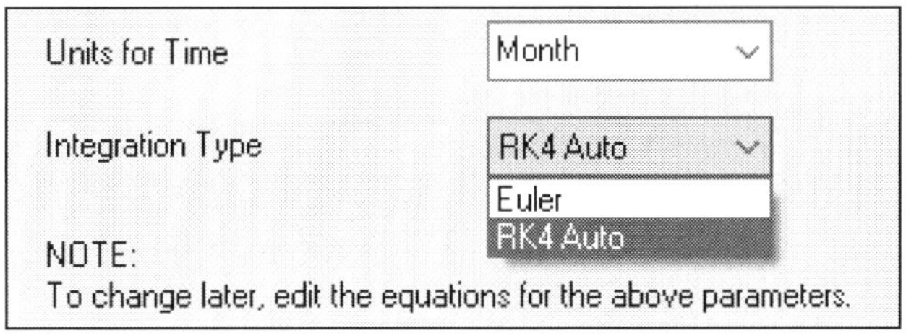

Now you can simulate the model. When the simulation is done, very complex behaviour is observed in the state variables or Stocks.

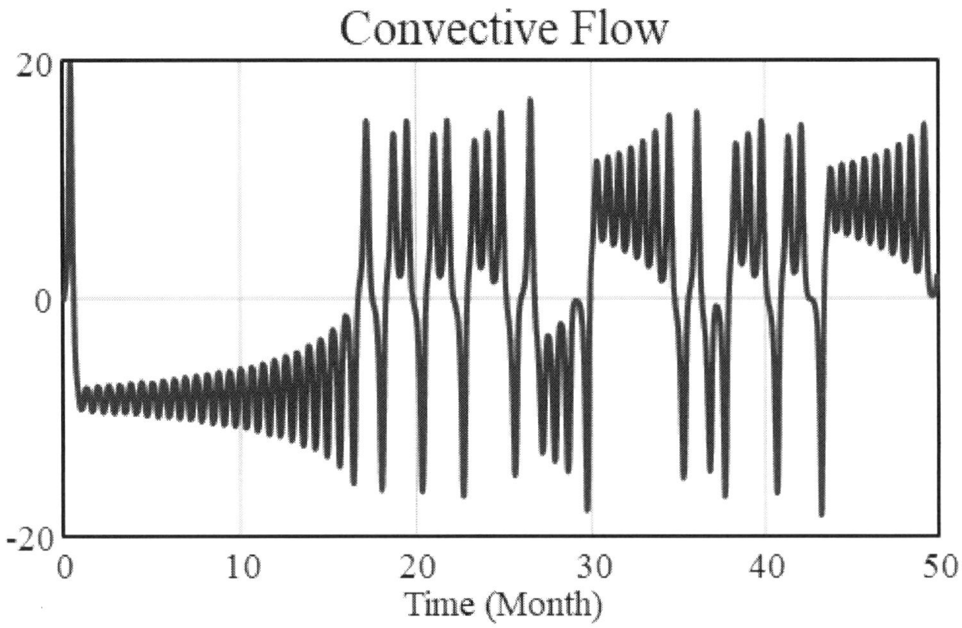

We can make an XY graph to compare the joint evolution of two variables, for example, the Convective Flow graph (on the X axis) with respect to the Vertical Temperature Difference (on the Y axis). To make it, click the Control Panel icon and select the variables as indicated in the following figure.

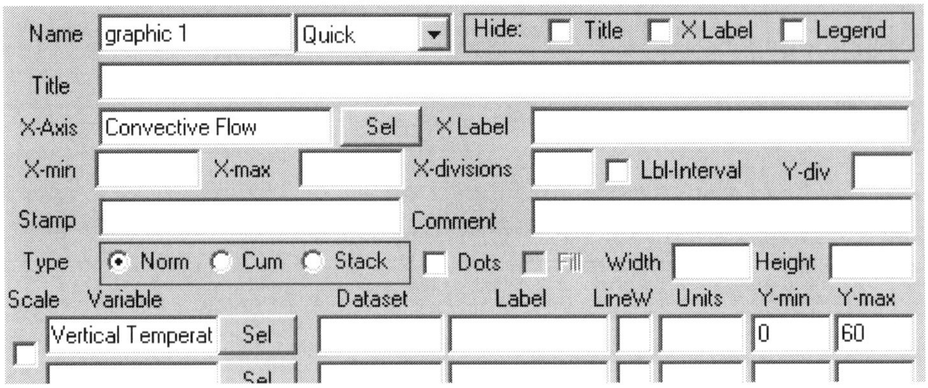

Strictly speaking, what we are doing is making a graph of phase space. Phase space is the mathematical space created by the variables that describe a dynamic system. Each point of the phase space represents a possible state of the system. The system's evolution in time is represented with a trajectory in the phase space.

Phase space study is of special interest. The dissipative systems have regions of phase space toward which the trajectories which originate from a specific region called the 'attractor basin' converge. There are predictable attractors with a simple structure, such as the point or the limit cycle. But there are other attractors known as 'strange attractors' in which small differences in the initial positions lead to positions which completely diverge. This is precisely the case of the Lorenz attractor with its peculiar form similar to that of a butterfly.

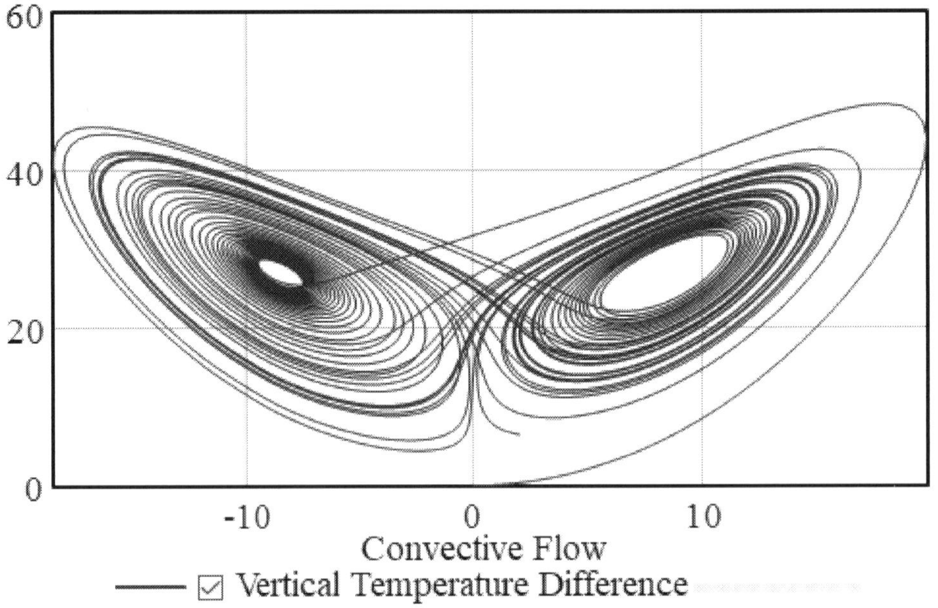

Variable time evolutions, generally complex, probably cannot lead to quick conclusions nor valid predictions, but phase space analysis allows us to see where the system's state converges and between which maximum and minimum values of its variables it evolves. All this is valuable information when the study of the complex situations that characterize the world in which we live is considered.

Edward Lorenz discovered this unexpected behaviour when he made the first simulations with the model he was studying. Without meaning to, he developed a valuable practical example of chaotic behaviour, which affirmed what had already been proposed in theory many years before. In 1890, Henri Poincaré published an article describing the fact that the sun-earth-moon system cannot be explained under traditional mechanics. In his words- 'small differences in the initial conditions produce very great ones in the final phenomena. A small error in the former will produce an enormous error in the latter. Prediction becomes impossible.' Subsequent research in this topic has made it possible to develop many examples of physical, chemical, biological and mathematical systems that present this phenomenon of unpredictability, which is called 'deterministic chaos'.

There is a good example in Julien Sprott's book, 'Chaos and Time-Series Analysis', which proposes a simple model with three variables, represented by the following equations:

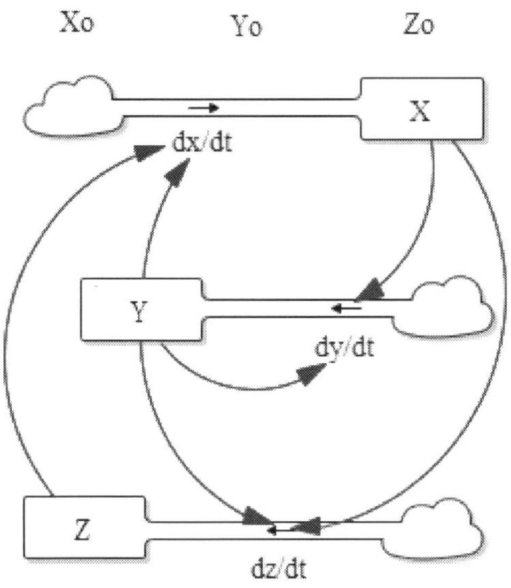

*dx/dt = y*z*
dy/dt = x-y
*dz/dt = 1-x*y*

The model's equations are omitted because they are very similar to those described in Edward Lorenz's model. Test with the initial values of X, Y and Z = 1. You will observe complete stability and then take the initial value of Y=0.99 and you will obtain the following graph.

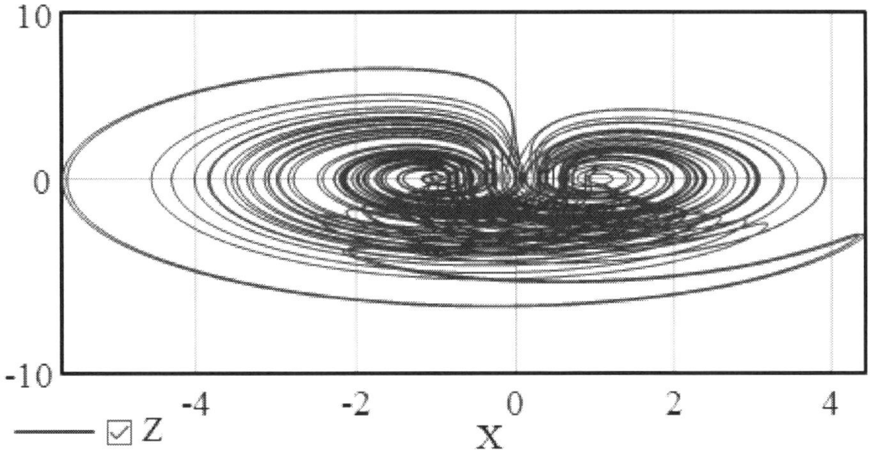

The systemic view and the powerful calculation tools of today make it possible to analyze chaotic behaviour problems very quickly with the advantage of having a clear visual image of the system's structure and its interrelationships. Likewise, the construction of phase space and the search for attractors is instantaneous, but should not be considered as a simple mathematical exercise.

3.27. The mysterious lamp

With the aim of eradicating widespread ancestral superstitions, scientists study and attempt to explain strange phenomena.

This investigation studies an electrical phenomenon with no logical explanation - a street lamp that starts to behave oddly when night falls. The street lamp's bulb begins to turn itself on and off at two-second intervals and a few hours later, always before daybreak, the light bulb burns out. In this way, a succession of light bulbs is destroyed in what seems to be an assault of the street lamp on the light bulbs.

No anomalies are observed in the street lamp's electrical and mechanical components and all the tests carried out by various technicians have demonstrated an absence of manufacturing faults.

Different brands of light bulbs have been tested and the wiring has been checked on numerous occasions. Furthermore, there are similar street lamps in the surrounding area that work perfectly well. So far, no plausible explanation for this strange phenomenon has been found.

Research teams

First research team

The first research team was made up of an electrical engineer specialising in underground conductivity, a mechanical engineer with widespread experience in industrial assembly and a well-known urban planner.

This first team drew up a detailed document describing the street lamp's location and its electrical and physical components. They were unable to find the cause of the strange behaviour within the specified time limit.

However, when a strange electrical problem affected the computer where it had been stored, this valuable document vanished.

Second research team

A second research team with a very different profile was set up. The collaborators included a clairvoyant specialising in paranormal phenomena, a retired gardener and his assistant, an estate agent and a municipal police officer acquainted with the area, as well as the tenants and former doormen of the block of flats in front of which the street lamp is situated who offered their disinterested contributions!

The meetings held by this second group were fraught with difficulties from the start, due to the aggressive attitude of several of the tenants, and the police officer's insistence on directing the investigation.

In an attempt to reach some form of consensus, the team was supplemented with a group of psychologists specialising in behavioural disorders, a famous Argentinean psychiatrist, an American journalist, a biologist and a landscape technician.

This second team never managed to draw up a report due to the persistent differences of opinion between several of its members, which made it impossible to settle on a final document.

Third research team

Finally, a team of lecturers and students of System Dynamics was entrusted with the research, which they were to use as fieldwork in their study of counterintuitive phenomena.

Their results, shown below, allowed them to understand why the phenomenon had occurred and in their conclusions they were able to propose ways in which the problem could easily be solved.

The third research team's final report

Location

The street lamp which was the subject of study is located in front of a block of flats at number 18 of the Doctor Francesc Darder Street in Barcelona (Spain). The street is named after an explorer and biologist who returned from Namibia with the desiccated body of an enigmatic black sorcerer. The body was exhibited to the public in the town of Banyoles for more than fifty years.

The building is surrounded by a leafy area and there are two identical street lamps in front of it. Photographs of the building can be viewed at http://www.qdq.com. The street lamp exhibiting the enigmatic behaviour is the one situated next to the steps leading to the main entrance.

It might be worth noting that there is a chapel directly in front of the street lamp studied, and that the USA consulate in Barcelona is less than a hundred metres away.

Description of the street lamp

The street lamp is made of metal and the post is painted white. At the top end of the post, there is a metal cylinder of the same colour with vents on either side, inside of which is the light bulb.

It is a 250-watt light bulb, receiving 220 volts of electric current. High-quality, energy-efficient, long-life Philips light bulbs are used for the street lamps in the building's immediate surroundings. Light bulbs of various types, brands and models were used in the investigation.

When daylight fades, the street lamp turns itself on automatically by means of a device that is fixed to the wall.

The research team prepared the following diagram. The top part of the document shows the sequence in which the light bulb turns itself on and off, including the time when it first switches itself on (because the bulb is controlled by a light sensor, this time may vary) and the intervals between switching on and off. After a long wait, the hour at which the light bulb burns out was recorded.

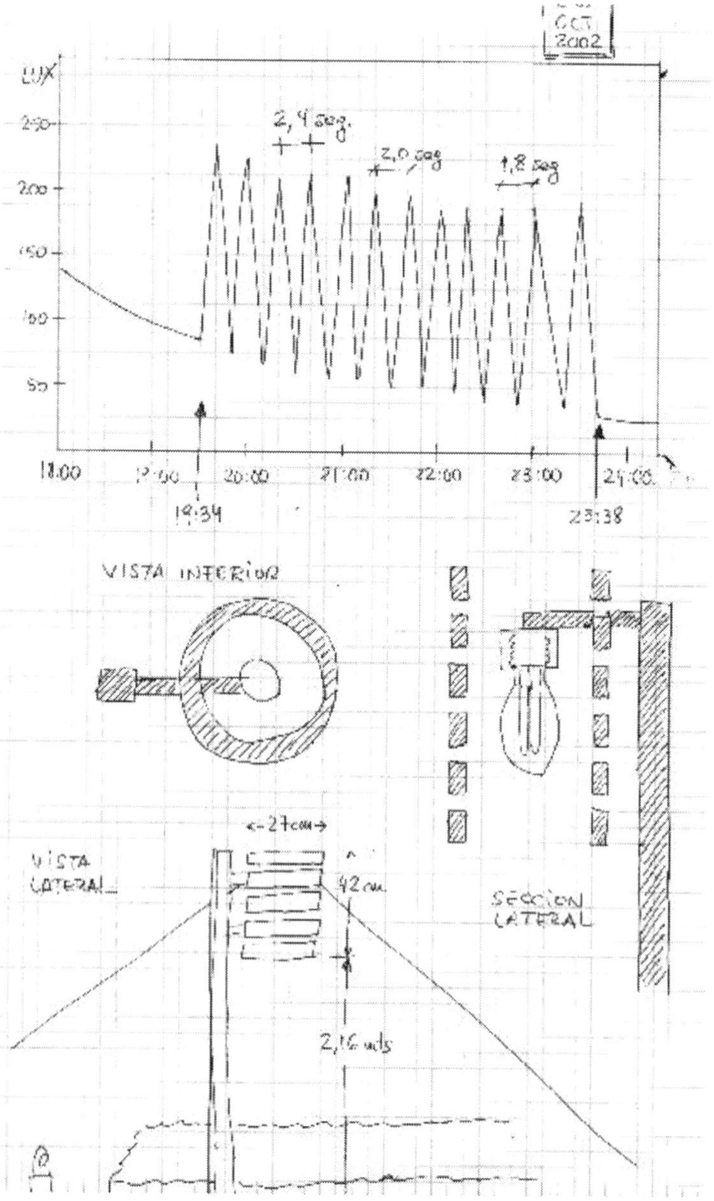

The document's middle section features cross-sections of the street lamp and the drawing at the bottom shows its measurements.

Hypotheses

The third research team began by gathering all the hypotheses propounded.

The tenants attributed the behaviour to strong lateral winds that penetrate through the side vents, banging the light bulb against the lamp and causing it to burn out. This theory fails to explain why the light bulbs burn out when there is no wind or why only the light bulbs in one street lamp burn out.

Other hypotheses involved the prayers spoken in the nearby chapel at night and the sorcerer brought by Mr. Darder, the biologist, who might be seeking revenge.

There was also speculation surrounding a magnetic field that might be caused by a secret device installed in the nearby USA consulate, but the fact that all the other electrical equipment in the building functioned correctly failed to endorse this hypothesis.

Analysis of the problem with a dynamic simulation model

The third research team chose to disregard these theories due to their lack of rigour and because they do not in any way explain the phenomenon. Instead, they decided to carry out a dynamic simulation model.

They structured the model according to a causal diagram and then drew up a Forrester Diagram, which they introduced in a computer. The following diagram shows the final model, which includes the basic elements of the system studied.

In short, the model demonstrates that the sensor which switches the light bulb on when the natural light is less than the desired value and switches it off when it exceeds this value.

DESIRED SYSTEM

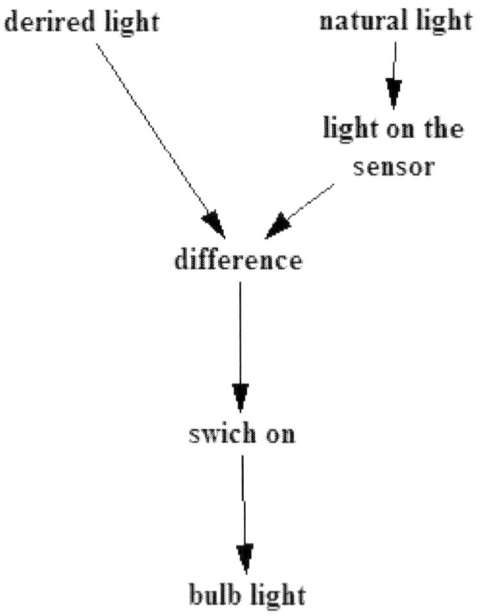

REAL SYSYTEM

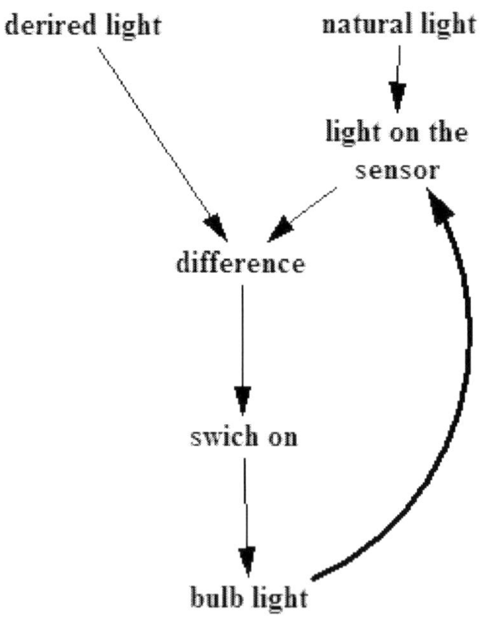

When the light bulb switches itself on because there is little natural light, it shines on the light sensor, which reacts as if there was plenty of light and consequently turns the light bulb off. The system's behaviour clearly demonstrates that the oscillations occur as soon as the natural light descends to a certain value.

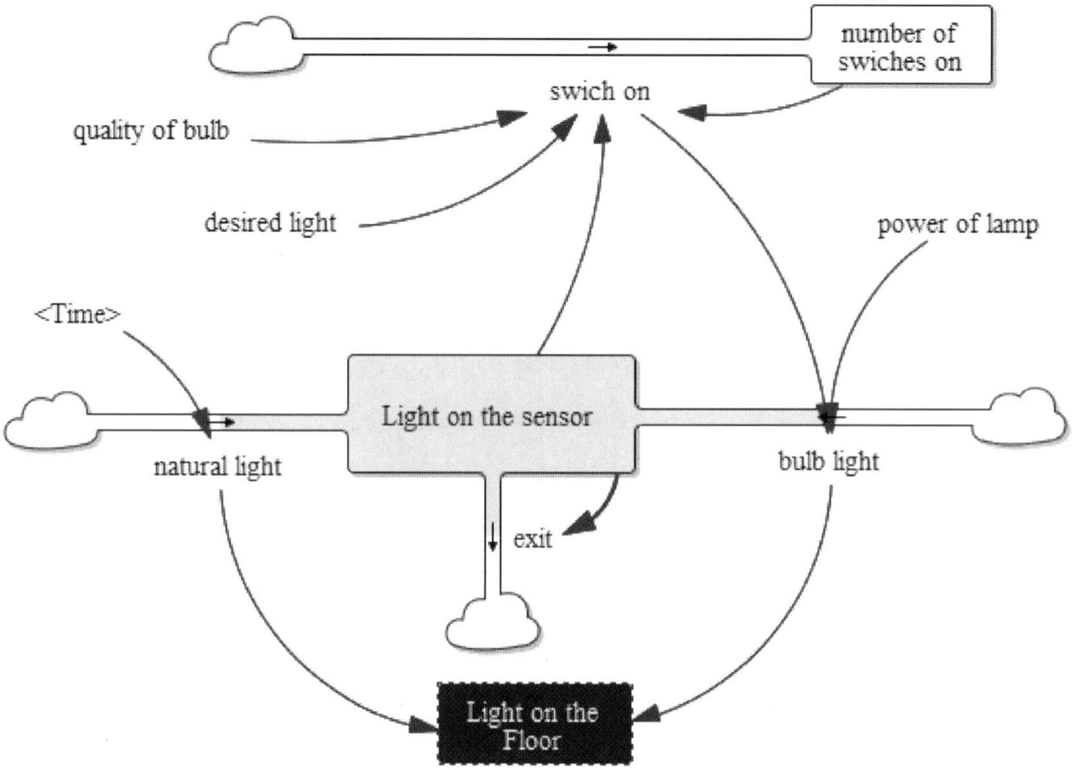

Equations

This is an advanced exercise. Could you build the equations of this model? It may be necessary to use the functions IF THEN ELSE, RAMP, and RANDOM NORMAL. Be patient.

File - New model or Model - Settings
Model - Settings: INITIAL TIME = 1800 FINAL TIME = 2400 Units: Hour

(01) bulb light=IF THEN ELSE(swich on=1 ,power of lamp , 0)

(02) desired light= 100

(03) exit=Light on the sensor

(04) Light on the Floor=MAX(0,bulb light+natural light)

(05) Light on the sensor= INTEG (bulb light+natural light-exit
Initial value: natural light

(06) natural light= 160-RAMP(0.22,1800,2400)*(2400/Time)^3

(07) number of swiches on= INTEG (swich on
Initial value: 0

(08) power of lamp=RANDOM NORMAL(90, 110 , 100, 20, 777)

(09) quality of bulb=200

(10) swich on=IF THEN ELSE(Light on the sensor<=desired light, 1 , 0)*IF THEN ELSE(number of swiches on >=quality of bulb, 0 , 1)

Obtained behaviour

The model reproduces the light bulb's switching-on, switching-off sequence and one can conclude that it is this that causes the light bulb to burn out after a few hours.

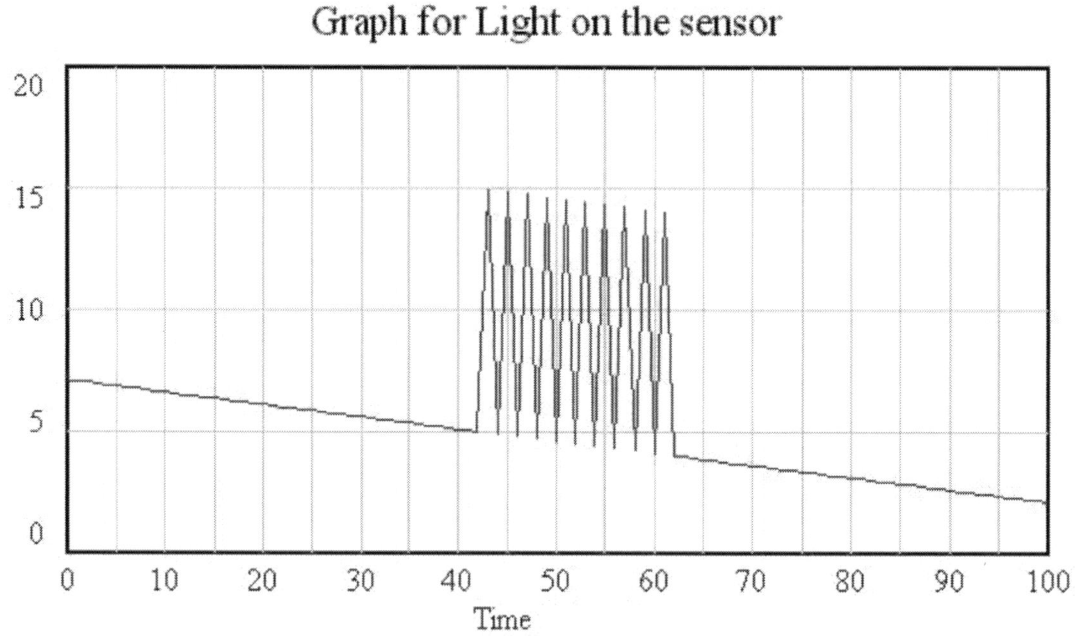

4. GUIDE TO CREATING A MODEL

Once the theory and exercises have been completed, the reader should be able to deal with the creation of a simulation model based on a theme of interest.

It is normal at this point to feel an uncertainty and concern. Therefore it is necessary to recommend calm. To create a model applying System Dynamics requires some theoretical knowledge and experience that the reader may well have acquired in previous chapters and the perception that it is a craft project that requires a lot of patience.

It is necessary to continue a certain order in this process. The following scheme shows the stages to follow and describes the essential characteristics of each stage. Even though there is a lineal structure, it is possible that on some occasions it will be necessary to repeat some stages in light of any new perceptions that we may have acquired from the theme that we are analysing.

SCHEMA

A – CREATING A CAUSAL LOOP DIAGRAM (CLD)

1. DEFINE THE PROBLEM
2. IDENTIFY THE MOST IMPORTANT ELEMENTS
3. IDENTIFY THE ELEMENTS OF A SECONDARY IMPORTANCE
4. IDENTIFY THE ELEMENTS OF A TERTIARY IMPORTANCE
5. DEFINE THE RELATIONS
6. IDENTIFY THE FEEDBACKS
7. ELIMINATE THE IRRELEVANT ELEMENTS
8. IMAGINE POSSIBLE SOLUTIONS TO SOLVE THE PROBLEM

B- CREATING A STOCK AND FLOW DIAGRAM (SFD)

9. CHARACTERISE THE ELEMENTS
10. WRITE THE EQUATIONS
11. ASSIGN VALUES TO THE PARAMETERS
12. CREATE A FIRST VERSION OF THE MODEL
13. STABILIZE THE MODEL
14. IDENTIFY THE KEY ELEMENTS
15. SIMULATE

C- WRITING THE CONCLUSIONS

4.1. Creating a Causal Loop Diagram (CLD)

There are certain reputed authors who reject the need to create a causal diagram now that even the slightest complicated system has a number of curves that make it impossible to analyse its behaviour and therefore do not the attempt to estimate or find solutions that have a minimal guarantee of success. They conclude arguing that the flow diagram is much more explicit and useful as it shows the curves that exist in the system with clarity, which definitively are the regulated elements over which we should have control.

Even though these arguments are understandable, we cannot forget some of the undoubted value of the causal diagram. The most important of these values is that it is a simple method to order ideas that can seem unclear at the beginning of the study. Secondly, seen as a simple preliminary stage, it allows the (creator) to easily pass the elements and relations from the system to the flow diagram. Thirdly, it allows a clear and fluent communication with the final user, something of which the flow diagrams are not capable.

1. Define the problem

Without a doubt this is the key stage in the project.

Frequently the customer doesn't explain clearly to the consultant who is going to create the model, the final purpose of the study. It is necessary to press as much as possible for a precise definition of the problem that we have to analyse. If possible, it is a good idea to have the definition in writing signed by the costumer. We need to concentrate our efforts in one direction. If this is not the right direction or has to be modified, our efforts will have been, in best case scenario, useless and frequently frustrating.

It is convenient to define the problem in terms that we can clearly appreciate when it improves or deteriorates. These are valid quantitative definitions (customer waiting minutes) and qualitative definitions (fear of flying), but we need to make an effort to put the definition in concrete terms, an effort that will be soon rewarded.

It is not at all useful to define the problem in terms similar to these – 'the problem is poor decision making due to a lack of communication between the retailers as a result of the divergence in the methodological aspects and the consequences that derive from the development and improvement of the client's knowledge.'

It is advisable to describe the problem in the centre of an A4 white sheet of paper.

2. Identify the most important elements

In this stage it is necessary to write the names of all the elements that we think have an influence in the problem. Again, they could be qualitative or quantitative elements but we always need to be able to evaluate any improvement or increase or a decrease or deterioration.

It is very convenient to compile information about scientific or technical studies that guarantee this causal relation or an expert opinion on the matter we have to tackle. In this stage it is not necessary to worry about the magnitude of this relation or how we are going to quantify or shape it.

The names of these elements should be written in brainstorming style around the problem in the centre of the page.

3. Identify the elements of a secondary importance

Once we have localised the elements that directly influence the state of the problem in the terms that we have defined, it is necessary to identify the elements of a secondary importance that also have a secondary influence.

These elements are not directly related with the problem but they do condition elements that have a decisive influence. Therefore we need to bear in mind the state and evolution of these elements. The names of these elements should be written around the elements of primary importance.

4. Identify the elements of a tertiary importance

It is necessary to repeat the previous process with new elements that have an influence to a lesser degree and so on. This process can be repeated as many times as necessary.

Definitively, we will take up again the definition of the System to build a model formed by all the interrelated elements so that if we modify the state of one of the elements, it will result in the modification of another element.

The inevitable question is to know when to stop. It is said that God is the ultimate reason behind all things and that we must reach Him using the number 7. In reality, it is not necessary to reach the seventh order of the causal relation but simply to try to limit the quantity of elements to the size of the page. In other words, the elements that can't fit on the page do not have a significant influence in the problem that we want to analyse.

5. Define the relations

The following stage consists of drawing the arrows or influences that we believe exist between the elements of the system. If the definition of the elements is correct, it will not be difficult to assign a positive or negative sign to each one of the relations. In the case that it is not possible to clearly establish the sign of the relations, it is necessary to define again the implicated elements.

To understand the meaning of the causal relation and its sign shouldn't represent any great difficulty. However, there are phenomena where the meaning of the causal relation is not so evident, in the way that it could be difficult for an alien to identify whether it is the rain that causes people to open their umbrellas or the opening of the umbrellas that causes the rain to fall.

6. Identify the feedbacks

The curves will give us signs about the possible behaviour of the system and also the possible measures to increase its effects or to minimise them. For this we have to identify the curves that exist and their corresponding signs. From this we can see in the positive curves the motors of change and in the negative curves, the causes of stability in the system.

This is a good moment to identify those relations where there are significant delays, be they material delays or from information, and we will mark them in the diagram so that this aspect creates its own dynamic in the system.

7. Eliminate the irrelevant elements

It is necessary to eliminate from the system those elements initially included but which have since been shown to have a relation that isn't relevant. This could be simply because the effects that these relations have are produced outside the planned time scale of the model.

In some ways the construction of a model is like the action of an accordion in that there are stages of expansion of the model, adding new elements and stages of simplification, eliminating unnecessary elements. It is advisable that the final format is as small as possible.

8. Imagine possible solutions to solve the problem

In view of the causal diagram that we have, with the causal relations identified, the curves with their respective signs, the material and information delays and the elimination of unnecessary elements, we can start to try to identify, if it is possible, some of the behaviour patterns of the systems. If this is possible, we can start to come up with some solutions for the problem.

On many occasions this means the end of the job, so that we have acquired a profound knowledge of the causes that precipitate the problem and we are able to propose solutions based on this knowledge and in the systems own dynamic.

The most effective solutions always come from the modification of the relations between the elements more than an attempt to modify the nature of the elements.

4.2. Creating a Stock And Flow Diagram (SFD)

The creation of the flow diagram takes place directly on the computer screen with the simulation software that we use and should not represent any special difficulty if we already have the causal diagram. In general, it is composed of the same elements although it is usually necessary to add some auxiliary elements.

9. Characterise the elements

Briefly returning to the indications that were given in the chapter 'The Creation of a Model', we can say that it is necessary to firstly identify the stocks of the system and for this we can take a mental photo of the system and those elements that appear in each stock of the system. The variations in these elements are the fluxes. They have to have the same units as well as a temporal component. The rest of the elements are auxiliary Variables.

The fluxes don't usually appear in the causal diagram in an explicit form and should be added in the creation of the flow diagram.

10. Write the equations

In this stage we need to make concrete the relations that exist between the elements. For this we need to use simple arithmetic formulae, to make use of the functions that the software has to offer, or use the tables when it is difficult to establish an equation.

11. Assign values to the parameters

Some elements of the model are constants on the horizon of defined simulation and we should assign them a value. In some instances we have this information and in others we must assign a reasonable value. In these types of models, precision does not usually contribute great advantages. Although we know precisely the value that a constant has had in the past, it is more useful without a doubt, to know whether or not this value will remain a constant in the future. It is possible to calculate with precision the life expectancy of the past however, it would be much more useful to know the trend or the changes that occur in this trend after making changes to the structure of the model.

The equations usually incorporate parameters to which we need to assign a value. It is important to ensure that the value is as explicit and well documented as possible. Unlike the constants that are very visible, the parameters in an equation can not be perceived by the reader or the final user and can decisively influence the behaviour of the model.

12. Create a first version of the model

It is impossible to create a complete model at the first attempt but it is very useful to have a simple model that works. It is necessary to keep creating different versions that incorporate improvement.

13. Stabilise the model

The first versions of the model are usually unstable because we haven't assigned the correct values to some of the variables. It is very useful to have a model that functions with all its variables stable.

14. Identify the key elements

In this stage we have to localise the key elements in the behaviour of the system. These will be the elements upon which we must concentrate to improve the state of the system and solve the problem.

15. Simulate

The generation of proposals must be based on the introduction of modifications in the model that afterwards can be put into practice so that we can select the modifications that produce the best results.

4.3. Template for creating a SFD

Write the problem in five or six words in the center of the sheet. Define it through a concept that may vary over time and indicate the units of measurement for it. If the problem focuses on a non-quantifiable parameter (such as health) use a scale from 0 to 100 to assess the status of the problem at all times.

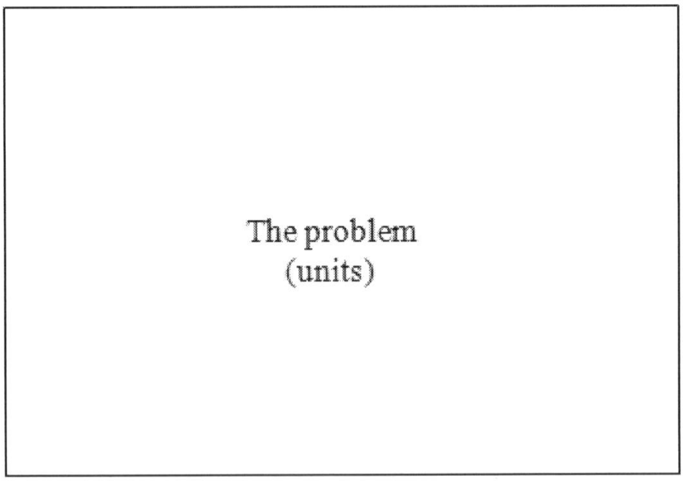

Then write the elements directly or indirectly related to the problem and the relationships that you think exist between them, in a Causal Diagram. This is a first version; it is necessary to select the most relevant factors that influence the problem. Later, you will reconsider if it is necessary to include new elements or delete some. At this point you can use Vensim PLE software (free for personal use at vensim.com)

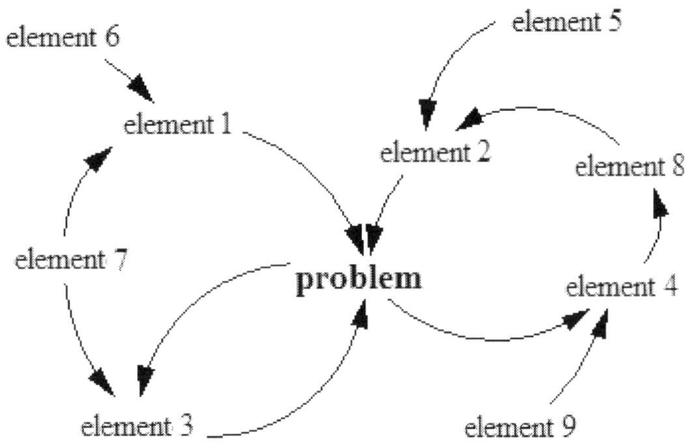

Indicate the loops and the signs of these loops, as they can give you the clue to the current behavior of the system and allow you to manipulate it according to your purpose. So you may want to promote some positive loops to quickly change the current situation, or you might want to activate negative loops to stabilize the current behavior of the system.

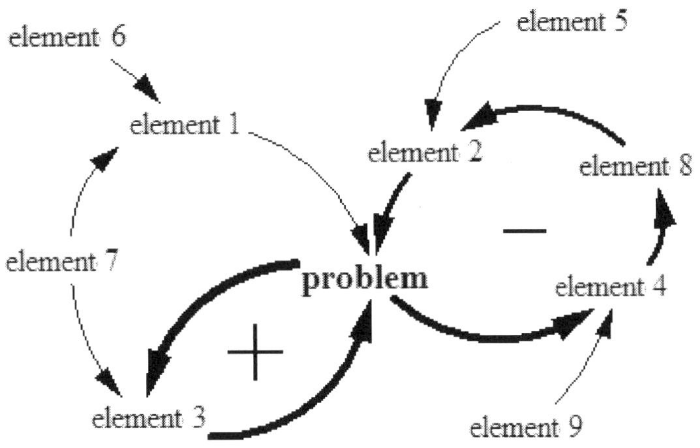

It is possible that at this point you can offer some conclusions about the actions that should be taken to manipulate the system according to your purposes. If so, do so and do not continue with the following steps. If the analysis based on the Causal Diagram still does not allow you to draw conclusions then you must construct the Stock and Flow Diagram. This step may help you rethink some relationships defined in the Causal Loop Diagram.

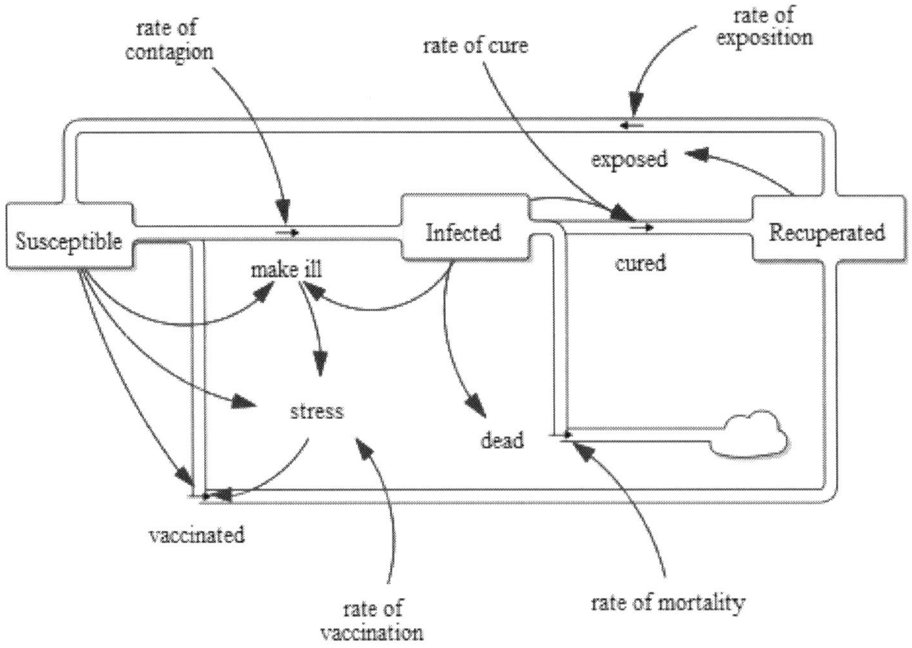

The next step is to define mathematically or with the help of tables the relationships between the elements. It is convenient to define the variables reproducing a stable situation of the system in order to be able to see more easily the impact of possible actions that we wish to study.

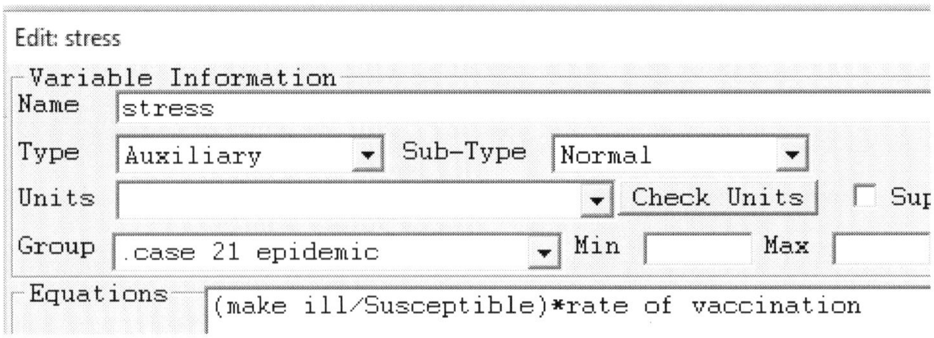

Finally, it is necessary to show the conclusions to our client, and for this the best thing is to use the Causal Diagram and the graphs with the simulations obtained with the model, without showing the Flow Diagram or the equations if they are not expressly requested.

4.4. Writing the conclusions

The final stage consists of the elaboration of the conclusions once we consider that we have completed the simulation process. These conclusions have to be precise, clearly indicating the proposals. We can also include a causal diagram which doesn't necessarily need to be the complete model. It can be a simplified version that must be put in an annex.

It is better to avoid titles for the conclusions like 'The construction of a simulation model for the study…' and instead, use titles like 'The study of the problem…' because the final user is not interested in whether we have created a simulation model, a spreadsheet or an IT programme.

In short, the final user wants to understand the proposals that we can offer and we have to convince the user about its advantages. If the model helps us with this last matter we will use it but in general, to explain the model requires the user to make an effort to be very attentive, something that is not usually popular.

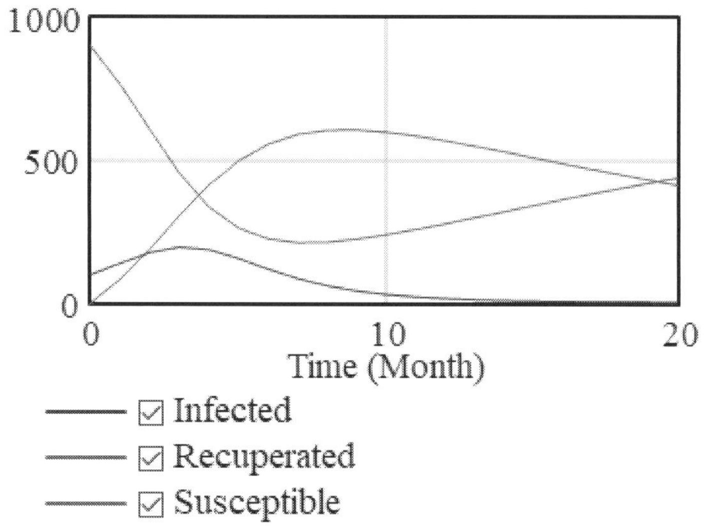

5. CONCLUSION

To build a simulation model you will need The Force with you. The road is long, dangerous, but above all… exciting. Don't be afraid.

If you have acquired the Knowledge from this book, as I hope, put it to good use. Be patient and generous and always act in good faith. In this way, you will be a Knight of the XXIst century.

ANNEX

I. History and basic concepts

Systems

Definitions:

a) 'A system is a set of interconnected elements (Von Bertalanffy, 1968)
b) 'A system can be defined as any set of variables selected by the observer from the whole found in a *real machine*' (Ashby, 1952)
c) 'So far, it will be enough to think of a system as a set of physical objects in a limited part of space that remain identifiable as a group throughout a significant amount of time' (Bergmann, 1957)
d) 'A whole composed of many parts. A set of attributes' (Cherry, 1957)
e) 'A set of parts put together to achieve a common goal' (Forrester)
f) 'A complex unity formed by many different facts subject to a common plan or aiming at the same purpose' (Webster's new international dictionary)

The existence of a common goal as one main feature of a system should not hide the fact that within the same system there may be conflicts of interest. For example, we might think that in a football match, each player tries to minimize effort in order to avoid exhaustion and injuries, thus taking advantage of his teammates. Overall, however, they act with the purpose of achieving the same objective, which is a final victory. Hence, the role of the team's coach consists of balancing this apparent conflict of interests so that the team works toward achieving the common goal.

Open systems

The analytic-reductionist approach aims at reducing a system to its elements in order to study them, and the systemic approach combines the elements to better understand the types of interactions between them. Bertalanffy knew that many systems, by nature, were not closed. For example, if we separate a living organism from its normal environment, it is likely to die due to an absence of oxygen, water and food. In fact, organisms are open systems that can not survive without constantly exchanging matter and energy with their environments.

In the late 1920s, Bertalanffy wrote:

'As the basic characteristic of a living being is its organization, a normal evaluation of its parts and isolated processes cannot give us a plausible explanation of vital phenomena. This evaluation does not give us information about the coordination between parts and process. Thus, the elemental task of biology should be to discover the laws of biological systems (at all levels of organization). We think that the attempts to find a basis for theoretical biology point to an essential change in the conception of the world. We will call this conception, which

is considered a research method, organismic biology, and since it attempts to be explanatory, a theory for organism systems'

The systemic program was the starting point of the General System Theory, replacing the term *organism* with *organized entities* such as social groups, personalities or technological devices.

As stated by Bertalanffy (1942):

'In certain conditions, open systems are near an independent state of time called uniform state'

This uniform state is characterized by a relatively high order, represented by the existence of marked differences in the components of a system.

General System Theory

General System Theory (GST) was developed in 1940 by the biologist Ludwig von Bertalanffy. When formulated, this theory focused neither on regulation phenomena nor informational process. However, it was closer to scientific consensus than cybernetics. In fact, Bertalanffy was particularly interested in open systems.

The systemic approach is devoted to the study of the different interactions between the elements of a system and their environment. Common relationships are given in a variety of systems describing a distinctive nature. This leads to the creation of general systems. Then, it is plausible to consider a general system as a type of particular system with the same relational structure, so that any of them can function as a model of the others. This creates the need to build different theories for different systems depending on the formal context where the author is conducting his research. All the same, it is also reasonable to create a General System Theory in order to describe the specific features found in any system - not specific content but a formal mathematic theory.

A General System Theory which is successfully applicable to both real and imaginary systems should be capable of describing systems with any number of discrete or continuous variables. Thus, as Mesarovic stated, *'a system is any subset of a cartesian generalized product'*.

The importance of the interactions in a systemic approach will serve the purpose of distinguishing between input variables generated by the environment and output variables generated by the system itself. In turn, we will have to take into consideration the temporal transition in complex systems with different inner states, whether in deterministic or probabilistic processes. In cases of high systemic interest, the output in a system reacts over the input by using a feedback loop causing a non-linear process. Consequently, processes derived from regulation and balance, usually given in open systems both live and electronic, are of interest to the General System Theory.

System Dynamics

System Dynamics is a methodology used to understand change over time through equations in finite differences or differential equations. Once the representation of these processes is modeled, we can study the dynamic of the total number of available states in a system. System Dynamics has its origin in the 1930s, when the servomechanism theory emerged. Servomechanisms are instruments which produce feedback from the output level to the input level.

In the 1950s, Jay Forrester developed Industrial Dynamics at MIT (Massachusetts Institute of Technology), assisted by the initial advances of computerized systems. By putting together the systemic approach and computer simulation, Forrester's efforts resulted in better solutions to problems in the industrial area. In the beginning of the 50s, he published Industrial Dynamics (1961).

In the 1960s, these tools were also used in social contexts. Later this decade, an array of books was published: *Principles of Systems* (1968), *Urban Dynamics* (1969), *World Dynamics* (1971) for the Club of Roma, *Counterintuitive Behavior of Social Systems* (1971) and *The Life Cycle of Economic Development* (1973) are only examples. A special mention must be given to the book by D.L. Meadows, *Dynamics of Growth in a Finite World* (1972), as it was published one year before the energy and raw materials crisis in 1973 and partially predicted the consequences of this crisis.

Since then, applications of System Dynamics have extended to several different areas, such as ecology, for which it is an effective resource for studying complex phenomena in nature. In recent years, we have seen an increasing public interest in this application, and various articles have been published citing the feedback among a variety of elements within an environment: its complexity, the existence of leverage-points, no return points, etc.

The study of social and economic processes based on mental models is complex and must take into account the existence of multiple links of feedback. Extrapolation of System Dynamics to this field is very difficult. However, it is accurate to say that this methodology fills the gap which exists between instruments used to analyze these processes.

The application of Systems to social economy is based on the fact that in this area it is also possible to define systems integrated by elements which are stably interconnected. These elements obey a group of rules, such as being connected by logic, market, demography, etc.

In 1977, Jaime Terceiro made the following comment-

> *The main disadvantage in dynamic programming, except for relatively simple cases in which is possible to obtain an analytic relation to the recurrent relation, is that it requires a memory capacity that surpasses the practical limits of modern computers*

Although this assertion was true in 1977, nowadays, the amazing evolution of software and hardware in recent years allows us to create these types of models with just a PC and some basic computer skills. As a result, the door has been opened for a generalized application of this methodology.

As this tool gains popularity and is applied to more fields, those who frequently use it can, to some extent, adapt it according to their needs.

Feedback

Cybernetics introduces the idea of circularity through the concept of feedback, breaking away from Newtonian classical science, where the effects are bound linearly.

The use of this concept might explain the evolution of social systems where two different types of feedback are found.

Cybernetics and Social Sciences

Norbert Wiener, the father of cybernetics, was a firm defender of cybernetics as a way to approach social sciences and society. Wiener was also convinced that any type of behaviour can be explained by the principles of cybernetics such as communication, control of entropy through learning by using feedback loops, etc (cf. *The Human Use of Human Beings* and *Cybernetics*. Also, *The study of control and communication in the animal and machine*)

Cyberneticians studied nervous systems with the purpose of understanding the human condition. They concluded that observations independent from the observer are not possible. For example, when a person writes, in any language, he uses a structure in his nervous system which is the result of a history of previous interactions with languages since he was a child.

Apart from the obvious disadvantages of the researcher's subjective influence, it can sometimes act as a catalyst for change processes. Some aspects of psychology such as family therapy trace their origins back to cybernetics (Watzlawick, 1967), considering that unusual behaviors may be the result of interactions within the family. As Watzlawick states-

> *We basically affirm that interpersonal systems - groups of strangers, couples, families, psychotherapeutic or even international relations - can be understood as feedback circuits, since a person's behavior affects and is affected by others.*

In the following pages an approach which can be used to interpret reality will be described. The *correct* or *the best* way to observe reality does not exist, given that it is impossible to define a specific path as the best or most accurate.

Cybernetics

The word cybernetics originates from the Greek word 'Kybernetes' which appeared for the first time in the writings of Plato and later, in the 19th century, André Marie Ampere used the word to refer to different forms of government.

By 1943, a group of scientists, led by the mathematician Nobert Wiener, had recognized the need to choose an appropriate word to describe this array of theories and concepts. In 1947, the group adopted the word 'Cybernetics' which gained popularity in Weiner's book, '*Cybernetics, or the Study of Control and Communication in the Animal and Machine'* (1948). Since then, the discipline has received an increasing degree of interest. Cybernetics has expanded both as an interdisciplinary science which aims to study control or self-control (Wiener), and also as science which seeks the efficacy of action (Couffignal).

State of Space

Cybernetics addresses the difference between the presence and absence of specific features, also referred to as dimensions or attributes. For example, a system called billiard ball may have features such as a particular colour, weight, position or speed. The presence or absence of each feature can be represented with Boolean variables that take on two values: 'yes' when the system has the feature, or 'no' when the system does not have the feature.

The binary representation can be generalized to a subsequent feature with multiple, discrete or continuous values. The set of all possible states of a system is referred to as State of Space. An essential component of cybernetic modeling is the quantitative measure of State of Space's dimension, or the number of different states. This measure is called variety. The variety is defined by the number of elements in a Space of State: V=log ([S])

Entropy

The concept of entropy was introduced by Clausius during the 19th century. It was intended to act as a measure of disorder in gas molecules in order to balance thermodynamic accountability.

Statistically, disorder exists because of the number of distinct states a system is capable of adopting. If one compares two systems, system A will have more disorder than system B whenever the number of distinct states found in A is higher than in B.

In a closed system, entropy increases according to Clausius equation-

$$dS>0$$

whereas in an open system the total change of entropy can be expressed as specified by Prigogine-

$$dS=dSi+dSe$$

where dSe denotes the change of entropy due to importation that can be either negative or positive, and dSi represents the production of entropy due to irreversible processes, which is always positive.

FUNCTIONS

A function FUNCTION (#, A,B,C...) shows the relationship that exists between the elements.

In this way, the function Y=2 X indicates that Y will always have a value twice that of X, without any other restriction.
Example: Each chair costs 100 €, I want to know how much it will cost to buy 1,2,3,4... chairs. The formula would be:

Total Cost = Cost of each chair x the number of chairs
Or Total Cost = 100 x number of chairs

The software Vensim uses the points '.' to signal the decimal points and ',' to separate the elements of the formula. More information can be found typing the key F1 (Search Index for a Topic)

ABS(A)

ABS calculate the absolute value of A, this means the positive value of a figure. ABS (5.00) is equal to 5.00 and ABS (-5.00) is equal 5.00. This acts as the function IF THEN ELSE (X<0, -X, X) in the way that if X is negative it changes its sign, therefore the result is always positive.

Example: In the door of a factory there is a sensor that counts the number of people who pass through the door. It counts + 1 if a person enters and -1 if a person leaves. We want to know the total number of people that have passed through the door regardless of whether they are entering or leaving. One possible way to do this is:

Total = Entrances + ABS (Exits)

EXP (X)

EXP calculate e (2.718...) elevated to X

Example: This expression is used in some system models. For example e^2 is equal to 7,3875

IF THEN ELSE (Cond,X,Y)

The result is X if the condition is met, if the condition is not met the result is Y.
Example: When the monthly orders are more than 100 I want the price applied in the monthly bill to be 30, and if not I want the applied price to be 50.

Price = IF THEN ELSE (Orders>100,30,50)

INTEGER OF X

Give as a result the whole part of the value X if it has decimals.
Example: a cash dispenser can only pay euro notes, not coins. If the client types that they want to receive a factionary amount like 25.5, the cash dispenser programme eliminates the decimals:

Amount to pay = INTEGER (25.5)

And the result would be 25 € (remember that '.' indicates decimal).

LN(X)

Calculate the natural logarithm of X
Example: In some physical systems this expression is used. For example we have that the value of ln (50) is equal to 3,912.

MAX (A,B)

Calculate the maximum of A and B. The result will be B when B>A, and will be equal to A when B<A
Example: In a race between two cars A and B, we always want to know the velocity of the car that completes the circuit faster.

Velocity of the fastest car = MAX(Velocity of A, Velocity of B)

When A goes at 100 and B goes at 80, the velocity of the fastest will be 100, and when A goes at 100 and B goes at 120, the velocity of the fastest will be 120.

MIN(A,B)

Calculate the minimum of A and B. The result will be A when B>A, and it will be equal to B when B<A
Example: In a race between two cars A and B we always want to know the velocity of the car that circulates slower.

Velocity of the slower car = MIN (Velocity of A, Velocity of B)

When A goes at 100 and B goes at 80, the velocity of the slower will be 80 and when A goes at 100 and B goes at 120, the velocity of the slowest will be 100.

PULSE (A,B)

This function is worth 1 after the period A during B periods. Before and after it is worth 0. For example, a trader opens at 8 o'clock in the morning and stays open for 12 hours without interruption. Equation: State of the trader = PULSE (8,12) and the State is worth 1 when it is open and 0 when it is closed.

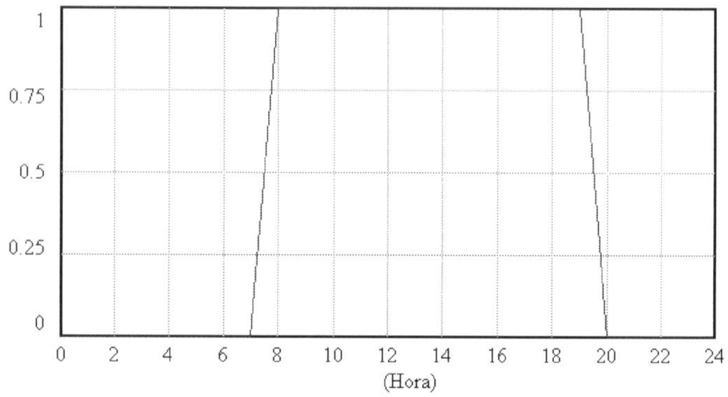

PULSE TRAIN (A,B,C,D)

Equal to the function PULSE but starting in the period A, with a duration of B periods, that repeats every C periods and letting it repeat itself after period D.
Example: A patient has to subject himself to daily medical treatment sessions. He starts everyday at 4.00 in the morning, the treatment lasts 2 hours and he has to start a session every 6 hours. At 18.00 he can't start a new session. The variable Treatment will be worth 1 in the periods where the patient has a session of treatment and 0 if he doesn't have it.

Treatment = PULSE TRAIN (4,2,6,18)

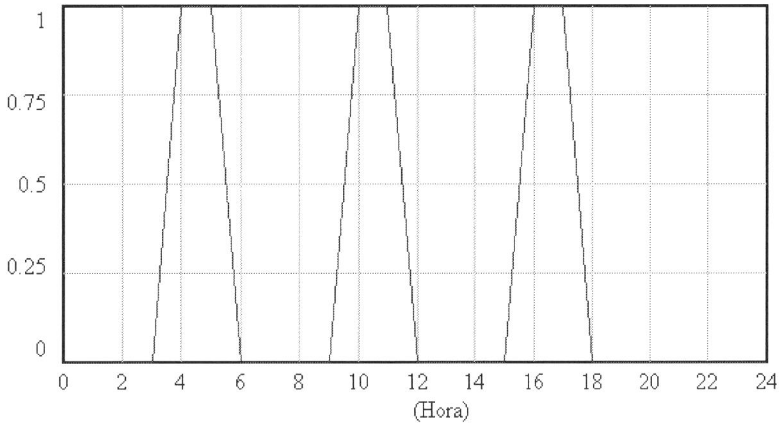

RAMP (S, T1,T2)

Value 0 until the period T1, from this instant increases S units every period until the period T2, and later remains constant.
Example: We want to open the floodgates to a reservoir in a progressive way from 2.00 in the morning until 12.00 the next day. The value of the floodgate is 0 if it is closed and 100 if it is completely open.

Floodgates = RAMP (10,2,12)

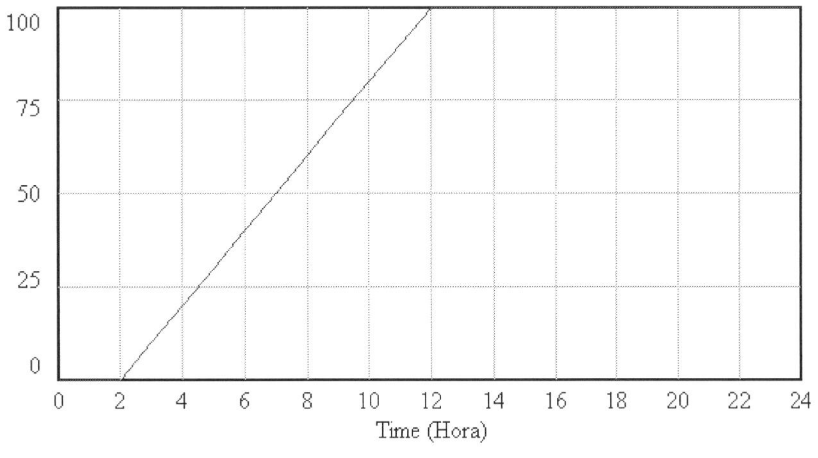

RANDOM UNIFORM (m,x,s)

Return a series of random values with a minimum of 'm' and a maximum of 'x', 's' is the parameter of the calculation of the random numbers, and can be any figure. If 's' is changed the series of random number is modified.
Example: We want to simulate the behaviour of body temperature that we know oscillates between 36 and 38 degrees during the day.

Temperature = RANDOM UNIFORM (36,38,99)

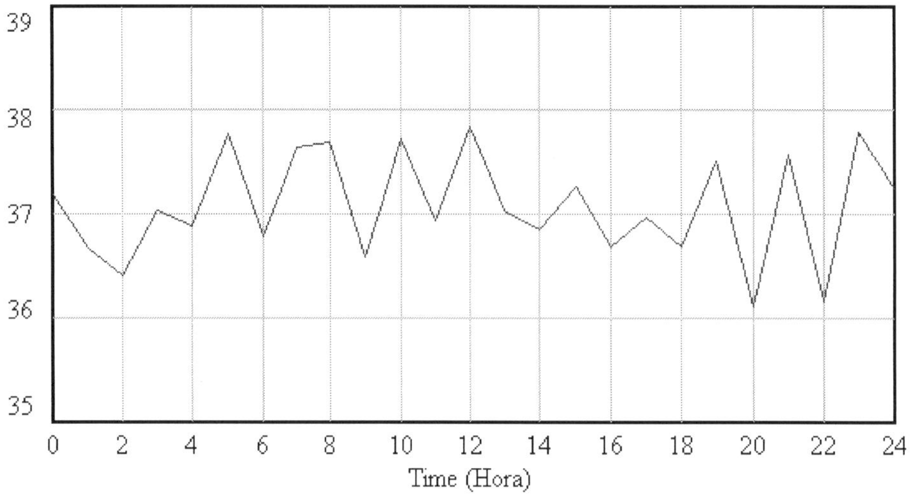

RANDOM NORMAL (m,n,p,x,s)

Simulate the function RANDOM NORMAL, return a series of random values with a minimum value of 36º, a maximum value 38º, an average of 36,5º and a standard deviation equal to 0,5º. We will use:

Temperature = RANDOM NORMAL(36, 38, 36.5,0.5,99)

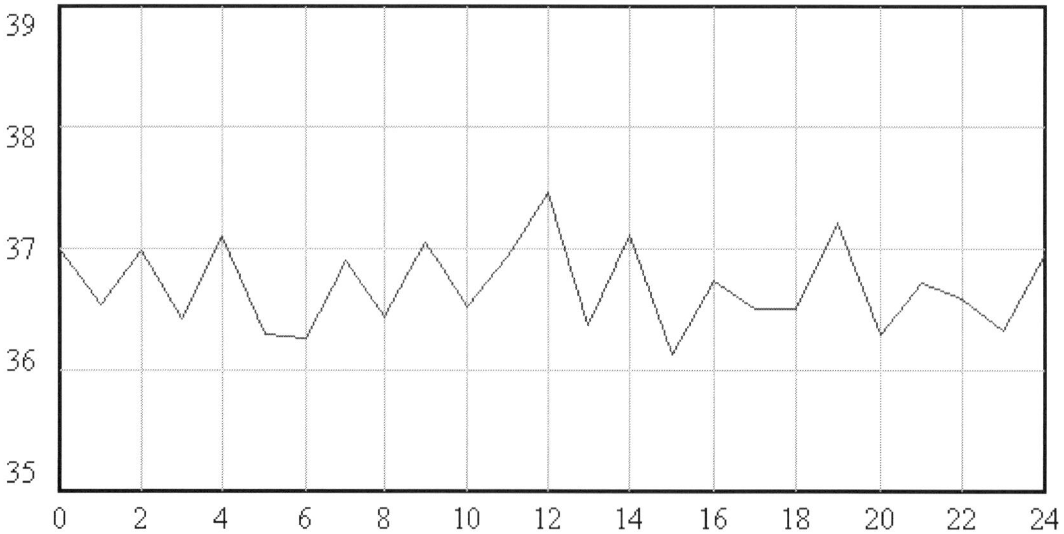

SIN(X)

Calculate the sine of X in radians.

SQRT(X)

Calculate the square root of X.

STEP(H,T)

The result is 0 until the moment T from then onwards it is H
Example: A variable values 0 until period 15 and after this stage it is worth 5.
Variable = Step(5,15)

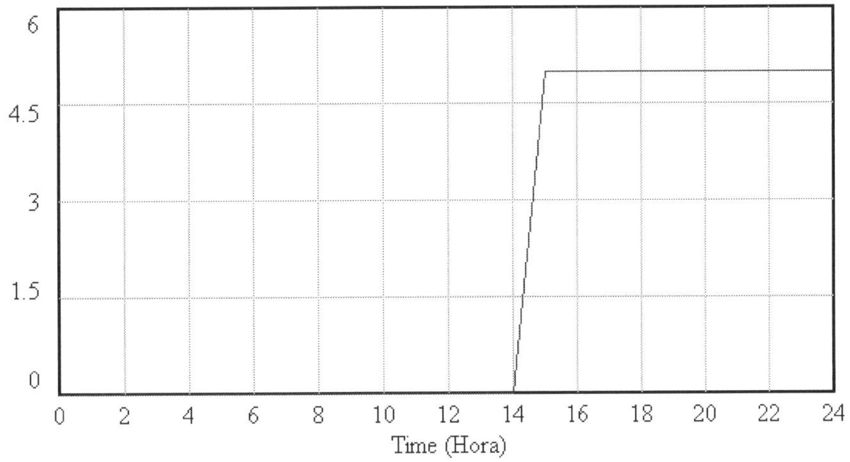

XIDZ (A,B,X)

The result is A/B, except when B = 0 then the result is X. This is used when we have to make the division A/B and at any moment B can be zero. This would give the quotient an infinite value and would cause the collapse of the model. In this case, if B is equal to 0, the variable takes the value X.

TABLES

Let us create a micro-model of table in order to see what a table is and how they are created. Suppose that we have an element 'b' and we want it to depend on another element 'a', in other words, element 'a' influences element 'b'. The relationship between the two elements is complicated and cannot be defined with an arithmetic relation (of the type (b = 4a+2). In these cases, tables are used.

'a' is for example the quantity of rain that falls and 'b' is the percentage of people with umbrellas. We don't know the arithmetic relation between both variables but we know that when there is more rain, there is a higher percentage of people with umbrellas. In a previous study we have found the following results:

-When it doesn't rain the percentage of people with umbrellas is nil (a=0, b=0)
-When it has rained 30 litres of water, the percentage is 10% (a=30, b=10)
-When it has rained 50 litres of water, the percentage is 100% (a=50, b=100)
If we draw on an axis X,Y, we would have the following values:
Point 1: (0,0) Point 2: (30,10) Point 3: (50,100)

Let us create a small model that can calculate the percentage of people who carry umbrellas depending on the quantity of rain that falls. It is necessary to follow these steps:
-Open Vensim and click on File, New Model, OK
-Create an auxiliary variable called 'rain', another called 'umbrellas' and another called 'table'
-Draw an arrow from the rain to the umbrellas and another from the table to the umbrellas.

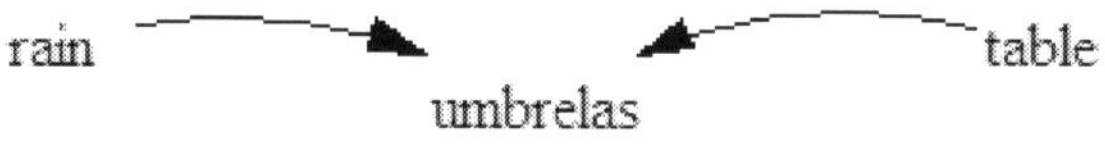

- Write the equation rain = RAMP(1,1,50) with this we simulate that the rain has increased 1 litre every period from period 1 to period 50.
- Write the equation umbrellas = table (rain)

To write the table equation do this:
- Click the name of the table
- Choose the option Type: Lookup
- Click on the icon AsGraph, then click the icon 

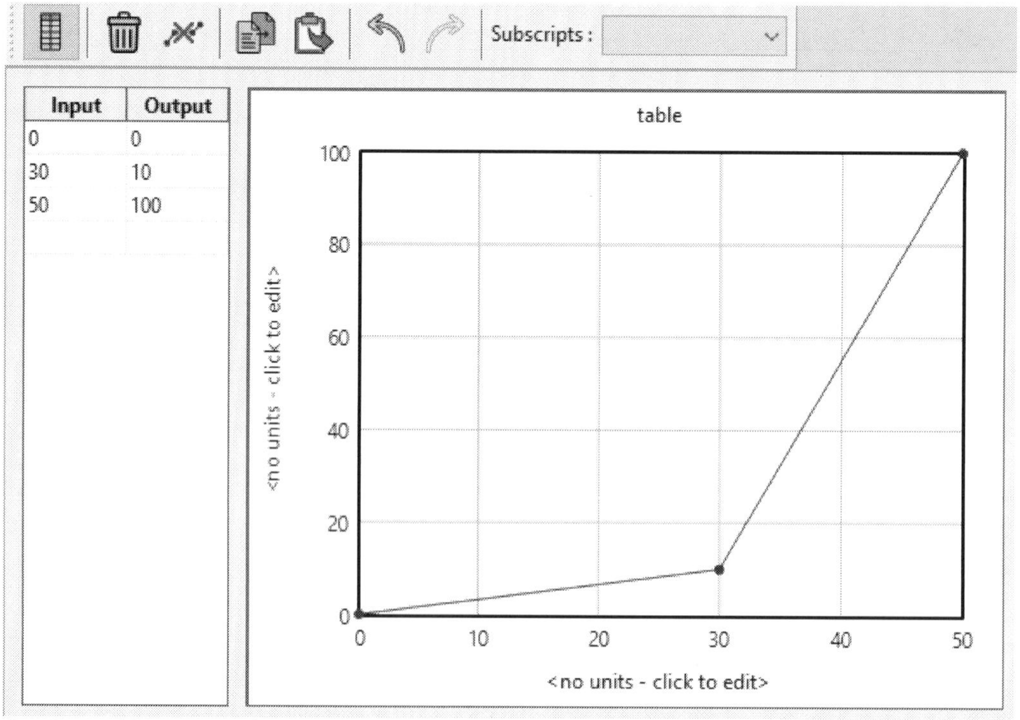

- In the Input/Output columns enter the three rows of pairs of values: (0.0), (30,10) and (50,100)
- Click OK and OK again.
- Click the icon Simulate 

- Look at the behaviour of the rain and the umbrellas

In this case we have created a table to obtain the result of the dependent variable (umbrellas) based on the values that take the independent variable (rain)

DELAYS

In the systems we frequently find that the responses of a variable in relation to another are not instantaneous but are often delayed. These variables can be shaped depending on whether they are information delays or material delays. It is also possible to shape the variables considering that the response is strong at the beginning (first order) or the response presents a significant delay (third order).

INFORMATION DELAYS (used in causal links)

DELAY 1 (I,T) Exponential delay of the first order for the variable I and the period T

DELAY 1 I(I,T,N) The same as the DELAY 1 but starting the simulation in the value N instead of I

SMOOTH 3 (X,T) Exponential delay of the third order for the value X and in the period T

SMOOTH 3I (X,T,N) The same as SMOOTH 3 but starting with the simulation in the value of N instead of X.

DELAY FIXED (X,T,N) Delay the X value at the T period starting the simulation in N instead of X

MATERIAL DELAYS (used in flows)

SMOOTH (X,T) Exponential delay of the first order for the variable X and the period T

SMOOTH (X,T,N) The same as SMOOTH but starting the simulation in the value N instead of X

DELAY3 (I,T) Exponential delay of the third order for the value I and the period T

DELAY 3I (I,T,N) The same as DELAY 3 starting the simulation in the value N instead of I.

To check the effects of the different delays, it is useful to create a very simple model. Place the different delays that we want to compare.
Example: Imagine that we want to simulate a relation between the moment that it starts to rain (minute 10) and the percentage of people that carry umbrellas. Wc scc the small model and the equations with a delay of the third order.

rain = STEP (100,10)
umbrellas = DELAY3 (rain, delay)
delay = 8

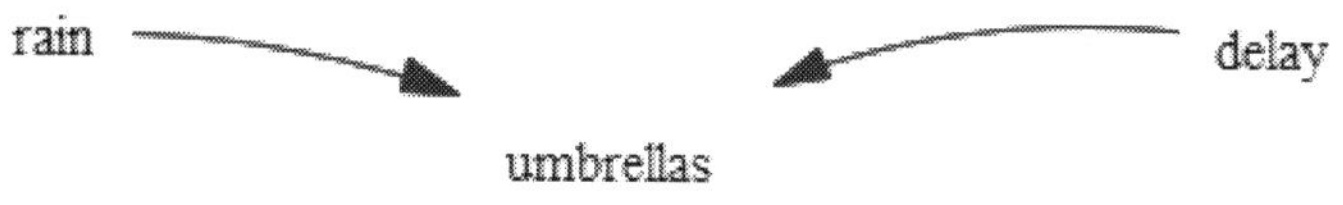

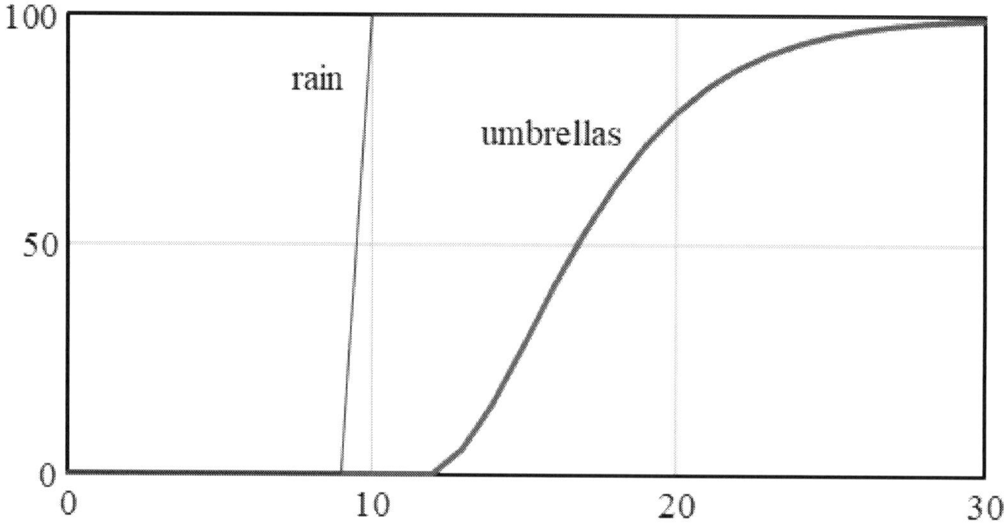

And now using different delay times for a function of the first order (DELAY1).

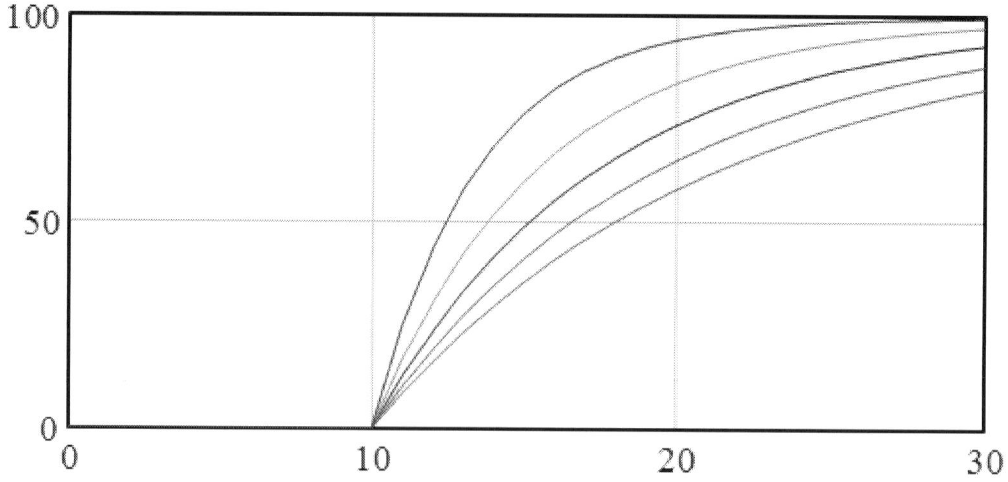

III. Frequently Asked Questions

How does a function with a temporal delay work?

We will see the practical functioning of a function with a temporal delay like SMOOTH using a very simple model.

We create the following model:

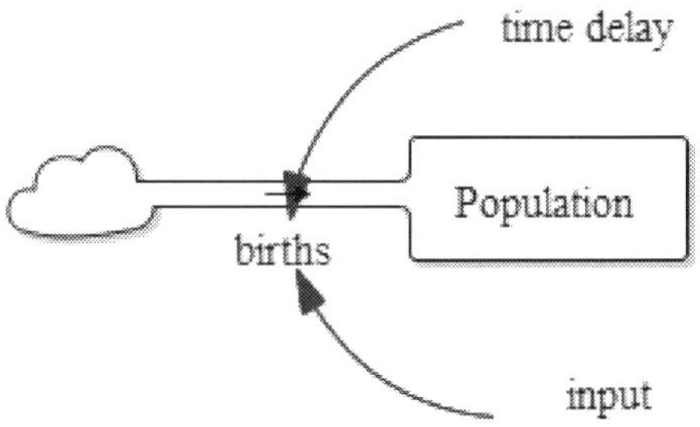

Births = SMOOTH (input, time delay)

input = step (100,10)

time delay= 25

The variable 'input' has a value of 0 until the period 10, in this period it takes a value of 100 and maintains this value indefinitely. The variable 'Births' takes the same value as the 'input' with a temporal delay value of 25. We can see the result of the model in the following table (left image). In a spreadsheet (right image) we can calculate the same values and see the formula that it produces.

Time	Births
0	0
1	0
2	0
3	0
4	0
5	0
6	0
7	0
8	0
9	0
10	0
11	4
12	7.84
13	11.5264
14	15.0653
15	18.4627
16	21.7242
17	24.8553
18	27.861
19	30.7466
20	33.5167

C18 = =+C17+(B17-C17)/25

	A	B	C
1			
2			
3	Time	input	**calc**
4	0	0	0
5	1	0	0
6	2	0	0
7	3	0	0
8	4	0	0
9	5	0	0
10	6	0	0
11	7	0	0
12	8	0	0
13	9	0	0
14	10	100	0
15	11	100	4,0000
16	12	100	7,8400
17	13	100	11,5264
18	14	100	15,0653
19	15	100	18,4627
20	16	100	21,7242
21	17	100	24,8553
22	18	100	27,8610
23	19	100	30,7466
24	20	100	33,5167

What is the difference between a process of Addiction and Shifting the burden?

This question is interesting because of the significance it implies. In both situations the system manages to equalise the Real State with the Desired State with external help.

We talk about Addiction when an object - a thing - intervenes and by Shifting the burden when another system intervenes with its own objectives.

The consequences of this detail are important because the object of an Addiction would never leave us and therefore we don't have to expect any change if we don't want one. On the contrary, the system that supports our charge today can decide tomorrow to stop supporting us and provoke a crisis.

For example, we can be addicted to tobacco and in this case, if we manage to reduce our stress with this practice we can be sure that we will always be able to do it as long as we don't decide to give up smoking. On the contrary, if we have passed the charge of our low incomes to our father, it's possible that one unexpected day the subject of our charge decides that he has already been patient with us and he stops helping us.

Are these forecast models?

Forecast models are those that given some initial conditions, we are interested in finding out about the state of the system after some time, with the characteristic that we can not intervene in a significant way. The most well know models of this type are the weather forecast models. To work with these models it is essential to have as much data as possible about the initial situation. System Dynamics is not usually used to make predictions as firstly, we can and want to manipulate the system and secondly, in general we don't have much data about the situation.

With the data available to us we will see the state of the system and study the different alternatives that can improve it based on what we want to achieve. Using this system, we are able to foresee the consequences of our actions on the model. This is done to select the most efficient action as the system is not left to evolve freely.

We could also use this model to foresee what would happen if we did nothing, however, in general, this type of forecast is not very precise because of a lack of data. This lack of precision does not stop us comparing different alternatives of behaviour in the system and to make a classification of the results from best to worst.

When is a delay of the first order and when is it of the third order?

We will consider that a variable has a delay of the first order when it reacts quickly to an impulse. For example, there is a delay that exists between the time that a switch is flicked on and the moment the room lights up. It happens very quickly but the delay exists. The most important factor is that the light bulb gives 90% of its light in the instant the switch is flicked and that the following 10% follows in a few seconds.

A delay of the third order is produced when the response to an impulse is delayed considerably at the time and at the beginning the response is slow. For example, today the price of a product increases and the customers continue to consume the same quantity until they find a substitute product.

Delays have a decisive influence in the behaviour of many systems. For example, let us take air conditioners. If now the temperature in the room is 40º and we switch the thermostat to 25º, at first the machine functions at high performance and in the first five minutes the temperature falls 10º to 30º. In the following five minutes, the temperature falls 4º and afterwards it takes half an hour before the temperature drops another 1º to reach the objective of 25º because the system is working at low performance. This is a system with a delay in the first order. At the beginning it adjusts its state rapidly to the desired state based on the difference that exists between the two.

The same system with a delay of infinite order with an adjusting time of 10 minutes maintains itself at 40º for 10 minutes and afterwards the temperature is reduced suddenly to 15º. The lower the order of the delay, the faster it will start to respond. The higher the order of delay, the longer it will take to respond.

To get a visual image, we can imagine that the delay is a set of stocks that separate the entrance or input of the exit or output. The impulses pass from one stock to another in each period. If the delay is of the first order, there is only one stock between the entrance and the exit, if the delay is of the third order there are three stocks between the exits.

What is the duplication period of a variable?

Let us suppose that we are making a model of the evolution of the balance of a current account with a fixed interest. In other words, there is a stock that is the balance in the account, a flux that is the interest and an auxiliary variable that acts as a fixed interest. The flux is calculated as the balance for the type of interest. We want to know how many years are necessary to duplicate the interest.

We know that the duplication period of the balance is equal to 0.7 / i, and i is the type of interest. How is this demonstrated?

We have $(1+i)^t=2$, in other words one unit more than the interest in t years has to be equal to 2, being t the period of duplication, t is the number of years it takes for the capital of 1 to turn into 2.

Also, $\ln(1+i)^t = \ln 2$ applying logarithms

And therefore, $t \ln (1+i) = \ln 2$

We isolate the t that will be the duplication time $t = \ln 2 / \ln(1+i)$

And we have $\ln 2=0.699$ and that $\ln (1+i)$ is always very close to equal to i and from there we have approx $t=0.7/i$

What is the difference between the limiting factors and the key factors?

The key factors are elements of the system that are very sensitive. They are always the same. Anyone would be sensitive if somebody put a finger in their eye and would probably react with violence. But in reality we have two eyes and it would not be life threatening to lose one of them.

Each system has its own key factors and in order to discover them we need to invest a certain amount of time and effort. It is important to understand these key factors if we want to manipulate the system without altering any factors that could provoke a violent response. On the other hand, we need to try to take advantage of the factors that produce a positive reaction in the system. It is very important to remember that in general they are hidden and that they are always the same.

The limiting factors, on the other hand, are usually very visible and they usually change with time. They are the elements which will condition the state of the system now or in the immediate future. However, tomorrow they could be other different elements. For example, I'm hungry so I don't work - I go to have something to eat. Once I have eaten, the limiting factor is that I don't have paper, so I go to get paper. When I have paper, I don't have any ideas. In other words the limiting factors are constantly changing.

Which time interval should I calculate?

Often in a simulation we want to demonstrate the results of the simulation in a temporal scale or period. Meanwhile the calculations we want are realised in a shorter period of time. For example, in a variable we want to simulate the temporal evolution of a worker's salary throughout his life that will start when he is 18 and finish when he is 65. The unit of time with which we want to see the results is logically a year. We want the model to produce data monthly, as the worker is paid monthly. In this case we will use the option time step to define the period of calculation 1/12, 0.083333

The software works in binary codes and can't manage a period number with precision therefore we have to be conscious that this error exists. That, in general, will be very small and almost certainly less than those we introduce in some of the constants that we are going to use in this model

For example, is in the previous example we define Time Step as 0.83, by the end of a year we would have 0.083x12=0.996. This implies an annual error of 0.4% and by the end of 10 years the error will be 4%

As long as it is possible we should use potentials of 2 so that we have options for Time Step:1,0.5,0.25,0.125,0.0625… As is logical we have to use in all of the model, units that are coherent with the definitions that we make in Time Step in the way that if it corresponds to one month, the variable have to take this period as a reference (monthly salary, monthly taxes, monthly bills) instead of the instead of the period of time that we will see appearing in the graphs (years).

Which temporal horizon should we define?

This is an essential aspect that requires special attention in each model. We need to be generous in the definition of the time limit of the simulation. Restrictions from the point of view of the hardware or the software don't exist. The existing software executes simulations in just a few seconds.

We need to avoid focusing too much on the temporal horizon that the client or user suggests. Sometimes, certain phenomena can manifest in the model shortly after the temporal horizon chosen, that can also show themselves shortly before inside the horizon that we have chosen.

A wide temporal horizon allows us to have the security that certain phenomena are really what they seem to be, in the way that a system with stable oscillations doesn't start to grow - and are therefore unstable - after a certain determined period.

What practical use does the introduction of NOISE to the model have?

Actually it is almost impossible to observe the natural, business or social processes in a single parameter that evolves in a lineal way over a long period of time. In general, we can see that it follows a determined evolution with frequent small variations.

The cause of these small variations of the variable is owed to the existence of stationary factors that act in an occasional way, from external factors that have modified the state of the system in a determined moment and may also be due to the inevitable mistakes in the measure of the state of the system. If we build a model with the aim of understanding the natural dynamic of the studied system or to perceive better the structure that defines its behaviour, we should not be worried about the factors that modify the state of the system in a small magnitude.

The most important factor is to define if these small variations that we observe are actually of importance or not. If these variations are not important to the model we can omit them. If these variations are seen to be useful, they would require a more detailed investigation.

In the software, we have the function NOISE and it is convenient to have an idea of its possible uses. If we have an historical series and a model that reproduces the average values of the historical series and we add the function NOISE to establish behaviour that is similar to real behaviour, we are obliged to define a certain magnitude for the statistic that defines the dispersion of the values (for example the typical deviation).

The magnitude of this statistic parameter is the way to quantify the random and exact aspects that are unknown. The noise in a system also tells us about its capacity to stabilise itself in front of small disturbances. If the system finds itself dominated by a positive loop, it will enter into a phase of instability as soon as it is altered by small fluctuations coming from a NOISE function. On the other hand, if the structure of the system has negative loops it will be capable of rapidly compensating these fluctuations.

IV. Download all the models

The models explained in this book are available in:

http://atc-innova.com/zip_one.htm

V. Acknowledgements

Without a doubt this book is indebted to those who were my teachers, especially Pere Escorsa for having signalled the importance of this discipline as an instrument of analysis. I am also indebted to some friends, José Alfonso Delgado who convinced me of the need to have books available on this subject and encouraged me to dedicate the effort needed to give form to the book.

Secondly, this book is the product of many years of teaching and therefore **my greatest thanks are to all my students** who, with their continuous questions, have made me reflect on the theoretical concepts and make the examples clearer and simpler.

Those who have collaborated in this book:

- Mario Guido Pérez (Chemical Engineer, Argentina) author of the models of the Chemical Reactor, Ingestion of Toxics, Golden Number and Butterfly Effect.

- Claudio M. Enrique (UNL, Santa Fe, Argentina) author of the model Study of the Oscillatory Movements.

- Gustavo Adolfo Juarez (University de Catamarca, Argentina) for his collaboration in the model Development of an Epidemic.

- José Ignacio Fernandez Mendez (UNAM, Mexico) and Michael Frenchman (Consultant, USA) in the model of the fishery of shrimp in Campeche.

- Josep Maria Banyeres (Engineer, Spain) in the Barays of Angkor model.

- Mohamed Nemiche (Doctor in Physical Sciences; Morocco) in the part of history and basic concepts.

- Antoni Lacasa Ruiz (Artist, Spain) who has offered his experience in the drawings that illustrate and make more agreeable the text.

... and all the translators of the original book, specially to the reviewers Terry Walker and John Lowry.

Last but not least, sincerely thank John Sterman who kindly accepted to write the preface of this book.

VI. Next step: Agent- Based Modeling?

Simulation models founded on the concept of System Dynamics assume all of the elements defined in the system are homogeneous. It is for this reason when using the concept of "young people" the assumption is the model will express the same characteristics and behavior. Similarly, using the concept of "received orders" it is assumed characteristics and behavior are all the same for modeling purposes.

Under circumstances when it is not possible to determine the homogeneity assumption for example, 10 types of young people in terms of their buyer profile and 10 types of products requiring analyzes. Agent-Based Models or ABM are best suited to conduct the analyzes than System Dynamics models.

Agent-Based Model are ideal to simulate actions and interactions of many agents or entities in order to evaluate the effects on the whole system. ABM-type models are used in multi-disciplinary fields such as, business, biology, ecology, and the social sciences just to name a few.

What is an Agent-Based Model?

An ABM is a methodology that enables the creation of simulation models based on identifying individual behavior of the entities (people, animals, things, etc.) that make up a system, and is used to analyze what changes can be introduced to the real system to achieve the desired behavior.

In an ABM individual entity represent the most basic elements of the system. The software simulates how those individual entities interact, and the resulting evolution of the system can be studied both: from the aggregate perspective and with respect to individual behavior.

When using an ABM

Prior to creating a model with the ABM methodology, we have to brainstorm the following options:

1. Often times the problem can be analyzed well with the help of an Excel. If so, without hesitation use Excel.
2. If there's a lot of available historical data and you want to make a forecast, consider using statistical software. For example, we want to make a forecast for the next month of traffic on a highway, of which there are daily historical traffic data for the last five years. No construction or any other specific event is scheduled for next month. In this case, software such as SPSS or Mathlab is best suited for this option.
3. Under the notion the main elements of the system are homogeneous, in this case the ideal tool is to create a System Dynamics model. For example,

you want to establish the limit of tuna catches in a certain area. It can be assumed that all tunas are the same, and the fisheries are classified into 3 types. A software with the capability to create System Dynamics models such as Vensim is the best option.

Discarding the previous options, we already have the certainty that an ABM may be just what we need. An ABM model is used when we have a system composed of elements that, although similar, have characteristics that make them unique and critical to understand the system as a whole to decide policies that must be applied to manage it and achieve the proposed objective.

For example, let's say you want to plan the activities of a taxi fleet, and you cannot assume that they are the same, due to different vehicle brands, different purchase dates and daily accumulated kilometers, etc. Therefore, although they are all similar, the characteristics of the taxies cannot be based on an average value, the same for all, because that average taxi does not exist.

What is ABM useful for?

ABM simulation models are used across multiple industries and fields. The following are a few examples on the usage of ABMs: In the academic field ABMs are useful for research projects, final projects and doctoral theses e.g., since it allows the concepts to be clearly ordered, rigorous calculation made to finally offer convincing results. ABMs are useful as a planning tool for production and managing warehouses. In economics useful to manage multiple public resources, especially investments. The environment to analyze the environmental impact of human activity on natural ecosystems. The multiple disciplines of the social sciences can integrate ABM models as an analysis tool allowing analysis of small groups to groups that encompass millions of people.

Software used

There are several available generic tools which allows you to create ABMs, such as Java, Anylogic, NetLogo, Mathlab, C++ or Python nonetheless, these tools require existing or prior knowledge of the software, in order to be able to build an ABM-type model. To avoid user frustration it is recommended to use Ventity software which is designed specifically for creating ABMs. The advantage of Ventity is prior knowledge in programming is not required. Additionally, Ventity software is free for personal and educational usage.

Previous knowledge

Using Ventity for ABM does not require any programming knowledge, which is a great advantage. Nor are special mathematical knowledge necessary, an average user of Excel can approach this book with complete confidence and peace of mind. The person who has prior programming knowledge will be able to directly manipulate the instructions of the model, since it is ultimately a txt file, and the individual will also be able to find some equivalence with the existing knowledge from programming.

Book: Agent-Based Modeling and Simulation
ISBN: 979-8599111306

Starting the usage of a new methodology always requires effort. This book shows in an orderly fashion the concepts as well as the steps to follow to build an ABM. The step-by-step creation of various models of increasing complexity are shown, in order for the reader to progressively gain mastery of this tool.

The themes of the models are diverse reflecting the multiple applications of this tool. The reason for this diversity of themes is not so much to show possible applications, but instead to make reading the book an enjoyable experience.

Focus of the book

It is assumed that the reader is completely unaware of the ABM models, beyond having an idea of their possible utility. Therefore, the book begins by explaining the most basic concepts and the terms used and then progressively explaining in each of the practical exercises the benefits the software offers. Each exercise brings some news and also helps to consolidate what has been seen up to that moment.

Obviously, a certain effort of concentration is required from the reader, but the content of the book is designed so that the reader can advance at a good pace. The approach is therefore to start from the most basic level and progress until the reader can make their own model applying the ABM methodology. This is the goal of this book, but it is not the end of the road thus, a follow-up book introduces the reader to additional features such as creating entities, defining cohorts, and doing sensitivity analysis, optimization, and calibration.

The book as basic guide

Creating an ABM entails knowledge of the Ventity software, this book offers a detail guide for the creation of ABM simulation models. Your own experiences will enhance and provide the skills to build more complex models applicable to your particular area of work interest.

Videos

Build a model in one minute https://youtu.be/EVOjzILbTIc
Simulate covid-19 in a city https://youtu.be/_fTb_Drxl34
Simulate a production process https://youtu.be/2dgoKxRhFdQ
Simulate multiple agents (airplanes) moving https://youtu.be/IzYgxDgMOfA
Simulate Christmas gifts distribution https://youtu.be/UxHFCPb0oQ8

VII. Recommended books

Includes examples of the most common mistakes in building a simulation model.
ISBN 979-8662618657

A book for Vensim PLE PLUS users
ISBN 979-8655650183

A books to explore other fields…Feedbacks.
From Causal Diagrams to System Thinking
ISBN: 978-1790616015

VIII. Recommended papers

Collection of books
Selected papers on System Dynamics

Modeling and Simulating Business Dynamics

ISBN: 9781686997556

Paper 1. New Technologies and Employment
Paper 2. Dynamic Balanced Scorecard
Paper 3. The Procurement Process
Paper 4. Scenario Planning Workshop
Paper 5. Risk Analysis Methods
Paper 6. Stereotypes in Socio-Economic Systems
Paper 7. Enterprise Resource Planning
Paper 8. Marketing Research
Paper 9. Group Model Building
Paper 10. Business Dynamics Simulator
Paper 11. Strategic Decision Support
Paper 12. Rare Earth Elements
Paper 13. Building a Learning Lab
Paper 14. Human Resource Planning

http://atc-innova.com/09192.htm

Modeling in Ecology and the Environment

ISBN: 9781687000323

Paper 1. Ecosystem Services Management
Paper 2. Marine Protected Areas
Paper 3. Sustainable Tourism Management
Paper 4. Great Apes and Land Cover
Paper 5. Drivers of Change in Arid Environments
Paper 6. Greenhouse Gases and Carbon Footprint
Paper 7. Facilitating Green Innovation
Paper 8. Water Quality Modeling
Paper 9. Wastewater Treatment
Paper 10. Green Growth and Ecotax
Paper 11. Economic Valuation of Natural Resources
Paper 12. Participatory Methods in Environment Projects
Paper 13. Recoverers and Recycling Organization
Paper 14. Gamification and project management

http://atc-innova.com/09193.htm

Modeling the Economy: Money and Finances

ISBN: 9781687003133

Paper 1. Modeling National Economies
Paper 2. Modeling Ecological-Economic Systems
Paper 3. Teaching Economics with a Simulator
Paper 4. Modeling and Simulation the Financial Sector
Paper 5. The Neoclassical Growth Modeled
Paper 6. Social Security Funds Sustainability
Paper 7. A Two-region Model
Paper 8. Simulation of S-shaped Growth
Paper 9. Public Policies Decision Making
Paper 10. Dynamic Balanced Scorecard
Paper 11. A Case Study for Business Schools
Paper 12. FAO's Model for Policy Guidelines
Paper 13. Scenario Planning and Implement. Challenges
Paper 14. Sustainable Finance Through Ecotax
Paper 15. Economic Valuation of Natural Resources
Paper 16. Impact of New Technologies on Employment
Paper 17. Bass Diffusion Model

http://atc-innova.com/09194.htm

Modeling and Simulation in Energy Management

ISBN: 9781687004932

Paper 1. Modeling and Simulating Energy Policies
Paper 2. Water-Energy-Food Nexus
Paper 3. Environmental-Social Pressures in Mining
Paper 4. Impacts of Electric Vehicle Diffusion
Paper 5. Forecasting Electricity Demand Market
Paper 6. Rare Earths Production Forecasting
Paper 7. Risk Analysis of Offshore Fire
Paper 8. Scenario Planning Implementation
Paper 9. Energy and Environmental Protection
Paper 10. Simulating Petroleum Peak Curve
Paper 11. Participative Group Model Building
Paper 12. Green Growth and Ecotax
Paper 13. Enterprise Resource Planning Implementat.
Paper 14. Occupational Accident Prevention
Paper 15. Gamification and Project Management

http://atc-innova.com/09195.htm

Modeling and Simulation of Healthcare Systems

ISBN: 9781687006745

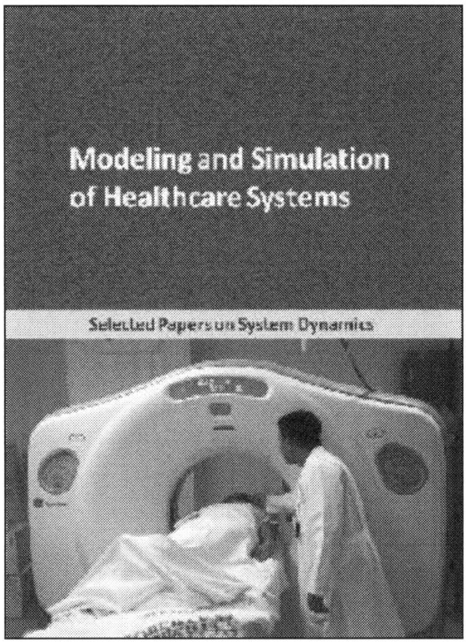

Paper 1. Patient Flow and Hospital Staffing Requirem.
Paper 2. Forecasting the Shortage of Neurosurgeons
Paper 3. Simulating Childhood Obesity
Paper 4. Modeling Vector Borne Disease
Paper 5. Generic Model of Contagious Disease
Paper 6. Increasing Patient's Safety Level
Paper 7. Prevention Rabies and Vaccination Coverage
Paper 8. Deceased Donors for Organ Transplantation
Paper 9. Antimicrobial and Antibiotic Resistance
Paper 10. Cost Treatment of Hepatitis C
Paper 11. Model for Suicide Prevention
Paper 12. Modeling For Human Resources Planning
Paper 13. Scenario Planning Workshop
Paper 14. Stereotypes in Socio-Economic Systems
Paper 15. The Bass Diffusion Model
Paper 16. Evaluate Future Businesses Performance
Paper 17 Group Model Building
Paper 18. Infectious Disease Model

http://atc-innova.com/09196.htm

Modeling the Housing and Urban Dynamics

ISBN: 9781687008367

Paper 1. Evaluating Electric Vehicle Diffusion
Paper 2. Modeling and Simulating Traffic Jams
Paper 3. Road Infrastructure Investment
Paper 4. Mgment of the Gravel Road Maintenance
Paper 5. Critical Infrastructure Protection
Paper 6. Modeling Waste from Home Devices
Paper 7. Scenario Planning Workshop
Paper 8. Group Model Building
Paper 9. Global Container Multimodal Transport
Paper 10. Gamification and Project Management
Paper 11. Business Simulator for Business Schools
Paper 12. Economic Simulations of Ordinary Waste
Paper 13. Perception and Utility

http://atc-innova.com/09197.htm

Modeling Supply Chains and Industrial Dynamics

ISBN: 9781687009975

Paper 1. Basic Logistic Model
Paper 2. Variability of Demand and Bullwhip Effect
Paper 3. Effectiveness of Lean Manufacturing
Paper 4. Life Cycle of Computers
Paper 5. Influence of 3D Printing
Paper 6. Forecasting Retail Electricity Market
Paper 7. Machine Learning Simulation
Paper 8. Improving the Productivity
Paper 9. Increasing Port Security and Performance
Paper 10. A Work Accident Simulation
Paper 11. Behavior of Procurement Process
Paper 12. Impact of New Technologies on Employment
Paper 13. Knowledge Mnt in Small and Medium Enterprises
Paper 14. Modeling the Intensive Farming Production
Paper 15. Scenario Planning Workshop
Paper 16. Environmental Issues in Mining Activities
Paper 17. Enterprise Resource Planning Implementation
Paper 18. Business Dynamics Simulator

http://atc-innova.com/09198.htm

Modeling Labor, Human Resources, and Social Dynamics

ISBN: 9781687015389

Paper 1. Gamification to elaborate a Business Model
Paper 2. Stress-Injury Pathway to Military Suicide
Paper 3. Modeling For Human Resources Planning
Paper 4. Conditions for Recoverers' Well-being
Paper 5. Utility perceived by Individuals
Paper 6. Occupational Accident Prevention
Paper 7. Planning & Mainten. of Corporate Knowledge
Paper 8. Strategy Simulator for Newspapers
Paper 9. Emerging-State Actor Model
Paper 10. Disinformation and Rumor Maker
Paper 11. Stereotypes in Socio-Economic Systems
Paper 12. Risk of Fire Based on Human Factors
Paper 13. Enhancing Services Mgent Through Modelling
Paper 14. Impact of New Technologies on Employment
Paper 15. The Bass Diffusion Model
Paper 16. Business Dynamics Simulator

http://atc-innova.com/09199.htm

Modeling Sustainable Development

ISBN: 9781700341600

Paper 1. Model for Evaluation of Reuse of Electronic Waste
Paper 2. Addressing the Sustainability Issue in Smart Cities
Paper 3. Sustainability Keys in Socio-Economic Systems
Paper 4. Participatory Methods in Group Model Building
Paper 5. Sustainability of Marine Protected Areas
Paper 6. Waste remanufacturing modeling from household
Paper 7. Sustainability of ecological arid environments
Paper 8. Model for Facilitating Green Innovation
Paper 9. Sustainability Assessment of Asset Mgnt Decisions
Paper 10. Green Growth and Ecotax
Paper 11. Sustainable Development in Intensive Farming
Paper 12. Environmental and Social Pressures in Mining
Paper 13. Input-Output for Ecological-Economic Analysis
Paper 14. Gamification and Innovation Management

http://atc-innova.com/091910.htm

Modeling Agriculture and Food Production

ISBN: 9781686984570

Modeling in Agriculture
and Food Production
Selected papers based on System Dynamics

Paper 1. Sustainable Food Availability
Paper 2. A FAO's Model of Agro-food Systems
Paper 3. Crop Modelling in a Rainfed
Paper 4. Fishing in a Natural Environment
Paper 5. A Growing Farm of Chicken
Paper 6. Agro-food Industry Modeled
Paper 7. Livestock Production Modeled
Paper 8. Dormice and Hazelnuts Production
Paper 9. Constraints in Piggery Industry
Paper 10. Sheep Producers and Consumers
Paper 11. Poultry Supply Chain
Paper 12. The Winery Industry
Paper 13. Group Model Building
Paper 14. Dynamic Input-Output
Paper 15. Greenhouse Gases and Carbon Footprint
Paper 16. Water-Energy-Food Nexus

http://atc-innova.com/09191.htm

Made in United States
Orlando, FL
26 April 2023

32499675R10161